AF385820

Advances in Geographical and Environmental Sciences

Series Editors

Yukio Himiyama, Hokkaido University of Education, Asahikawa, Japan

Subhash Anand, Department of Geography, University of Delhi, New Delhi, India

Advances in Geographical and Environmental Sciences synthesizes series diagnostigation and prognostication of earth environment, incorporating challenging interactive areas within ecological envelope of geosphere, biosphere, hydrosphere, atmosphere and cryosphere. It deals with land use land cover change (LUCC), urbanization, energy flux, land-ocean fluxes, climate, food security, ecohydrology, biodiversity, natural hazards and disasters, human health and their mutual interaction and feedback mechanism in order to contribute towards sustainable future. The geosciences methods range from traditional field techniques and conventional data collection, use of remote sensing and geographical information system, computer aided technique to advance geostatistical and dynamic modeling.

The series integrate past, present and future of geospheric attributes incorporating biophysical and human dimensions in spatio-temporal perspectives. The geosciences, encompassing land-ocean-atmosphere interaction is considered as a vital component in the context of environmental issues, especially in observation and prediction of air and water pollution, global warming and urban heat islands. It is important to communicate the advances in geosciences to increase resilience of society through capacity building for mitigating the impact of natural hazards and disasters. Sustainability of human society depends strongly on the earth environment, and thus the development of geosciences is critical for a better understanding of our living environment, and its sustainable development.

Geoscience also has the responsibility to not confine itself to addressing current problems but it is also developing a framework to address future issues. In order to build a 'Future Earth Model' for understanding and predicting the functioning of the whole climatic system, collaboration of experts in the traditional earth disciplines as well as in ecology, information technology, instrumentation and complex system is essential, through initiatives from human geoscientists. Thus human geoscience is emerging as key policy science for contributing towards sustainability/survivality science together with future earth initiative.

Advances in Geographical and Environmental Sciences series publishes books that contain novel approaches in tackling issues of human geoscience in its broadest sense-books in the series should focus on true progress in a particular area or region. The series includes monographs and edited volumes without any limitations in the page numbers.

All submitted proposals are carefully reviewed by the series editor.

Publication Ethics and Publication Malpractice Statement: https://www.springern ature.com/gp/policies/book-publishing-policies

Ethical, Conflict-of-Interest, and Editorial Policy Statement: https://www.spr inger.com/us/authors-editors/journal-author/journal-author-helpdesk/before-you-start/before-you-start/1330#c14214

Aspassia D. Chatziefthimiou · Alex Amato ·
Gonzalo Castro de la Mata · Mona Al-Kuwari
Editors

Pathways to Nature Conservation and Resilience in Hot and Arid Lands

The Case of Qatar

 Springer

Editors
Aspassia D. Chatziefthimiou
Earthna, Qatar Foundation
Doha, Qatar

Gonzalo Castro de la Mata
Earthna, Qatar Foundation
Doha, Qatar

Alex Amato
Earthna, Qatar Foundation
Doha, Qatar

Mona Al-Kuwari
Earthna, Qatar Foundation
Doha, Qatar

ISSN 2198-3542   ISSN 2198-3550 (electronic)
Advances in Geographical and Environmental Sciences
ISBN 978-3-032-11045-9   ISBN 978-3-032-11046-6 (eBook)
https://doi.org/10.1007/978-3-032-11046-6

This work was supported by Qatar National Library.

© The Editor(s) (if applicable) and The Author(s) 2026. This book is an open access publication.

Open Access This book is licensed under the terms of the Creative Commons Attribution 4.0 International License (http://creativecommons.org/licenses/by/4.0/), which permits use, sharing, adaptation, distribution and reproduction in any medium or format, as long as you give appropriate credit to the original author(s) and the source, provide a link to the Creative Commons license and indicate if changes were made.

The images or other third party material in this book are included in the book's Creative Commons license, unless indicated otherwise in a credit line to the material. If material is not included in the book's Creative Commons license and your intended use is not permitted by statutory regulation or exceeds the permitted use, you will need to obtain permission directly from the copyright holder.

The use of general descriptive names, registered names, trademarks, service marks, etc. in this publication does not imply, even in the absence of a specific statement, that such names are exempt from the relevant protective laws and regulations and therefore free for general use.

The publisher, the authors and the editors are safe to assume that the advice and information in this book are believed to be true and accurate at the date of publication. Neither the publisher nor the authors or the editors give a warranty, expressed or implied, with respect to the material contained herein or for any errors or omissions that may have been made. The publisher remains neutral with regard to jurisdictional claims in published maps and institutional affiliations.

This Springer imprint is published by the registered company Springer Nature Switzerland AG
The registered company address is: Gewerbestrasse 11, 6330 Cham, Switzerland

If disposing of this product, please recycle the paper.

Acknowledgements

Open access funding provided by the Qatar National Library. We extend our sincere gratitude to the Earthna Center for Sustainability at Qatar Foundation for its valuable support throughout the development of this book. We are also thankful to the editorial team for their dedication in preparing the manuscript for publication. Special thanks go to all chapter authors and peer reviewers for their thoughtful contributions and rigorous efforts. We would also like to thank Sara Abdul Majid and Aisha Yusuf for their valuable support and assistance throughout the entire process.

Contents

Editors and Contributors

About the Editors

Dr. Aspassia D. Chatziefthimiou is a researcher and consultant on nature conservation and restoration, and on the environmental impacts of development in hot desert climates. Dr. Aspassia received her Ph.D. in Ecology and Evolution from Rutgers University, USA, and is a Certified Senior Ecologist by the Ecological Society of America.

Dr. Aspassia incorporates traditional ecological knowledge, ethnography, as well as tracking and trailing in her research and fieldwork. She develops educational programs, incorporating elements from all the above topics, to catalyze citizen science participation, and to deepen nature connection and appreciation in the population at large.

Currently, Dr. Aspassia serves as the Chief Scientific Officer for Earthna's Center for a Sustainable Future, Mangrove Restoration Program, and as the Ecological Research Consultant to the National Museum of Qatar, and to Virginia Commonwealth University's (IN)tangible Lab. e-mail: a.d.chatziefthimiou@gmail.com

Dr. Alex Amato has over 40 years' experience in the built environment and sustainability research. He has worked in both the private and public sectors in the UK, SE Asia, and now in the Middle East. His work covers a wide gamut, from design, to construction, product development, and research. He has worked in architectural practice and academy. He has also worked in industry for both the steel and concrete sectors, where he was engaged in the development of new sustainable manufacturing and construction systems.

Since gaining his Ph.D. at Oxford Brookes University over 25 years ago, he has sought to resolve the perplexing problem of integrating sustainability into the built environment and more widely afield. He is now part of the Earthna team, formed in 2022, where he is their Sustainability Research and Policy Advisor. Here he leads several research projects intended to provide Qatar with a clearer picture of its sustainability challenges.

These projects range from mangrove restoration and ecosystem protection to implementation of the circular economy where all of Qatar's economic sectors are examined in terms of their sustainability, both existing and potential. This recently has extended to modeling Qatar's domestic economy as a means for evidence-based policy formulation. His contention is that only by using quantitative methods will effective evidence-based policy emerge. e-mail: aamato@qf.org.qa

Dr. Gonzalo Castro de la Mata is an ecologist recognized as a global leader in the promotion of sustainability, with an emphasis on innovative free market solutions to environmental issues. As Executive Director of Earthna, he guides the center's mandate of enhancing Qatar's role within the global sustainability policy ecosystem.

His previous positions including Managing Director for External Affairs at Pluspetrol, Latin America's largest private oil and gas company; Chairman of the World Bank's Inspection Panel in Washington D.C.; Managing Director of Sustainable Forestry Management for the Americas; Head of Biodiversity at the Global Environmental Facility, also in Washington D.C.; Lead Environmental Specialist at the World Bank; and Director of the World Wildlife Fund's Latin American and Caribbean Program.

Gonzalo founded Ecosystem Services LLC, which generates carbon offsets through avoiding deforestation of the Amazon rainforest, and Wetlands for the Americas (now Wetlands International). He has acted as a consultant for multilateral and bilateral agencies, governments, and NGOs, and published over 250 articles in scientific journals, magazines, and books, as well as authoring two books himself.

He received a Ph.D. in Ecology and Population Biology from the University of Pennsylvania and holds a Master of Science in Biophysics and a Bachelor of Science in Biology from Cayetano Heredia in Lima, Peru. e-mail: gcastrodelamata@ qf.org.qa

Dr. Mona Al-Kuwari is the Editor and Publication Manager at Earthna, a center under Qatar Foundation dedicated to sustainability policy and research. She holds a Ph.D. in Sustainability Studies from Hamad Bin Khalifa University, with a research focus on integrating social and environmental dimensions into sustainable development frameworks. She also holds a Master's degree in Information Systems from the College of Engineering and a Bachelor of Science in Statistics from the College of Science, both from Qatar University.

Dr. Mona has played a leading role in advancing sustainability education, research dissemination, and policy outreach. Her scholarly work includes contributions to international journals, editorial roles, and presentations at global conferences. Her primary research interests lie in the development and implementation of sustainability performance assessment strategies, particularly within culturally grounded and policy-driven contexts.

She began her career at Qatar Foundation with the Qatar National Research Fund as a statistician. She later moved into the Office of Strategy, Management, and Partnerships in Higher Education, where she served as Senior Data Analytics and Impact Assessment Specialist. In parallel, she has significantly contributed to the Foundation's strategic sustainability initiatives, bringing data-driven insights and policy relevance to the forefront of institutional planning. e-mail: moalkuwari@qf.org.qa

Contributors

D. Abdelwahed UNESCO Office for Gulf States & Yemen, UNESCO, Doha, Qatar

S. Abdul Majid Eartha Centre for a Sustainable Future, Doha, Qatar

A. S. Al-Ansari Qatar Meteorology Department, Civil Aviation Authority, Doha, Qatar

S. AlHajri Ministry of Environment and Climate Change, Doha, Qatar

M. Ali Earthna, Center for a Sustainable Future, Qatar Foundation, Doha, Qatar

A. M. Al-Mannai Qatar Meteorology Department, Civil Aviation Authority, Doha, Qatar

N. Al-Mohannadi Qatar Meteorology Department, Civil Aviation Authority, Doha, Qatar

N. Al Salem National Museum of Qatar, Doha, Qatar

R. Al-Sehlawi Co-founder Atlas Bookstore for the Natural and Built Environment, Doha, Qatar

A. Amato Earthna, Center for a Sustainable Future, Qatar Foundation, Doha, Qatar

E. Athwal UNESCO Chair of Environmental Law and Sustainable Development, College of Law, Hamad Bin Khalifa University, Doha, Qatar

M. Baeuerle Sustainability Hive, Doha, Qatar

P. K. Bal Qatar Meteorology Department, Civil Aviation Authority, Doha, Qatar

R. Ben-Hamadou Earthna, Center for a Sustainable Future, Qatar Foundation, Doha, Qatar

B. Böer UNESCO South-Asia Office in New Delhi, New Delhi, India

T. R. R. Bontognali Space Exploration Institute, Neuchâtel, Switzerland; Department of Environmental Sciences, University of Basel, Basel, Switzerland

K. T. Bron GeoTree Consulting, Melbourne, Australia

J. Bruder Carnegie Mellon University in Qatar, Doha, Qatar

A. D. Chatziefthimiou Earthna, Center for a Sustainable Future, Qatar Foundation, Doha, Qatar

C. C. H. Lim Dadu Children's Museum of Qatar, Doha, Qatar

A. T. Conkey Independent Researcher, Texas, USA

M. Contestabile Earthna, Center for a Sustainable Future, Qatar Foundation, Doha, Qatar

G. Dimitropoulos UNESCO Chair of Environmental Law and Sustainable Development, College of Law, Hamad Bin Khalifa University, Doha, Qatar

B. W. Giraldes Environmental Science Center, Qatar University, Doha, Qatar

G. Haddad Arab International Academy—Lusail, Doha, Qatar

S. Hameed Snow Leopard Foundation, Islamabad, Pakistan

N. Hamwy Carnegie Mellon University in Qatar, Doha, Qatar

F. Hassan Friends of the Environment, Doha, Qatar

R. Hinnawi Earthna, Center for a Sustainable Future, Qatar Foundation, Doha, Qatar

K. K. Kanikicharla Qatar Meteorology Department, Civil Aviation Authority, Doha, Qatar

T. Karanisa Ministry of Rural Development and Food, Athens, Greece

J. Lari Ministry of Environment and Climate Change, Doha, Qatar

J. Lawler Qatar Environment and Energy Research Institute, Hamad Bin Khalifa University, Doha, Qatar

A. Leitão Environmental Sciences Center, Qatar University, Doha, Qatar

V. L. Miller Qatar Academy Msheireb, Qatar Foundation, Doha, Qatar

A. M. Dheen Mohamed Education Above All Foundation, Doha, Qatar

M. A. Nawaz Department of Biological and Environmental Sciences, Qatar University, Doha, Qatar

D. Olawuyi UNESCO Chair of Environmental Law and Sustainable Development, College of Law, Hamad Bin Khalifa University, Doha, Qatar

I. Petrovska University American College Skopje, Skopje, Republic of Macedonia;
Formerly, ARIU—University of Derby, Doha, Qatar

P. Range Environmental Science Center, Qatar University, Doha, Qatar

C. Reeves Sustainable Oceans Management Limited, Ripon, North Yorkshire, UK

R. Richer Duke Kunshan University, Kunshan, China

R. Riera BIOCON (Biodiversity and Conservation), ECOAQUA, University of Las Palmas de Gran Canaria, Canary Islands, Las Palmas de Gran Canaria, Spain

S. D. Sever Carnegie Mellon University in Qatar, Doha, Qatar

S. Tull Qatar Bird Records Committee, Doha, Qatar

A. S. Weber Weill Cornell Medicine Qatar, Doha, Qatar

J. MK. Wong Ministry of Environment and Climate Change, Doha, Qatar

R. Yousif Department of Geological Sciences & Geological Engineering, Queen's University Kingston, Kingston, Canada

List of Figures

List of Tables

Chapter 1
Geology, Hydrology, and Water Resources

How Geology and Hydrology Shape the Past, Present, and Future Environment in Qatar

R. Yousif, J. Lawler, and A. D. Chatziefthimiou

Abstract This chapter explores the intricate interplay between geology, hydrology, and environmental evolution in Qatar, emphasizing how these natural systems have shaped the region's past, present, and future. Qatar's surface geology is dominated by Eocene and Miocene carbonate formations, influenced by tectonic uplift and evaporative sedimentation in the Arabian Gulf. The peninsula's aquifers, primarily composed of Paleogene carbonates and evaporites, reflect a complex depositional history shaped by marine incursions and tectonic structures. Quaternary sea-level fluctuations and climatic shifts have significantly influenced coastal geomorphology, with evidence of ancient shorelines and paleo-lakes. Key landforms such as sand dunes, wadis, sabkhas, and caves illustrate the dynamic interaction of wind, water, and geological substrates in shaping Qatar's arid landscape. The chapter also highlights the critical role of geosystem services in supporting biodiversity and sustaining life in desert ecosystems. These include water regulation, erosion control, and nutrient cycling, with particular attention to the ecological value of mangroves, wetlands, and coral reefs. Qatar's water resources, heavily reliant on desalination and limited groundwater, face challenges from over-extraction, salinization, and brine disposal. The chapter underscores the importance of sustainable water management, including treated sewage re-use and managed aquifer recharge, to mitigate environmental impacts. Recommendations include promoting geological research, understanding

R. Yousif
Department of Geological Sciences & Geological Engineering, Queen's University Kingston, Kingston, Canada
e-mail: ruqaiya_a@hotmail.com

J. Lawler (✉)
Qatar Environment and Energy Research Institute, Hamad Bin Khalifa University, Doha, Qatar
e-mail: jenny.lawler@dcu.ie

A. D. Chatziefthimiou
Earthna, Center for a Sustainable Future, Qatar Foundation, Doha, Qatar
e-mail: a.d.chatziefthimiou@gmail.com

© The Author(s) 2026

A. D. Chatziefthimiou et al. (eds.), *Pathways to Nature Conservation and Resilience in Hot and Arid Lands*, Advances in Geographical and Environmental Sciences,
https://doi.org/10.1007/978-3-032-11046-6_1

past sea-level changes, improving water governance, and assessing the ecological consequences of water-related practices. These insights are vital for guiding Qatar's environmental policy and ensuring resilience in the face of climate change and rapid development.

1.1 Surface Geology

The Arabian Gulf is a foreland basin south of the Zagros Mountains. It is a relatively shallow basin (mostly <60 m deep), and its southern waters are the site of extensive photozoan carbonate sedimentation. This depositional realm includes the northern Oman and the United Arab Emirates, where the "Great Pearl Bank" is located. The southern Arabian Gulf is considered an important modern analog for evaporative carbonate ramp settings in which many Middle East hydrocarbon reservoir rocks are interpreted to have formed (Shinn 1969). Uplift of the Qatar Arch produced the Qatar Peninsula, which protrudes north into the southern Arabian Gulf. The surface rocks of Qatar are predominantly Eocene-age (55.8–33.9 million years) dolostones, and Miocene-age (23.03–5.3 million year) outcrops occur only in the southwestern part of the peninsula (Fig. 1.1) (Rivers et al. 2019a, b; Akbar 1995).

During the early Rus Formation, a large amount of evaporite and clay-rich siliciclastic material was deposited in the southern basin, which was smaller and less open. In time-equivalent strata of the northern basin, comparable evaporites and fine siliciclastic deposits are not found. Fine-grained siliciclastic deposits are interbedded with, but beneath, carbonate strata throughout Qatar, suggesting that the basins may have been connected during the subsequent deposition of the Al Khor Member of the Rus Formation and the Dammam Formation (Rivers et al. 2018, 2019b; Al Hajari 1990).

Early studies described the basic surface geology of the Qatar Peninsula as a succession of carbonate-dominated Cenozoic sedimentary rocks separable into two structural domains: (1) the broadly north-plunging Qatar-South Fars arch, which affects most of the peninsula and extends into the Gulf toward Iran and (2) the Dukhan "anticline," which is restricted to the western part of the country. Recent studies interpret the two antiformal structures as distinct, possibly even fault-separated structures, contrary to earlier interpretations that suggested they may be related and separated by a syncline (Al Hajari 1990).

The coastal regions of Qatar have been substantially influenced by recent sea-level changes (Rivers et al. 2020). Geochronological data of Quaternary paleo ocean conditions in arid regions, such as the Middle East, particularly the Arabian/Persian Gulf are still sparse and are limited to an understanding of pluvial lakes, shifts in the Intertropical Convergence Zone (ITCZ), and sea-level changes (uplift versus climatological changes) (Stevens et al. 2014; Wood et al. 2012). Mollusk-rich deposits of early Holocene (6000 year) and Pleistocene (125,000 year) age are exposed immediately landward of some coastal areas (Fig. 1.2; Rivers et al. 2018; Al Hajari 1990).

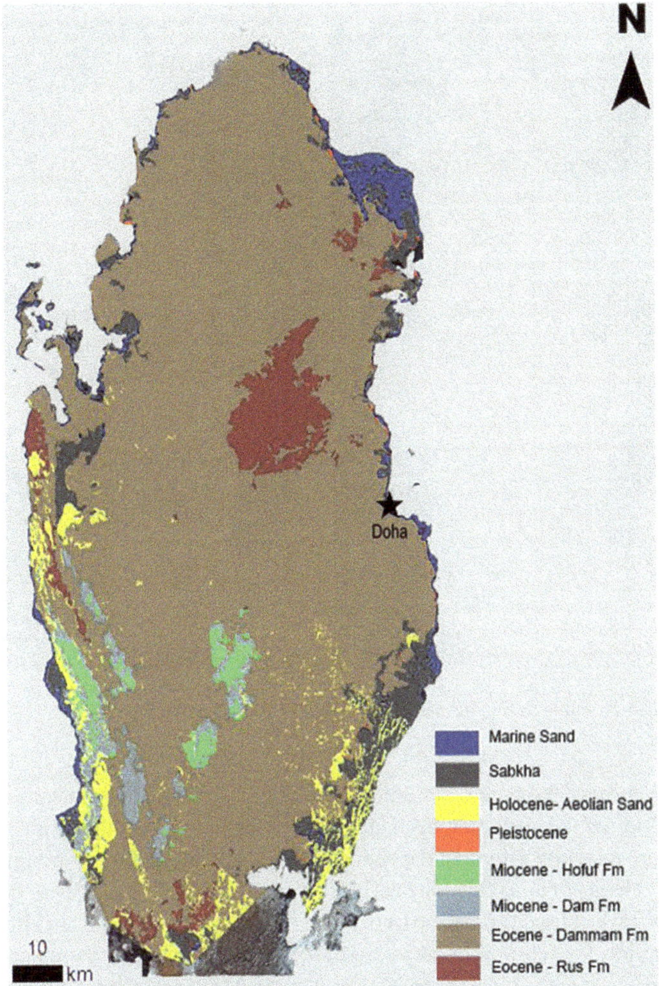

Fig. 1.1 Surface geological map of Qatar (Center for GIS). Most of Qatar's shallow-aquifer rocks, such as the Palecccocene and lower Eocene (Umm er Radhuma formation) and the lower to middle Eocene (Rus and Dammam formations), are composed of carbonates and evaporites that date back to the Paleogene era (Rivers et al. 2019b, 2020; Abu-Zeid 1991) Both the Traina Member of the Rus formation and the Umm er Radhuma formation were deposited in marine environments in two different basins. It is believed that a topographic high, whose location was determined by the presence of high-angle normal faults, once separated these basins, which extended to Qatar's south and north, respectively (Rivers et al. 2019a, 2019b)

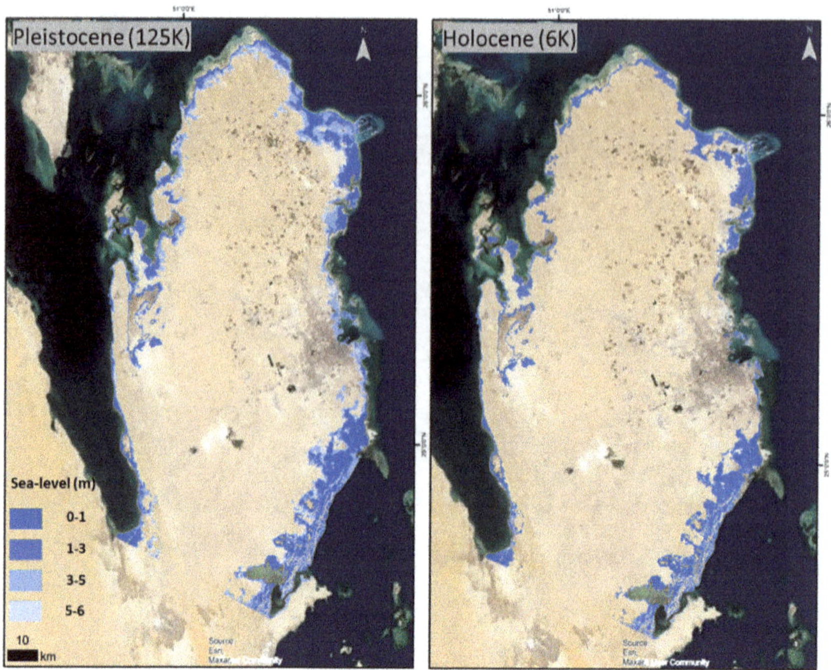

Fig. 1.2 Paleo coastline during the Pleistocene (125,000) and Holocene (6000)

Overall, most of the previous studies agree that the prominent influence on Middle to Late Quaternary environmental changes in Arabia was due to fluctuations in rainfall and moisture intensity triggered the changes of monsoon circulation strength (Michel et al. 2018; Heggy et al. 2022). The early Holocene was more humid than the present, which may have caused humans to bypass the region for occupation, and it represents the best analog for future climate changes. Figure 1.2 indicates the paleo coastline during the Pleistocene (125,000 ya) and Holocene (6,000 ya), which is the best model for 2 °C increase in average temperature (Jameson and Strohmenger 2012; Rivers et al. 2021). From this perspective, there is a need for further geochronological data in the context of understanding changes in this region and relating them to the main drivers of gradual alterations in the earth's ecosystems.

There are no published studies that describe the extent and environmental conditions of the paleoocean during the Quaternary. Purser and Seibold (1973), Lüning et al. (2018), and Rose (2010a, b) used coral terrace U/Th age dating at Qeshm Island (Iran), to demonstrate that the area underwent regional uplifting during the late Quaternary due to active salt doming, rather than sea-level fall (Mossadegh et al. 2013). Whereas Williams and Walkden (2002) studied Quaternary carbonate sediments in the southern Arabian/Persian Gulf and indicated that stratigraphic changes are mainly derived from sea-level fluctuations where marine deposits are found during the interglacial period, while aeolian sediments predominated during glacial periods (Williams and Walkden 2001, 2002; Williams 1999; Glenn et al. 2002).

1.2 Geological Landforms and Habitats

Sand dunes are a prominent geological feature in Qatar, mainly found in the southwestern part of the country. They are formed by the interaction of wind, sediment availability, and topography. The prevailing wind direction in Qatar is from the northwest, resulting in the accumulation of sand grains and the formation of dunes in the windward (upwind) side (Williams and Walkden 2001, 2002; Williams 1999, Glenn et al. 2002). These dunes are shaped and molded by the continuous action of wind, which transports sand grains and deposits them on the leeward (downwind) side, forming dunes with characteristic shapes and sizes (Rivers et al. 2020; Embabi and Ashour 1993).

Wadis and rawdas are typically dry riverbeds or channels found in arid regions like Qatar. They are formed by intermittent or ephemeral streams that carry water during periods of rainfall or flash floods. In Qatar, wadis and rawdas are often located in the northern and central parts of the country. These water channels form due to the combination of occasional heavy rainfall events and the underlying topography. Over time, the flowing water erodes the land, creating these distinct features (Williams and Walkden 2002).

Caves in Qatar are relatively rare, but they can be found in certain areas with suitable geological conditions. The formation of caves generally requires a combination of soluble rock, such as limestone or gypsum, and water action over an extended period (Rivers et al. 2018, 2019b, Al-Hajari 1990, Alsharhan et al 2001). In Qatar, limestone formations are present in some parts of the country, particularly in the northern region. Groundwater percolates through these limestone formations, dissolving the rock over time and creating cavities and underground passages that can develop into caves (Gutiérrez et al. 2014).

Sabkhas are coastal salt flats or salt pans that form in arid regions where evaporation rates are high and there is limited drainage (Kinsman 1964, 1969). In Qatar, both inland and coastal sabkhas can be found. Inland sabkhas develop in low-lying depressions where water accumulates during the rainy season. As the water evaporates, it leaves behind salt and mineral deposits, forming a crust on the surface (Ashour 1987). Coastal sabkhas, on the other hand, are formed by the interaction of seawater and shallow coastal areas. The high evaporation rates in these regions cause the precipitation of salts, leading to the formation of sabkhas along the coast (Strohmenger and Jameson 2018).

Overall, the formation of these geological features in Qatar is a result of the interplay between climate, topography, sediment availability, and various geological processes such as wind erosion, water erosion, and chemical weathering.

1.3 Coastal—Sea-Level Changes

Sea-level changes have been a natural part of the Earth's geological history, occurring over long periods of time due to various factors such as the Earth's orbit and the movement of tectonic plates (Fairbridge 1960). In the case of Qatar, the region has experienced several periods of sea-level changes over the past few million years (Murray-Wallace and Woodroffe 2014; Purkis et al. 2017). During the last Ice Age, which occurred between about 115,000 and 11,700 years ago, sea levels were much lower than they are today due to a large amount of water stored in glaciers and ice sheets (Rivers et al. 2021; Purkis et al. 2017). This caused the shoreline to be much further offshore than it is today, and Qatar would have been connected to the mainland by a large coastal plain (Jameson and Strohmenger 2012, 2014).

As the Ice Age ended and glaciers melted, sea levels began to rise rapidly, eventually reaching 2 m higher than their current levels around 7,000 years ago [8, 35]. During this time, the Qatari coast would have shifted inland, as the sea level rose and flooded the coastal plain (Al-Hajari 1990; Rivers et al. 2021; Purkis et al. 2017; Jameson and Strohmenger 2014). Since then, the sea level has fluctuated over time, influenced by factors such as global climate change, ocean currents, and tectonic activity (Heggy et al. 2022; Lambeck 1996). While the sea level has been relatively stable in recent history, the current rate of sea-level rise is much faster than the natural variability seen in the geological record (Milne et al. 2009).

1.4 Geology/Water Resources—Last Glaciation Maximum to 20,000–11,000 Ya

During the last glacial, the Qatar Peninsula, located in the Arabian Gulf area, has seen substantial changes in the distribution and availability of surface freshwater, notably in connection to lakes and precipitation patterns (Heggy et al. 2022). Qatar was part of a much bigger continent known as the Arabian Peninsula during the last glacial, which ended roughly 11,700 years ago (Lambeck 1996). The climate in the Qatar Peninsula was significantly colder and drier than it is now, and the region was most likely covered by sand dunes and little vegetation at the time. By the end of the glacial, the temperature had begun to warm, and the region had become more hospitable to life (Heggy et al. 2022). There have also been dramatic changes in the distribution of freshwater resources, including the development of lakes (Sowers et al. 2011).

There is evidence that many lakes existed in Qatar during the conclusion of the last glacial (Parker et al. 2006). Geological investigations, for example, have revealed sedimentary layers that imply the presence of a vast, shallow lake in the country's south, in a region known as the Al Kharrara Formation (Heggy et al. 2022). Water began to gather in low-lying places as the environment got wetter and more friendly to plant life, forming lakes such as Lake Dukan and Lake Tharthar in Iraq (Raza et al. 2011). These lakes were essential water supplies for the region's early human populations and supported a rich diversity of aquatic plant and animal life.

The lakes are assumed to have been supplied by rivers and streams that came from neighboring mountains, and they likely supported a rich range of aquatic life, including fish, water birds, and other species. However, as the climate began to warm and become drier following the last glaciation, the lakes began to dry up.

In addition to the changes in freshwater resources, the climate in Qatar has also undergone significant variations in precipitation patterns since the last glaciation (Raza et al. 2011). During the last glaciation, the region experienced relatively dry conditions, with limited rainfall. With the end of the glaciation, precipitation patterns became more variable, with alternating periods of drought and more abundant rainfall (Heggy et al. 2022). This variability in precipitation continues to the present day, with Qatar experiencing periodic droughts and flash floods. Today, the area is characterized by sand dunes and other desert landscapes, and there are no permanent surface water bodies in Qatar.

1.5 Geosystem Services

Geosystem services refer to the benefits that human societies derive from the Earth's natural systems, including the atmosphere, biosphere, geosphere, and hydrosphere (van Ree, van Beukering and Boekestijn 2017). In the context of the hot desert environment such as Qatar, geosystem services are particularly relevant. The geodiverse sub-surface as described already, with uniqueness in site-specific geological features, hydrological features, landforms, etc., sees Qatar's geodiversity intrinsically linked with biodiversity (Gray and Gordon 2020), not least through the medium of habitat provision and support for water and nutrient cycling for flora and fauna. Deserts are home to a variety of plant and animal species that have adapted to the harsh conditions of this environment. Geosystem services play a critical role in supporting biodiversity in the desert environment. For example, soil formation and nutrient cycling are essential for the growth and survival of plants, which in turn provide habitat and food for a wide range of desert animals. We often experience Shamal and other high winds in Qatar; however, plant life in these regions helps to stabilize the soil and reduce erosion (both in desert regions and urban planted areas). Native plants with deep root systems play an important role in absorbing and storing water. In desert ecosystems, limited rainfall and the presence of dry conditions can limit the availability of nutrients for plant growth. However, some desert plants are able to extract and store nutrients from the soil, providing a critical source of nutrients for other organisms in the ecosystem.

The regulation of water flow, erosion control, and carbon sequestration are other geosystem services that are essential for the maintenance of healthy desert ecosystems (Frisk et al. 2022).

Effective management of desert ecosystems requires an understanding of the geosystem services provided by these ecosystems. By recognizing the importance of these services, policymakers and land managers can make more informed decisions about how to protect and conserve these valuable natural resources. In the context of water, geosystem services can include water regulation and purification, as well as the provision of freshwater for irrigation and drinking. In Qatar's desert environment, some of these geosystem services need to be generated or encouraged through anthropogenic means, whereas others are readily available naturally. For example, wetlands can act as natural water filters, removing pollutants from water as it flows through the wetland. Qatar's coastal mangroves provide this service, where planted mangroves were established by researchers up to 30 years ago, and natural mangroves where fresh groundwater was said to occur (Al-Khayat, Vethamony and Nanajkar 2021). Artificial lagoons, which were used in Qatar for holding untreated wastewater for many years, have now been remediated and act as constructed wetlands (Draidia et al. 2022), supporting significant biodiversity.

Similarly, various natural Qatari habitats—characterized by water availability and sub-surface/geohydrological characteristics—marine environment, salt marshes, and mangrove forests; the arid lands (including dune communities); wetlands (including pond communities); Sabkhas (Abdelsamad et al. 2022) and Rawdahs (including the Ghaf tree communities) have been shown to support significant microbial and other biodiversity (Al-Thani and Yasseen 2021). Qatar's coral reefs provide habitat for a rich array of species and support fisheries, while the Al Shaheen region on the north coast of Qatar supports one of the largest populations of whale sharks in the world (Robinson et al. 2017).

While forests are not a natural feature of the Qatari geosystem, these can help to regulate water flow and reduce the impacts of flooding. Qatar has a plan to increase afforestation in the country by planting 10 million trees (Qatar News Agency 2022), and indeed succeeded in planting the millionth tree toward the end of 2022 (The Peninsula 2022a, b). Arid and desert regions also have important cultural and aesthetic values, providing opportunities for recreation, tourism, and artistic inspiration; geosystem services that are certainly provided by regions in Qatar such as the UNESCO heritage sites in Zekreet; the Inland Sea at Khor Al Udaid; and many other regions of beauty around the country. Qatar, like many arid regions, is facing significant threats, including habitat loss, over-extraction of water resources, and climate change, which can negatively impact these geosystem services. Conservation efforts are needed to protect and preserve these valuable ecosystems.

While history tells us about ancient water in Qatar and how the aquifers and the sub-surface changed over time, the landscape of modern water resources is critical in evaluating how the country can protect and support biodiversity. The State of Qatar is regarded as one of the most water-stressed countries in the world with virtually no conventional water resources—zero surface water resources, and mainly brackish groundwater resources which are over-abstracted, difficult to recharge, and cannot ever be expected to meet the needs of the growing population.

1.6 Modern Water Streams

1.6.1 Groundwater

Qatar's natural fresh groundwater resources are limited, and the depth to the water table varies from less than 1 m near the coastline to more than 70 m in some areas (Fig. 1.3).

Qatar's environment—like that of other GCC countries—is characterized by the lack of renewable freshwater resources such as rivers and lakes and by aquifers containing groundwater which is not replenished to the same extent to which it is abstracted, with corresponding deteriorating quality (Baalousha and Ouda 2017; Abdulrahman 2020), mainly seen as rising salinity (TDS, total dissolved solids). Qatar was classified as one of the most water-stressed countries in the world in

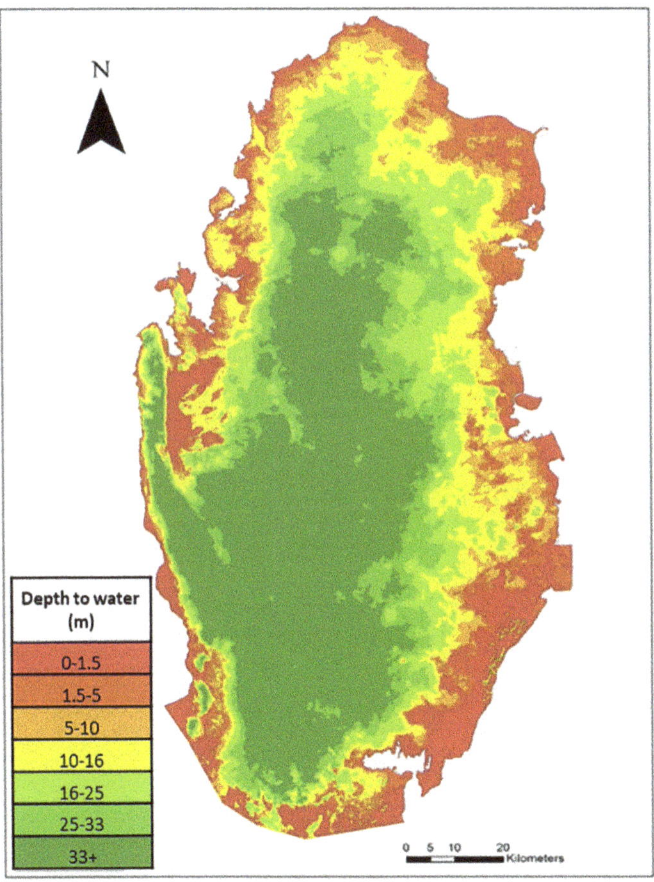

Fig. 1.3 Rated depth to water table (meters) (adapted from (Baalousha 2016b))

2019 (World Resources Institute 2019), due to the combination of low precipitation and low fresh groundwater levels, with natural renewable water resources (rainfall and groundwater) classified as far below the water poverty line (Darwish and Mohtar 2013) (Table 1.1). Abstraction of groundwater is estimated to be approximately three to five times the rate of natural recharge, in general, being used to meet agricultural requirements (Baalousha 2016a; Planning and Statistics Authority 2018b).

Qatar's groundwater can be divided into a number of geographical areas (Fig. 1.4), including the northern groundwater region (covering approximately 20% of Qatar and historically with good quality suitable for agricultural irrigation, at a depth of between 10 and 40 m), the southern groundwater basins (which cover approximately 50% of Qatar and have low quality not suitable for irrigation), a smaller area beneath Doha (where the water table is now high, mainly due to urban irrigation), and the Aruma deep groundwater basins, which at over 450 m deep are uneconomical for abstraction (Planning and Statistics Authority 2018b).

Prior to 1953 when the first seawater desalination plant was established, groundwater was the sole provider of water for drinking and irrigation. However, continuous deterioration in quality and availability of groundwater has occurred especially during the period of intense population increase and increased agriculture in Qatar. Irrigation hazard analysis performed on various wells located around Qatar indicates that up to 95% of samples are not suitable for agricultural irrigation, with over 60% having a specific electrical conductivity greater than 5000 μS/cm, which would likely cause harm to agriculture, and only salt-tolerant crops are suitable, while 34% of the irrigation water was found to be of very high salinity hazard and medium sodium hazard, and thus it is not suitable for irrigation in almost all soils, except soils with high permeability (Ahmad et al. 2020). The complex interrelationships between salt ions and chemical, biological, and geologic parameters and consequences on the natural, social, and built environment is called Freshwater Salinization Syndrome (FSS). FSS has direct and indirect effects on the mobilization of diverse chemical

Table 1.1 Natural water balance of Qatar's aquifers (annual average 1998–2016) (Planning and Statistics Authority 2018b)

Water balance	M m^3/ year	Data source
Recharge of aquifers from rainfall	71.6	Kahramaa and Ministry of Municipality and Environment (long-term annual average 1998–2017)
Inflow from Saudi Arabia	2.2	Al Dour Zone (long-term annual average 2006)
Total renewable water resources[1.1]	73.8	Sum of the rainfall recharge and inflow
Groundwater outflow into sea and deep saline aquifers	18.0	Kahramaa and Ministry of Municipality and Environment (long-term annual average 1998–2017)
Average annual water balance (groundwater safe yield)[2]	55.8	Renewable—Outflow

[1.1] According to FAO Aquastat
[2] without the returns from irrigation

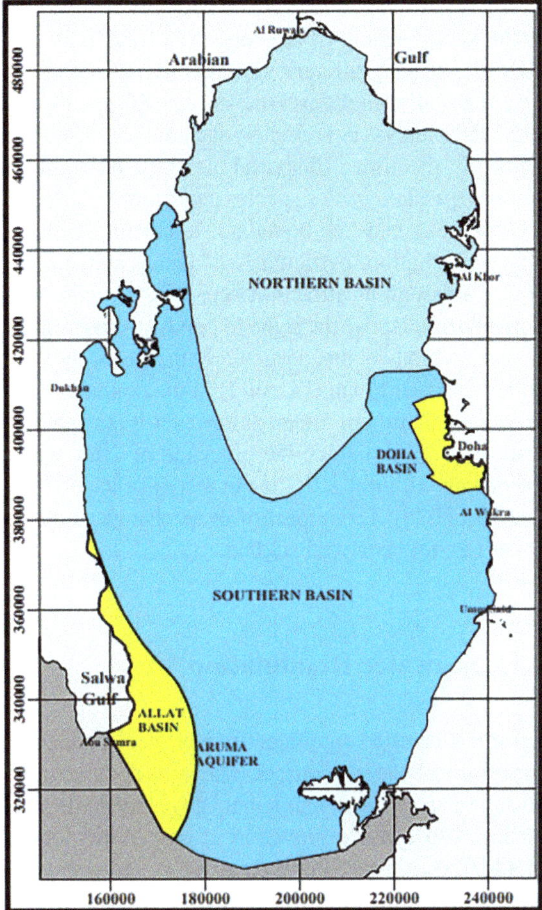

Fig. 1.4 Groundwater basins in the State of Qatar (Planning and Statistics Authority 2018b)

cocktails of ions, metals, nutrients, organics, and radionuclides in freshwaters with mounting impacts. This causes increasing risks to infrastructure, biodiversity, and critical ecosystem services (Kaushal et al. 2021).

An extensive survey by Schlumberger published in 2009 identified that the overall number of wells in Qatar with various uses was greater than 8500, with approximately 75% of the overall wells serving the needs of farms (Ahmad et al. 2020). Kahramaa, the General Electricity and Water Corporation of Qatar, has recently completed rehabilitating and activating over 300 existing recharge wells around Qatar, to increase the direct recharge to the aquifer system in the country from the rain and subsequently improve the groundwater condition and increase its availability for water security (The Peninsula, 2023), while at the same time extending efforts to drill and develop about 60 recharge/injection and observation wells to increase the direct recharge to storage zone and to increase the productivity of the aquifer system from the rain.

Managed aquifer recharge (MAR) is not easy or straightforward in Qatar due to the karstic nature of the sub-surface (Margat and van der Gun 2013; Baalousha 2016b). Virtually all agricultural activities using groundwater require local desalination, usually small reverse osmosis (RO) plants on the farm. The brine from these small RO plants is not typically well managed and is not well regulated or controlled. Typically, the brine is disposed of in trenches surrounding the farm boundary, where *Conocarpus lancifolius* species can grow well due to its halotolerance, and provide a wind break and tree boundary. Interestingly, this species is a good candidate for phytoremediation, particularly of heavy metals (Rasheed et al. 2020; Redha et al. 2021), although its attraction here is its drought resistance and fast growing nature. Some farms dispose the brine in ponds or evaporation ponds, which has concomitant impact on biodiversity, particularly insects or birds. The amount of agricultural brine being disposed to land or ponds in the country is not at present well known, due to lack of regulation in the area, however it is expected that a drastic increase in brine disposal occurred since the blockade of 2017 when it was necessary for Qatar to expand agricultural activities in response to food supply chain issues (Hussein and Lambert 2020). The impact of these changes in farming practices on biodiversity in Qatar has not been well studied.

1.7 Seawater Desalination

Qatar has invested significantly over the past decades in the exploitation of its non-conventional water resources. The most notable of these is the extensive desalination of seawater to meet municipal, governmental, and commercial needs. Qatar has managed to achieve provision of safe water for 100% of its resident population in addition to massively increasing its indigenous food security during the period 2017–2021. Sanitation, wastewater, and sewerage services have been achieved with extensive geographical penetration in the country, with concomitant reduction in environmental pollution, not only marine pollution due to direct sea discharge but also with the rehabilitation of lagoons and cessation of dumping of untreated water. This has also meant that vast amounts of water treated to a very high quality has become available in the State. To date this has not been fully utilized, with targets for up to 100% use for beneficial purposes yet to be realized under the country's strategic development plans.

Under Qatar's first National Development Strategy, NDS-1, it was identified that approximately 50% of the TSE volume (almost 2 M m^3/year) was being injected into the aquifer or discarded in surface reservoirs and was seen as a financial loss due to the high treatment costs, and the potential adverse impact of the injection of TSE on the aquifers. As such, the NDS-2 (Planning and Statistics Authority 2018a) for the period 2018–2022 was designed to build on the achievements of and lessons learned from the NDS-1 and continues to support implementation of the QNV 2030, in addition to improving economic performance and future aspects. The strategy pledges to rationalize water consumption and encourage the use of non-conventional water resources, among other key aspects. The NDS-2 has a number of targets which encouraged the expansion of TSE use in Qatar, including a target to provide infrastructure to use 70% of the TSE produced in different projects by 2022 and to establish a more holistic industrial zone water management plan. Under NDS-2 it was targeted to reduce loss of drinking water rate (actual + administrative) to 8% by 2022, which was achieved. Qatar National Development Strategy, QNDS-3, is under development during 2023, which will expand further the beneficial use of TSE, along with targeting further reduction in real losses. The usage of treated water for fodder irrigation has changed the balance of water use in the agricultural sector, helping to come some way toward alleviating the stress on the groundwater resources.

Treated sewage effluent is only one example of a non-conventional water resource in Qatar that can be a beneficial addition to the country's water balance. Produced water from the oil and gas industry is the water that is abstracted along with fossil fuels in a ratio of approximately 1.2:1 for gas fields. This water is treated and injected on-land, with estimates set at upward of tens of millions of m^3 per annum. Qatar has ongoing research initiatives in the treatment of this produced water for beneficial re-use, to supplement the use of treated sewage effluent. Along with this, various opportunities for water harvesting are underway. For example, the capture of humidity in large or commercial buildings using the building air handling units—capture, treatment, and re-use of air-conditioning condensate—has the potential to recover millions of m^3 of good quality water per year (although this water is typically sent to the sewer and as such, while it does not add to the water balance per se, it does allow for improvements in the energy efficiency of the water sector by allowing local treatment and re-use). Air to water generation units are nowadays commercially available at scales of up to 10,000 L per day freshwater generation, and the implementation of these type of units to support the agricultural sector are ongoing (Al-Ansari et al. 2022). Similarly, initiatives in greenhouse design which require less freshwater for cooling such as direct humidity capture and re-use in greenhouses (Tawalbeh et al. 2023) pave the way forward for reduction in water needs for the agriculture sector.

Water desalination today, in the GCC and elsewhere, still mainly relies on fossil-fuel-driven processes. On the water supply side, in Qatar, approximately 56% of all water available is produced via desalination of seawater (Kamal et al. 2021), while groundwater abstraction accounts for up to 25% (including abstraction of brackish groundwater that is locally desalinated on farms). The Qatar 30-year water masterplan initiated in 2009 includes major investment in desalination, water infrastructure, and wastewater treatment, with investment in water projects of more than 5 billion USD between 2010 and 2015. On the water demand side, the household sector, the agricultural industry, and the government sectors are the largest water users (Kamal et al. 2021).

Accelerated industrial development coupled with population growth has put pressure on Qatar's water resources. Water demand has grown considerably and continues to rise, and despite public awareness campaigns, Qatar has one of the highest domestic water consumption rates in the world (Mannan et al. 2019), at over 650 L/person/day

(Table 1.2), which is well over twice the European average—however it is important to note that this figure reflects potable water production, not all of which is for domestic use. As such the figure should be viewed as being somewhere between the water production rates and the wastewater treatment rates, which are approximately 220 L/person/day (Ashghal data, not published). As of 2021, desalination plants in Qatar generate 894 million m^3 per year of freshwater.

Thermal desalination today accounts for 65% of the market in Qatar (53% multi-stage flash (MSF), 12% multi-effect distillation (MED)), while 35% is provided by reverse osmosis (RO) (Mabrouk et al. 2017; Son et al. 2020). The country plans to transition toward > 55% of seawater desalination to be achieved using membrane or other sustainable technologies, supporting the transition toward integration of renewables such as solar PV into the energy mix. Although Qatar is classed as water stressed, the residents do not feel any shortage or lack of water due to the extensive desalination. However, desalination plants do have an impact on the environment and on biodiversity, in particular, the discharge of brines to the marine environment. While regulated, discharge of brine, with its often elevated temperature, salinity, and anti-scalant/anti-corrosion chemicals, can lead to ecological stresses by increasing water temperatures, salinities, and heavy metal concentrations, as well as decreasing dissolved oxygen levels. This leads to potentially serious and chronic threats to marine communities following exposure to desalination plant discharges, especially within the zoobenthos, echinodermata, seagrasses, and coral reefs (Sharifinia et al. 2019). However, there is also evidence that a well-managed brine discharge system—particularly from RO which does not suffer the temperature effects of thermal desalination—can have a positive effect particularly on the diversity and abundance of fish species, while having only a mild effect on other species such as polychaetes, bryozoans, sponges, barnacles, molluscs, and crustaceans (Kelaher et al. 2020).

Table 1.2 Water production and consumption in Qatar, 2017–2019 (adapted from (Kahramaa 2020)

Water production and consumption	2017	2018	2019
Average water per capita consumption (m^3/capita/year) [1]based on total water production	224	231	242
Water loss (%)	4.01	3.98	3.97
Water quality/Biological compliance (%)	99.7	99.7	99.4
Water production (Mm^3)	603	634	668
Water production growth (%)	+7.7	+5.1	+5.4
Maximum production (Mm^3/day)	1.78	1.84	1.98

1.8 Wastewater Management and Re-Use

At the same time, it is clear that anthropogenic activities, including the water management in the oil and gas industries, the discharge of industrial and domestic effluents, ballast water and brine wastes, accidental spillage of oil and radioactive wastes, and the modification of coastal structures, all have an impact on the marine biodiversity (Chitrakar et al. 2019). Produced water from the oil and gas industry, with the appropriate treatment, also has the potential to contribute in a beneficial manner to the water balance for the country and could supplement the TSE supply while also reducing discharge or ground injection.

Under QNV2030 and outlined in further details in the National Development Strategies (QNDS-1 and 2), Qatar has set an ambitious target for wastewater treatment and re-use. Qatar has continuously strived to upgrade its wastewater treatment plants (WWTPs) to produce treated effluents suitable for purposes such as landscaping, fodder crops irrigation, district cooling, and construction industry use. The vast majority of WWTPs in Qatar treat municipal wastewater at the very least secondary treatment status, with all of the larger plants including advanced and tertiary treatment. This has reduced the usage of desalinated freshwater for landscaping activities, although to date the quality of the irrigation water, in terms of non-regulated compounds such as micro-organic pollutants (pharmaceuticals, pesticides, etc.) which are sometimes not well removed in wastewater treatment plants, is not well studied. Extensive landscaping and beautification activities particularly around the greater Doha area are expected to have a significant impact on the biodiversity in the area, although this has not been well studied (Ministry of Environment 2014). Urban landscaping in particular requires further study to investigate the effectiveness of the current practices on biodiversity with a focus on irrigation levels, water retention, soil quality, and landscaping practices (Richer et al. 2016). Focus on native biodiversity can be encouraged with sustainable planting, with non-native trees and shrubs and planting of lawns often being seen as problematic (Lerman and Warren 2011; Rodriguezet al. 2017; Knudsen et al. 2022).

1.9 Recommendations

- **Encourage Geological Research**

  Geology is a fascinating and important topic of study, and encouraging geology research in Qatar may lead to a greater knowledge of the country's geology and natural resources, as well as a greater appreciation for the value of geosystem services. These are some ideas for promoting geological research in Qatar:

  1. Establish geological research centers: Establishing up geology research institutes in Qatar can give a platform for scholars to interact and undertake studies on various elements of geology in the nation. These centers can provide researchers

with cutting-edge technology and resources as well as financing for research initiatives. Partner with universities: partnering with universities in Qatar and other countries can facilitate research collaborations and provide access to expert faculty members and students. Universities can also provide funding opportunities for research projects and scholarships for students interested in pursuing geology research in Qatar.

2. Hold workshops and conferences on geology in Qatar: Holding workshops and conferences on geology in Qatar may bring together scholars, specialists, and professionals from many sectors to exchange ideas, present their research, and cooperate on future initiatives. These activities can also assist to raise awareness of the significance of geological research in Qatar.

3. Encourage students to seek postgraduate degrees and professions in geology by supporting graduate research programs. Scholarships, scholarships, and research assistantships can give students with the tools they need to do research and contribute to the growth of geology in Qatar.

4. Encourage international collaborations: Encouraging international relationships with research institutions can give possibilities for Qatari academics to work on global geological research initiatives, have access to international resources, and exchange information and skills.

Ultimately, encouraging geological research in Qatar can help to improve understanding of the country's natural resources, geological risks, and environmental conditions. By investing in research and collaborations, Qatar may become a regional and global leader in geological research.

- **Promote Understanding the Past Sea-Level Changes**

Knowing previous sea-level fluctuations in Qatar is significant for a variety of reasons. For starters, it may help us better understand the geology and environmental past of the area, as well as how it has evolved through time. Second, it can shed light on how Qatar's existing shoreline and terrain came to be, as well as how they could change in the future. Third, it can assist us in planning for the possible effects of future sea-level rise and climate change on the region.

Geological evidence such as sediment layers, fossils, and rock formations can be used to understand previous sea-level fluctuations in Qatar. Researchers can recreate the history of sea-level oscillations over thousands of years by examining these geological characteristics. For example, researchers have discovered evidence of marine fossils in layers several kilometers inland from the current beach, showing that sea level was previously significantly higher than it is now.

Another possibility would be to use satellite and remote sensing data to track present sea-level changes and forecast future trends. This can help us identify locations at danger of floods or erosion as sea levels rise, and devise measures to mitigate these risks.

Understanding historical sea-level fluctuations in Qatar is a crucial component of researching the region's geological and environmental history, and it can give vital insights into how it will evolve in the future.

- **Improve Water Management Strategies**

Qatar has built a robust water strategy which will impact directly on the biodiversity for the country upon implementation. In particular, groundwater abstraction reduction is crucial and proper metering and reporting with an aggressive timeline and a robust and transparent enforcement strategy needs to be implemented.

In addition, the brine management from agriculture and municipal (seawater) desalination must be carefully assessed for biodiversity impact, both terrestrial and marine. Current groundwater brine practices—land disposal and evaporative ponds—have an impact on biodiversity that should be assessed. There is a lack of studies in this area. The impact of seawater brine on the proposed marine protective areas also needs to be studied in light of the potential need to amend discharge regulations to promote biodiversity recovery.

Irrigation for urban landscaping and urban farming practices, including the 10 million trees project, should be studied and assessed for their impact on biodiversity. Included is the future implementation of bio-sludge for fertilization should it start to be allowed in Qatar along with expansion of TSE or treated industry water usage for urban irrigation.

- **Research on Environmental Contamination from the Water Sector**

There is a lack of investigative studies on the presence of contaminants at key sites in Qatar, such as the coastal mangroves. The dataset and potential need for remediation needs to be built. There is also a lack of information on agricultural and urban pesticide usage, its impact on groundwater, and impact on biodiversity. Increasing TSE usage and irrigation needs to be studied for its impact on the proliferation of contaminants of emerging concern in the environment such as antimicrobial resistance, pharmaceuticals and personal care products, plasticizers, and micro/nano-plastics, which can have an impact on both biodiversity and also environmental and human health.

References

Abdulrahman SA (2020) Water shortage in GCC countries: transferring water from Iraqi Kurdistan Region. Int J Environ Stud 77(2):191–207. https://doi.org/10.1080/00207233.2019.1690335

Abu-Zeid MM (1991) Lithostratigraphy and framework of sedimentation of the subsurface Paleogene succession in northern Qatar, Arabian Gulf. Neues Jahrbuch für Geologie und Paläontologie-Monatshefte, pp 191–204

Abdelsamad R, Disi ZA, Abu-Dieyeh M, Al-Ghouti MA, Zouari N (2022) Evidencing the role of carbonic anhydrase in the formation of carbonate minerals by bacterial strains isolated from extreme environments in Qatar. Heliyon 8(10):e11151

Ahmad AY, Al-Ghouti MA, Khraisheh M, Zouari N (2020) Hydrogeochemical characterization and quality evaluation of groundwater suitability for domestic and agricultural uses in the State of Qatar. Groundw Sustain Dev 11:100467. https://doi.org/10.1016/j.gsd.2020.100467

Al-Ansari EMAS, Husrevoglu YS, Yigiterhan O, Youssef N, Al-Maslamani IA, Abdel-Moati MA, Al-Mohamedi AJ, Aboobacker VM, Vethamony P (2022) Seasonal variability of hydrography off the east coast of Qatar, central Arabian Gulf. Arab J Geosci 15:1659. https://doi.org/10.1007/s12517-022-10927-4

Al-Hajari SA (1990) Geology of the tertiary and its influence on the aquifer system of Qatar and eastern Arabia. University of South Carolina

Ali Akbar A (1995) Application of remote sensing methods for discrimination of surficial sand types in Qatar Peninsula, the Arabian Gulf. University of Southampton

Al-Khayat JA, Vethamony P, Nanajkar M (2021) Molluscan diversity influenced by mangrove habitat in the Khors of Qatar. Wetlands 41(4):45. https://doi.org/10.1007/s13157-021-01441-6

Al-Thani RF, Yasseen BT (2021) Microbial ecology of Qatar, the Arabian Gulf: Possible roles of microorganisms. Front Mar Sci 8. https://doi.org/10.3389/fmars.2021.697269

Alsharhan A et al (2001) Hydrogeology of an arid region: the Arabian Gulf and adjoining areas. Elsevier

Ashour M (1987) Surficial deposits of Qatar Peninsula. Geological Society, London, Special Publications, vol 35, issue 1, pp 361–367

Baalousha HM (2016a) Development of a groundwater flow model for the highly parameterized Qatar aquifers. Model Earth Syst Environ 2(2):67

Baalousha HM (2016b) Groundwater vulnerability mapping of Qatar aquifers. J Afr Earth Sc 124:75–93. https://doi.org/10.1016/j.jafrearsci.2016.09.017

Baalousha HM, Ouda OKM (2017) Domestic water demand challenges in Qatar. Arab J Geosci 10(24):537. https://doi.org/10.1007/s12517-017-3330-4

Chitrakar P, Baawain MS, Sana A, Al-Mamun A (2019) Current status of marine pollution and mitigation strategies in arid region: a detailed review. Ocean Sci J 54(3):317–348. https://doi.org/10.1007/s12601-019-0027-5

Darwish MA, Mohtar R (2013) Qatar water challenges. Desalin Water Treat 51(1–3):75–86. https://doi.org/10.1080/19443994.2012.693582

Draidia A, Tareen M, Bayraktar N, Cramer ERA, Chen K-C (2022) Bird communities and the rehabilitation of Al Karaana Lagoons in Qatar. Birds 3(4), Article 4. https://doi.org/10.3390/birds3040022

Embabi N, Ashour M (1993) Barchan dunes in Qatar. J Arid Environ 25(1):49–69

Fairbridge RW (1960) The changing level of the sea. Sci Am 202(5):70–79

Frisk EL et al (2022) The geosystem services concept—what is it and can it support subsurface planning? Ecosyst Serv 58:101493. https://doi.org/10.1016/j.ecoser.2022.101493

Glennie K et al (2002) Quaternary climatic changes over southern Arabia and the Thar Desert, India. Geological Society, London, Special Publications, vol 195, issue 1, pp 301–316

Gray M, Gordon JE (2020) Geodiversity and the '8Gs': a response to Brocx & Semeniuk (2019). Aust J Earth Sci 67(3):437–444. https://doi.org/10.1080/08120099.2020.1722965

Gutiérrez F et al (2014) A review on natural and human-induced geohazards and impacts in karst. Earth Sci Rev 138:61–88

Heggy E et al (2022) Exploring the nature of buried linear features in the Qatar Peninsula: archaeological and paleoclimatic implications. ISPRS J Photogramm Remote Sens 183:210–227

Hussein H, Lambert LA (2020) A rentier state under blockade: Qatar's water-energy-food predicament from energy abundance and food insecurity to a silent water crisis. Water 12(1051):1051. https://doi.org/10.3390/w12041051

Jameson J, Strohmenger CJ (2012) Relative sea-level changes during the Late Pleistocene to Holocene of Qatar: implications for eustasy and tectonics. Search and discovery article, p 50704

Jameson J, Strohmenger CJ (2014) Resolving eustasy from neotectonics in the sea-level history of the Pliocene to Holocene of Qatar. In: IPTC 2014: International petroleum technology conference. European association of geoscientists & engineers

Kaushal SS et al (2021) Freshwater salinization syndrome: from emerging global problem to managing risks. Biogeochemistry 154(2):255–292. https://doi.org/10.1007/s10533-021-00784-w

Kamal A, Al-Ghamdi SG, Koç M (2021) Assessing the impact of water efficiency policies on Qatar's electricity and water sectors. Energies 14(14), Article 4348. https://doi.org/10.3390/en1 4144348

Kahramaa (2020) Kahramaa sustainability report 2018–2019. https://km.qa/Tarsheed/SavingWit hTarsheed/Kahramaa%20sustainability%20report_2020_.pdf

Kelaher BP, Clark GF, Johnston EL, Coleman MA (2020) Effect of desalination discharge on the abundance and diversity of reef fishes. Environ Sci Technol 54(2):735–744. https://doi.org/10. 1021/acs.est.9b03565

Kinsman D (1964) The recent carbonate sediments near Halat el Bahrani, Trucial Coast, Persian Gulf. In: Developments in sedimentology. Elsevier, pp 185–192

Kinsman DJ (1969) Modes of formation, sedimentary associations, and diagnostic features of shallow-water and supratidal evaporites. AAPG Bull 53(4):830–840

Knudsen BT, Stage C, Zandersen M (2022) Interspecies park life: participatory experiments and micro-utopian landscaping to increase urban biodiverse entanglement. Space Cult 25(4):720–742. https://doi.org/10.1177/1206331219863312

Lambeck K (1996) Shoreline reconstructions for the Persian Gulf since the last glacial maximum. Earth Planet Sci Lett 142(1–2):43–57

Lerman SB, Warren PS (2011) The conservation value of residential yards: linking birds and people. Ecol Appl 21(4):1327–1339

Lüning S et al (2018) Hydroclimate in Africa during the medieval climate anomaly. Palaeogeogr Palaeoclimatol Palaeoecol 495:309–322

Mabrouk A, Koç M, Abdala A (2017) Technoeconomic analysis of tri hybrid reverse osmosis–forward osmosis–multi stage flash desalination process. Desalin Water Treat 98:1–15. https:// doi.org/10.5004/dwt.2017.21564

Margat J, van der Gun J (2013) Groundwater around the world: a geographic synopsis (1st ed.). CRC Press

Mannan M, Alhaj M, Mabrouk AN, Al-Ghamdi SG (2019) Examining the life-cycle environmental impacts of desalination: a case study in the State of Qatar. Desalination 452:238–246. https:// doi.org/10.1016/j.desal.2018.11.017

Ministry of Environment (2014) Qatar national biodiversity strategy and action plan 2015–2025. https://www.fao.org/faolex/results/details/en/c/LEX-FAOC180906/

Milne GA et al (2009) Identifying the causes of sea-level change. Nat Geosci 2(7):471–478

Michel S et al (2018) Comparing dune migration measured from remote sensing with sand flux prediction based on weather data and model: a test case in Qatar. Earth Planet Sci Lett 497:12–21

Mossadegh ZK et al (2013) Palaeoecology of well-preserved coral communities in a siliciclastic environment from the Late Pleistocene (MIS 7), Kish Island, Persian Gulf (Iran): the development of low-relief reef frameworks (biostromes) in increasingly restricted environments. Int J Earth Sci 102:545–570

Murray-Wallace CV, Woodroffe CD (2014) Quaternary sea-level changes: a global perspective. Cambridge University Press

Parker AG et al (2006) A record of Holocene climate change from lake geochemical analyses in southeastern Arabia. Quatern Res 66(3):465–476

Planning and Statistics Authority (2018) National development strategy 2. https://www.psa.gov.qa/ en/knowledge/Documents/NDS2Final.pdf

Planning and Statistics Authority (2021) Water statistics in the State of Qatar 2017. https://www. psa.gov.qa/en/statistics/Statistical%20Releases/Environmental/Water/2017/Water-Statistics-2017-EN.pdf

Purser B, Seibold E (1973) The principal environmental factors influencing Holocene sedimentation and diagenesis in the Persian Gulf. In The Persian Gulf. Springer, pp 1–9

Purkis SJ et al (2017) Complex interplay between depositional and petrophysical environments on a Holocene tidal flat (Al Ruwais, Qatar). AAPG annual convention and exhibition

Purkis S et al (2017) Complex interplay between depositional and petrophysical environments in Holocene tidal carbonates (Al Ruwais, Qatar). Sedimentology 64(6):1646–1675

Qatar News Agency (2022) Qatar 2022/One million trees initiative, Qatars contribution to reduce carbon emissions-1. https://www.qna.org.qa/en/newsbulletins/2022-11/18/0038-qatar-2022-one-million-trees-initiative,-qatars-contribution-to-reduce-carbon-emissions-1

Rasheed F et al (2020) Phytoaccumulation of Zn, Pb, and Cd in Conocarpus lancifolius irrigated with wastewater: does physiological response influence heavy metal uptake? Int J Phytorem 22(3):287–294. https://doi.org/10.1080/15226514.2019.1658711

Raza HA et al (2011) Animal and bird trade and hunting in Iraq. Nature Iraq, Sulaimani, Iraq/Ministry of Environment, Baghdad

van Ree CCDF, van Beukering PJH, Boekestijn J (2017) Geosystem services: a hidden link in ecosystem management. Ecosyst Serv 26:58–69. https://doi.org/10.1016/j.ecoser.2017.05.013

Redha A, Al-Hasan R, Afzal M (2021) Synergistic and concentration-dependent toxicity of multiple heavy metals compared with single heavy metals in Conocarpus lancifolius. Environ Sci Pollut Res 28(18):23258–23272. https://doi.org/10.1007/s11356-020-12271-0

Richer R et al (2016) Native plant landscaping and species selection to promote sustainability and biodiversity in Qatar. Qatar Green Build Conf Proc 2016(3):41. https://doi.org/10.5339/qproc.2016.qgbc.41

Rivers JM et al (2019a) The depositional history of near-surface Qatar aquifer rocks and its impact on matrix flow and storage properties. Arab J Geosci 12(12):380

Rivers JM et al (2019b) The geochemistry of Qatar coastal waters and its impact on carbonate sediment chemistry and early marine diagenesis. Sed Geol 89(4):293–309

Rivers JM, Larson KP (2018) The Cenozoic kinematics of Qatar: evidence for high-angle faulting along the Dukhan 'anticline.' Mar Pet Geol 92:953–961

Rivers JM et al (2020) Mixed siliciclastic-carbonate-evaporite sedimentation in an arid eolian landscape: the Khor Al Adaid tide-dominated coastal embayment, Qatar. Sediment Geol 408:105730

Rivers JM et al (2021) Cenozoic coastal carbonate deposits of Qatar: evidence for dolomite preservation bias in highly arid systems. Sedimentology 68(2):771–787

Rodriguez SL, Peterson MN, Moorman CJ (2017) Does education influence wildlife-friendly landscaping preferences? Urban Ecosyst 20(2):489–496. https://doi.org/10.1007/s11252-016-0609-2

Rose J (2010a) Quaternary climates: a perspective for global warming. Proc Geol Assoc 121(3):334–341

Rose JI (2010b) New light on human prehistory in the Arabo-Persian Gulf Oasis. Curr Anthropol 51(6):849–883

Robinson DP et al (2017) Some like it hot: repeat migration and residency of whale sharks within an extreme natural environment. PLoS ONE 12(9):e0185360. https://doi.org/10.1371/journal.pone.0185360

Sharifinia M, Afshari Bahmanbeigloo Z, Smith WO Jr, Yap CK, Keshavarzifard M (2019) Prevention is better than cure: Persian Gulf biodiversity vulnerability to the impacts of desalination plants. Glob Change Biol 25(12):4022–4033. https://doi.org/10.1111/gcb.14808

Shinn EA (1969) Submarine lithification of Holocene carbonate sediments in the Persian Gulf. Sedimentology 12(1–2):109–144

Sowers J, Vengosh A, Weinthal E (2011) Climate change, water resources, and the politics of adaptation in the Middle East and North Africa. Clim Change 104(3–4):599–627

Son HS, Shahzad MW, Ghaffour N, Ng KC (2020) Pilot studies on synergetic impacts of energy utilization in hybrid desalination system: Multi-effect distillation and adsorption cycle (MED-AD). Desalination 477:114266. https://doi.org/10.1016/j.desal.2019.114266

Stevens T et al (2014) Eustatic control of late quaternary sea-level change in the Arabian/Persian Gulf. Quatern Res 82(1):175–184

Strohmenger CJ, Jameson J (2018) Gypsum stromatolites from Sawda Nathil: relicts from a southern coastline of Qatar. Carbonates Evaporites 33(2):169–186

Tawalbeh M, Javed RMN, Al-Othman A, Almomani F (2023) Salinity gradient solar ponds hybrid systems for power generation and water desalination. Energy Conv Manag 289:117180. https://doi.org/10.1016/j.enconman.2023.117180

The Peninsula (2022) Kahramaa completes strategic projects, eyes implementing 2023–2027 plan. https://thepeninsulaqatar.com/article/01/01/2023/kahramaa-completes-strategic-projects-eyes-implementing-2023-2027-plan

The Peninsula (2022) PM plants millionth tree as part of 'Plant Million Trees' initiative. https://thepeninsulaqatar.com/article/19/12/2022/pm-plants-millionth-tree-as-part-of-plant-million-trees-initiative

Williams AH, Walkden GM (2002) Late quaternary highstand deposits of the southern Arabian Gulf: A record of sea-level and climate change. Geological Society, London, Special Publications, vol 195, issue 1, pp 371–386

Williams AH (1999) Glacioeustatic cyclicity in Quaternary carbonates of the southern Arabian Gulf: Sedimentology, sequence stratigraphy, paleoenvironments and climatic record. University of Aberdeen

Williams AH, Walkden GM (2001) Carbonate eolianites from a eustatically influenced ramp-like setting: The Quaternary of the southern Arabian Gulf

Wood WW et al (2012) Rapid late Pleistocene/Holocene uplift and coastal evolution of the southern Arabian (Persian) Gulf. Quatern Res 77(2):215–220

World Resources Institute (2019) Aqueduct. https://www.wri.org/aqueduct

Open Access This chapter is licensed under the terms of the Creative Commons Attribution 4.0 International License (http://creativecommons.org/licenses/by/4.0/), which permits use, sharing, adaptation, distribution and reproduction in any medium or format, as long as you give appropriate credit to the original author(s) and the source, provide a link to the Creative Commons license and indicate if changes were made.

The images or other third party material in this chapter are included in the chapter's Creative Commons license, unless indicated otherwise in a credit line to the material. If material is not included in the chapter's Creative Commons license and your intended use is not permitted by statutory regulation or exceeds the permitted use, you will need to obtain permission directly from the copyright holder.

Chapter 2
Climate, Climate Modeling, and Forecasting

K. K. Kanikicharla, P. K. Bal, N. Al-Mohannadi, A. S. Al-Ansari,
A. M. Al-Mannai, A. Amato, and A. D. Chatziefthimiou

Observed and expected future changes in the climate over qatar and adjoining region.

Abstract This study assesses future climate change in Qatar under three IPCC greenhouse gas emission scenarios: SSP1-2.6, SSP2-4.5, and SSP5-8.5. Using multi-model simulations made under IPCC's CMIP6, projected changes in seasonal and annual mean temperature, heat stress index, and rainfall are analyzed for three future periods—the 2030s, 2050s, and 2080s—relative to the 1986–2005 baseline. Average temperature increases over Qatar are projected to range from 1.4 to 1.5 °C in the 2030s, 1.8 to 2.6 °C in the 2050s, and 2.0 to 5.3 °C by the 2080s, depending on the emission scenario. The heat stress index is expected to rise by 13.3 °C by the century's end, shifting conditions from "extreme caution" to "danger" levels during summer. Additionally, under the high-emission scenario (SSP5-8.5), annual rainfall could

K. K. Kanikicharla (✉) · P. K. Bal · N. Al-Mohannadi · A. S. Al-Ansari · A. M. Al-Mannai
Qatar Meteorology Department, Civil Aviation Authority, Doha, Qatar
e-mail: krishnakumar.kanikicharla@gmail.com

P. K. Bal
e-mail: Prasanta.Bal@caa.gov.qa

N. Al-Mohannadi
e-mail: naser.almohanadi@caa.gov.qa

A. S. Al-Ansari
e-mail: mjmmarri@mme.gov.qa

A. M. Al-Mannai
e-mail: abdulla.almannai@caa.gov.qa

A. Amato · A. D. Chatziefthimiou
Earthna, Center for a Sustainable Future, Qatar Foundation, Doha, Qatar
e-mail: aamato1uk@mac.com

A. D. Chatziefthimiou
e-mail: a.d.chatziefthimiou@gmail.com

© The Author(s) 2026

A. D. Chatziefthimiou et al. (eds.), *Pathways to Nature Conservation and Resilience in Hot and Arid Lands*, Advances in Geographical and Environmental Sciences, https://doi.org/10.1007/978-3-032-11046-6_2

increase by 44% by 2100. These projected changes in temperature, rainfall, and heat stress may have significant ecological and environmental management implications for Qatar and the surrounding region.

2.1 Introduction

The State of Qatar is dominated by subtropical dry, hot desert climate with low annual rainfall, very high temperatures in summer, and a big difference between maximum and minimum temperatures, especially in the inland areas. The coastal areas are influenced by the Arabian Gulf, and have lower maximum, but higher minimum temperatures and a higher moisture percentage in the air. Qatar with its very hot and dry climate, fragile ecosystems, and high dependence on energy intensive livelihood is more vulnerable to any adverse impacts of climate change. In view of this, 60 years of Meteorological Observations in Qatar and more than a century-long simulations from a large suite of global climate models made under WMO's Intergovernmental Panel on Climate Change (IPCC) and World Climate Research Program (WCRP) have been analyzed for assessing the present and future climate change for Qatar and adjacent regions.

The recently released Sixth Assessment Report (AR6) of the IPCC Working Group I (WGI) titled "Climate Change 2021: The Physical Science Basis" states that the "recent changes in the climate are widespread, rapid, and intensifying, and unprecedented in thousands of years. It is unequivocal that human influence has warmed the atmosphere, ocean and land. Widespread and rapid changes in the atmosphere, ocean, cryosphere and biosphere have occurred." It also indicates that "continued emissions of greenhouse gases will cause further warming and changes in all components of the climate system." Using the current generation of global coupled ocean-atmospheric models (CMIP6), the IPCC estimates that the likely increase in global mean temperatures by the end of twenty-first century range from 1.4 °C to 4.4 °C for different greenhouse gas emission scenarios in the future. It is also expected that global warming could manifest differently on a regional scale where the expected future changes might differ from place to place around the world. The IPCC predicts that Middle East and North African (MENA) region's climate difficulties may increase during the next century (Masson-Delmotte et al. 2021) and parts of the MENA region may be uninhabitable by end of the century, if the present trends continue (Pal et al. 2016). Therefore, having a good understanding on the expected future climate under different greenhouse gas (GHG) emission scenarios is vital for planning suitable mitigation and adaptation strategies. Hence, we try here to present observed changes in temperature and rainfall and possible future changes in the weather and climate over Qatar and adjacent regions.

2.2 Observed Changes

Climatological normals of temperature, rainfall, relative humidity, and wind speed computed based on 30 years (1987–2016) of meteorological data recorded at Doha International Airport are given in Table 2.1. Table also provides information on the frequency of different weather events that occur in Qatar such as fog, thunderstorm, Shamals, and sandstorms. Shamals are strong wind events that are experienced in this part of the world. Days with northerly to northwesterly winds exceeding a speed of 17 knots (31.5 km/h) and persisting at least for 3 h are considered as Shamal days. The duration of data used to compute these frequencies is stated in the table. As can be seen from the table, June to September is the peak summer season in Qatar with very hot temperatures and very low rainfall. During summers, the temperatures sometimes soar to about 50 °C. It is extremely hot during the day, while the nights are warm and not as extreme as the day conditions. Humidity is also quite high during the summer months. Rainfall occurs mostly during the winter months in sudden, short, and heavy cloudbursts. Over this region, the rainfall is contributed by thunderstorm activity due to occasional extra-tropical (synoptic scale) systems approaching the region from the west. North to northwesterly winds dominate Doha in all seasons (Fig. 2.1). It is also true with other stations in Qatar.

Before examining future changes in the climate over Qatar and the adjacent region, it is useful to understand how climate has been changing during the past 60 years since systematic observations began in Qatar. Doha International Airport (DIA) has been recording meteorological parameters since 1962 while observations began much later in other stations in Qatar. Hence, we use temperature and rainfall data at DIA to assess the long-term trends. During the last 60 years (1962–2021), Doha has experienced a significant warming with mean annual temperature increasing by 3.1 °C/60 yrs. However, the night-time temperatures are increasing more rapidly (5.2 °C/60 yrs) than the day-time temperatures (1.5 °C/60 yrs) leading to a significant reduction in the day–night temperature difference (−3.7 °C/60 yrs) (Fig. 2.2). Increased greenhouse gases and rapid urbanization are responsible for the pronounced temperature trends observed in Doha (Mohammed et al., 2021), and particularly in the night-time temperatures. However, it may be stated here that the temperature trends in other non-urban places in Qatar are not as pronounced (Fig. 2.3). On the other hand, the annual total rainfall showed a slight decreasing trend during the same period (1962–2021) (Fig. 2.4).

2.3 Future Projections

Estimates of expected future changes in the annual mean temperature and rainfall for Qatar and adjacent region are made using climate simulations from a large suite of global climate models made under IPCC's Coupled Model Intercomparison Project

Table 2.1 Monthly climatological normals of different meteorological variables and the frequency of occurrence of significant weather events at Doha international airport

S.No	Doha international airport Parameter	Period	JAN	FEB	MAR	APR	MAY	JUN	JUL	AUG	SEP	OCT	NOV	DEC	ANNUAL
1	Mean temperature (°C)	1987–2016	17.8	19.0	22.2	27.0	32.7	35.1	36.1	35.4	33.5	30.2	25.1	20.3	27.9
2	Mean maximum temperature (°C)	1987–2016	22.2	23.7	27.5	33.0	39.4	41.9	42.3	41.2	39.1	35.5	29.7	24.7	33.4
3	Mean minimum temperature (°C)	1987–2016	14.2	15.3	18.0	22.4	27.5	29.8	31.4	31.2	29.1	25.8	21.3	16.7	23.6
4	Mean rainfall (mm)	1987–2016	11.4	13.5	19.3	4.2	1.1	0.0	0.0	0.0	0.0	0.6	9.3	17.1	76.5
5	Mean relative humidity (%)	1987–2016	79	77	70	62	52	51	56	65	66	69	73	78	67
6	Mean wind speed (kt)	1987–2015	7.2	8.3	8.5	8.1	8.6	8.9	7.8	6.8	6.5	6.1	6.8	6.7	7.5
7	Mean number of days with fog	1962–2016	3.0	2.2	0.8	0.2	0.3	0.5	0.5	0.4	1.1	1.7	1.1	2.6	14.4
8	Mean number of days with thunderstorm	1962–2016	0.5	1.1	2.0	1.5	0.5	0.0	0.0	0.0	0.1	0.4	0.7	0.7	7.5
9	Mean number of days with shamal winds	1975–2004	2.7	4.2	5.6	5.9	8.0	10.9	6.7	5.0	1.6	1.9	3.0	1.7	57.2
10	Mean number of days with sand storm/thick dust haza (visibility < 1000 m)	1962–2016	0.3	0.2	0.4	0.5	0.3	0.8	0.9	0.3	0.1	0.1	0.2	0.1	4.2

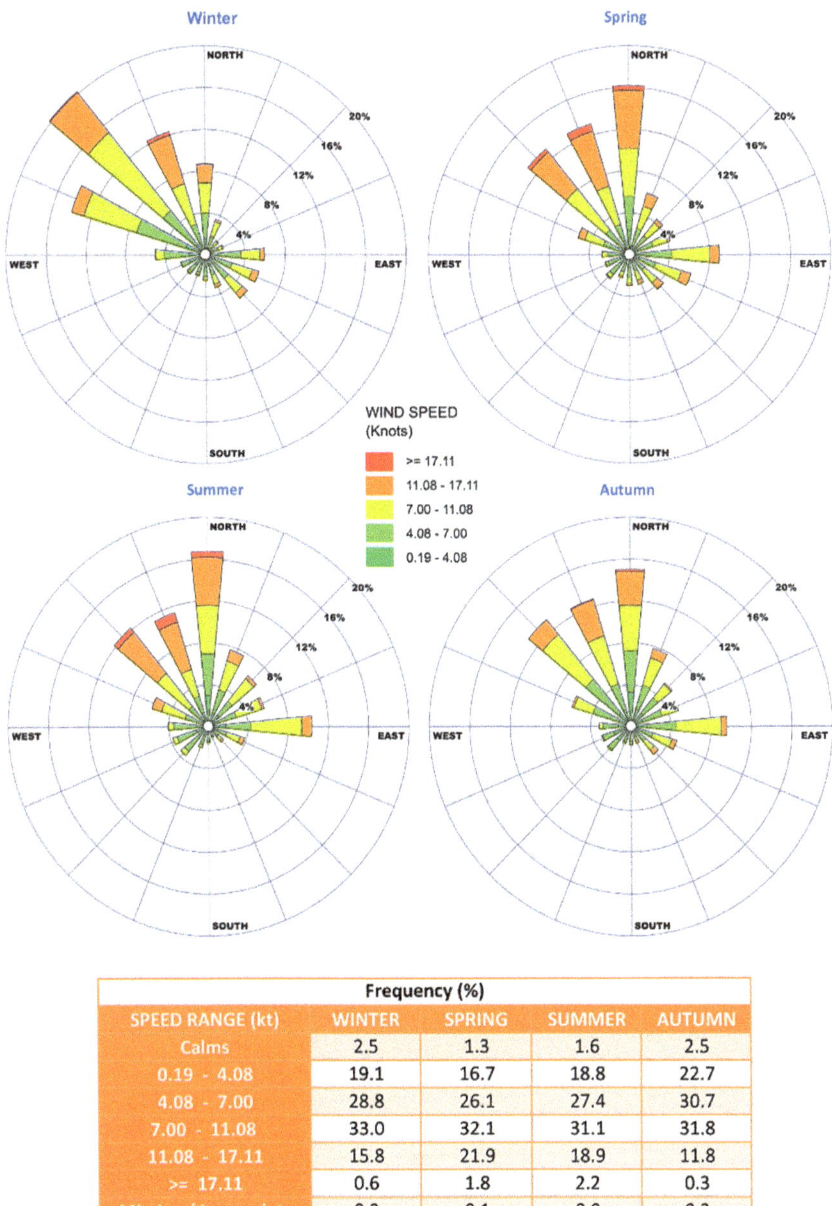

Frequency (%)				
SPEED RANGE (kt)	WINTER	SPRING	SUMMER	AUTUMN
Calms	2.5	1.3	1.6	2.5
0.19 - 4.08	19.1	16.7	18.8	22.7
4.08 - 7.00	28.8	26.1	27.4	30.7
7.00 - 11.08	33.0	32.1	31.1	31.8
11.08 - 17.11	15.8	21.9	18.9	11.8
>= 17.11	0.6	1.8	2.2	0.3
Missing / Incomplete	0.2	0.1	0.0	0.2

Fig. 2.1 Seasonal wind-rose diagrams made using hourly wind speed and direction data at Doha International Airport during 2007–2016

K. K. Kanikicharla et al.

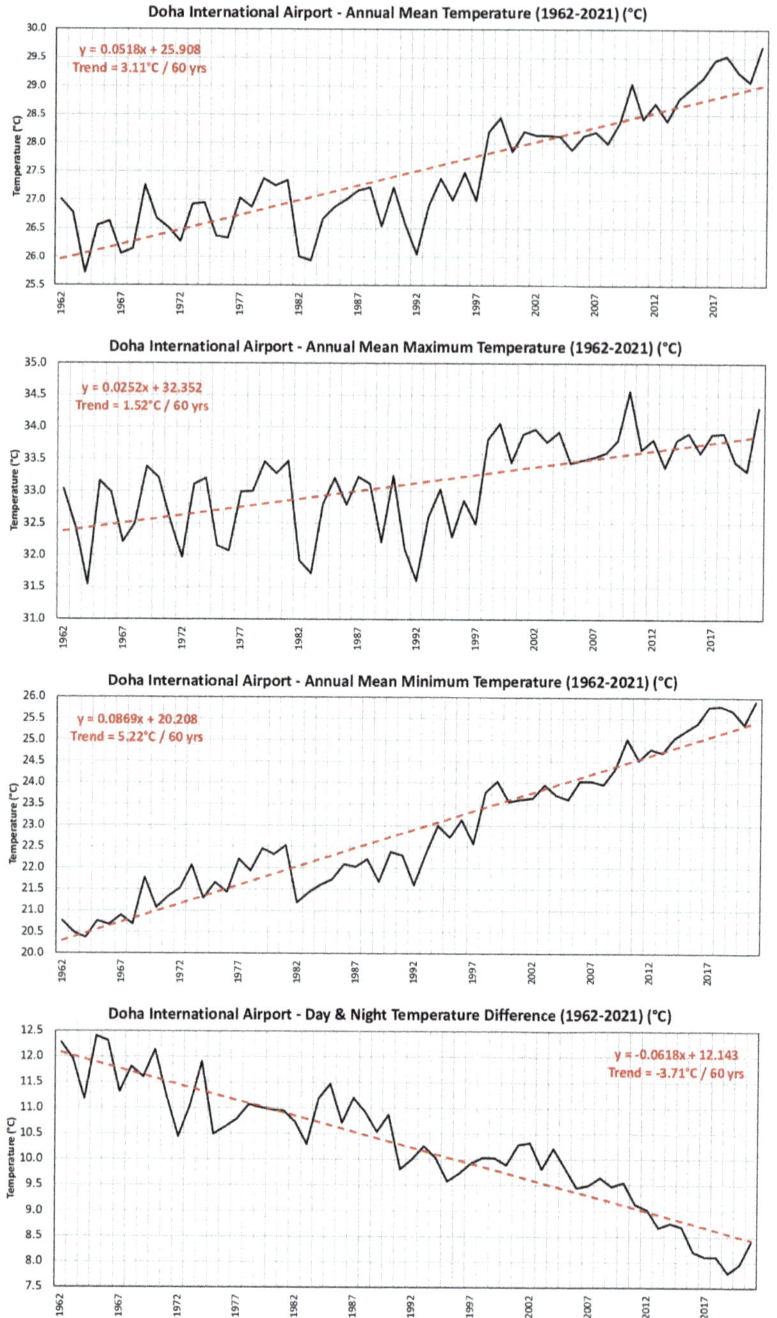

Fig. 2.2 Time series of annual (1) mean temperature, (2) maximum temperature, (3) minimum temperature, and (4) their difference during 1962–2021 at Doha International Airport

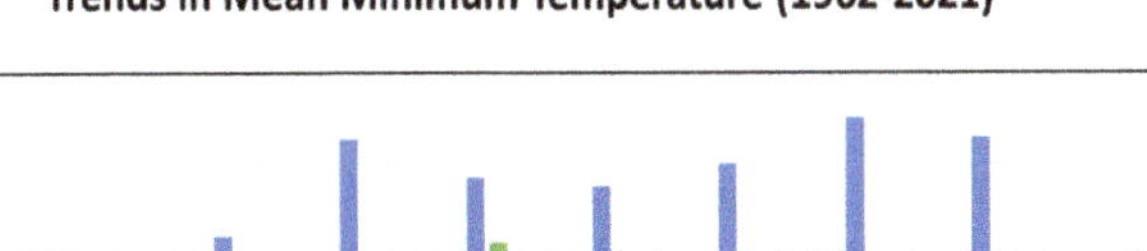

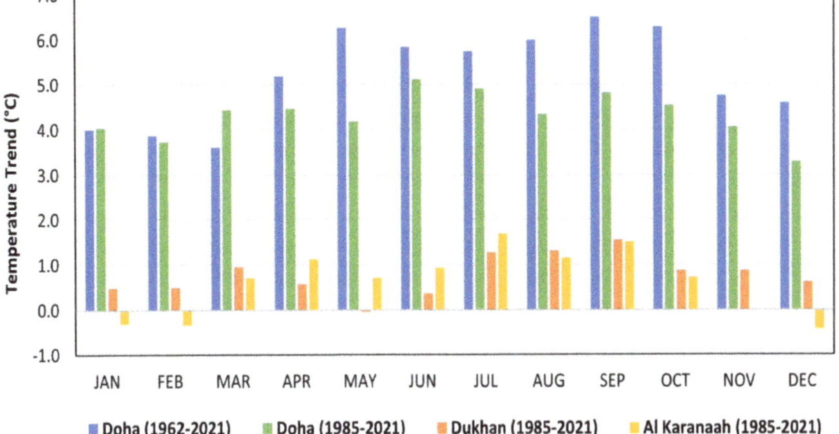

Fig. 2.3 Trends in monthly mean minimum temperature at Doha, Dukhan, and Al Karanaah Stations in Qatar during the periods 1962–2021 and 1985–2021

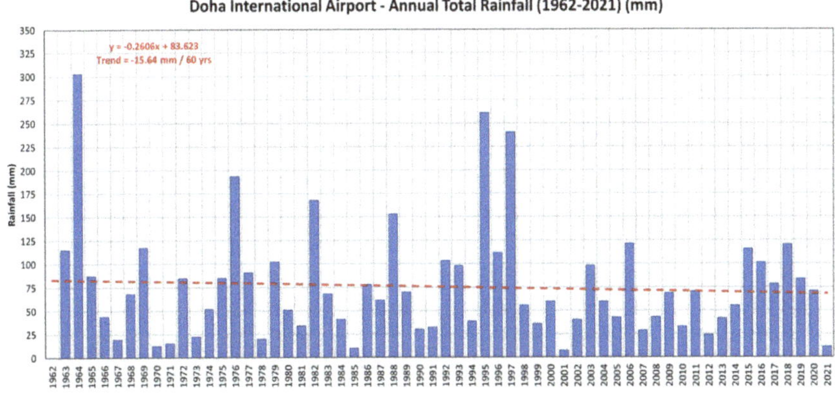

Fig. 2.4 Time series of annual total rainfall (mm) during 1962–2021 at doha international airport

(CMIP6) for three different greenhouse gas emission scenarios (SSP1-2.6, SSP2-4.5, and SSP5-8.5) known as Shared Socio-economic Pathways (SSPs) (Riahi et al. 2017). While SSP1 scenarios represent sustainability pathways, the SSP2 scenarios are akin to middle of the road pathways. The SSP5 scenarios follow fossil-fueled development pathway. These model simulations are therefore based on updated emission pathways, greater resolution, and enhanced dynamical processes that accurately represent earth's climate system (O'Neill et al. 2017; Tebaldi et al. 2021). Some recent studies have demonstrated the robustness of the CMIP6 models over the previous generation of climate models under CMIP5 in Asia (Zamani et al. 2020; Chen et al.

2021; Hamed et al. 2022). In this study, simulation data (1986–2100) of temperature, rainfall, and humidity on daily and monthly scales from 19 different CMIP6 global climate models (GCMs) have been utilized for estimating the future climate change over the Arabian Peninsula including the GCC and Qatar. Table 2.2 provides details on the model simulations used and the horizontal resolution of each model. Models are chosen as per the data availability. The horizontal resolution of these models ranges from ~100 × 100km to ~300 × 300km. Given this asymmetry in the resolution, first the data from different models have been converted to a uniform grid size of 0.5° × 0.5° (~50kmx50km). Some models have more than one simulation. However, we have considered only one simulation for each model. Future projections of temperature, rainfall, and heat stress index are presented by taking the mean of all the 19 models. The simulation data has been downloaded from the CMIP6 data portal via https://esgf-node.llnl.gov/. Future changes in different mean quantities are presented here for three time slices of near term (2021–2040, 2030s), mid-term (2041–2060, 2050s), and long term (2081–2100, 2080s) with respect to the baseline period of 1986–2005.

2.3.1 Temperature

While the IPCC assessments provide a general idea of global and continental scale expected future climate change, it is very important to assess the expected future change on a regional scale to generate policy relevant information. This can be understood from the fact that Doha City has warmed much more than that of the global mean temperature in the last 60 years. In view of this, we present here the expected future changes in annual mean temperatures for three different time slices and three different GHG emission scenarios as explained above. Figure 2.5 shows the spatial maps of changes in annual mean temperature (°C), computed by taking the mean of all the 19 global models for three different time slices centered around 2030s, 2050s, and 2080s, compared to the baseline period 1986–2005 over the broader Arabian Peninsula region including Qatar. It is found that projected temperature changes over Qatar as a whole are in the range of 1.4 °C to 1.5 °C for the 2030s, 1.8 °C to 2.6 °C for 2050s, and 2.0 °C to 5.3 °C for 2080s under three different (low-to-high emission) scenarios (Table 2.3). These changes in mean annual temperature are higher than the global mean temperature changes projected under different GHG emission scenarios. Table 2.3 also provides temperature values averaged over Qatar for each month to give an idea on how future changes are likely to be in different months. From these monthly temperature values, it appears that the temperature trends are expected to be uniform in all the months.

Table 2.2 Details of the CMIP6 global climate Models used in this study

Model name	Horizontal resolution (degrees)	Modeling center (group)
AWI-CM-1–1-MR	1.87 × 1.86	Alfred Wegener Institute (AWI)
FGOALS-g3	2.0 × 2.02	Institute of Atmospheric Physics, Chinese Academy of Sciences (CAS)
CanESM5	2.81 × 2.79	Canadian Centre for Climate Modelling and Analysis (CCCma)
CNRM-ESM2-1	1.40 × 1.40	Centre National de Recherches Meteorologiques (CNRM)
ACCESS-CM2	1.87 × 1.25	Commonwealth Scientific and Industrial Research Organization (CSIRO)
ACCESS-ESM1-5	1.87 × 1.25	Commonwealth Scientific and Industrial Research Organization (CSIRO)
EC-Earth3-Veg	0.70 × 0.70	Commonwealth Scientific and Industrial Research Organization, ARC Centre of Excellence for Climate System Science (CSIRO-ARCCSS) EC-Earth-Consortium
INM-CM4-8	2.0 × 1.50	Institute of Numerical Mathematics (INM)
INM-CM5-0	2.0 × 1.50	Institute of Numerical Mathematics (INM)
IPSL-CM5A2-INCA	3.75 × 1.89	Institut Pierre Simon Laplace (IPSL)
MIROC6	1.40 × 1.40	Japan Agency for Marine Earth Science and Technology, Atmosphere and Ocean Research Institute (MIROC)
HadGEM3-GC31-LL	1.87 × 1.25	Met Office Hadley Centre (MOHC)
KIOST-ESM	1.89 × 1.87	Korea Institute of Ocean Science and Technology (KIOST)
MPI-ESM1-2-LR	1.87 × 1.86	Max Planck Institute for Meteorology (MPI-M)
MRI-ESM2-0	1.12 × 1.12	Meteorological Research Institute (MRI)
NorESM2-LM	2.5 × 1.89	Norwegian Climate Centre (NCC)
NorESM2-MM	1.25 × 0.94	Norwegian Climate Centre (NCC)

(continued)

Table 2.2 (continued)

Model name	Horizontal resolution (degrees)	Modeling center (group)
KACE-1–0-G	1.87 × 1.25	National Institute of Meteorological Sciences—Korea Met. Administration (NIMS-KMA)
GFDL-ESM4	1.25 × 1.0	NOAA Geophysical Fluid Dynamics Laboratory (NOAA-GFDL)

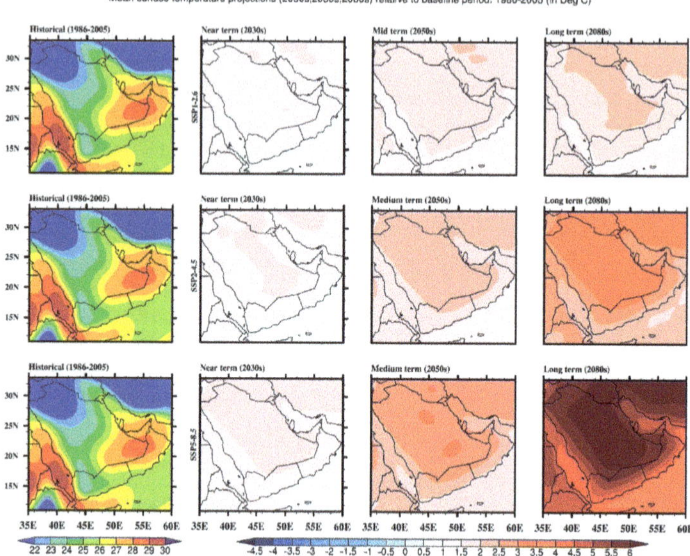

Fig. 2.5 Projected changes in mean annual surface air temperature for three future time slices (2030s, 2021–2040, 2050s, 2041–2060 and 2080s, 2081–2100) under three GHG emission scenarios—SSP1-2.6 (upper panel), SSP2-4.5 (middle panel), and SSP5-8.5 (lower panel). The changes (°C) are relative to the baseline period of 1986–2005 and mean of 19 models

2.3.2 Heat Stress Index

A heat stress index (HSI) (Steadman 1979) which takes into consideration mean temperature and relative humidity is used to assess the impact of climate change on human comfort and productivity in the future. Depending on the index exceeding certain threshold values, it is categorized into 1) caution (HSI: 27 °C to 32 °C), 2) extreme caution (HSI: 32 °C to 41 °C), 3) danger (HSI: 41 °C to 54 °C), and 4) extreme danger (HSI: >54 °C). Daily mean temperature and humidity data from 19 different CMIP6 global climate models are utilized to compute the heat stress index. Figure 2.6 shows the average summer (broadly defined season between May to October is considered here as summer) heat index over the GCC and its surrounding

Table 2.3 Monthly and annual surface mean temperature (°C) averaged over Qatar for the baseline and three different future time periods corresponding to two emission scenarios of SSP1-2.6 and SSP5-8.5

Monthly mean temperature (°C)	1986–2005	Near term (2021–2040)		Medium term (2041–2060)		Long term (2081–2100)	
Month	Historical	SSP1-2.6	SSP5-8.5	SSP1-2.6	SSP5-8.5	SSP1-2.6	SSP5-8.5
January	16.6	17.8	18.0	18.1	19.1	18.3	21.7
February	18.0	19.1	19.3	19.5	20.3	19.8	22.8
March	21.5	22.7	22.7	23.1	23.7	23.4	26.1
April	25.8	27.0	27.0	27.5	28.0	27.7	30.7
May	30.5	31.9	31.9	32.4	33.0	32.5	35.6
June	34.3	36.0	35.9	36.4	37.1	36.5	39.8
July	36.0	37.6	37.6	38.1	38.7	38.3	41.5
August	35.8	37.5	37.5	38.0	38.7	38.2	41.4
September	33.4	35.0	35.1	35.4	36.3	35.6	39.2
October	29.4	30.7	31.0	31.2	32.2	31.3	35.1
November	23.9	25.1	25.6	25.6	26.8	25.8	29.6
December	18.7	19.8	20.2	20.3	21.4	20.3	24.2
Annual mean temperature	**27.0**	**28.4**	**28.5**	**28.8**	**29.6**	**29.0**	**32.3**

region for the baseline period and for three future time periods representing 2030s, 2050s, and 2080s under two different emission scenarios (SSP1-2.6 and SSP5-8.5). Higher heat stress in the future is going to be a serious problem in the entire GCC region. It is found that heat stress index is likely to increase over Qatar by about 6.6 °C under low emission scenario (SSP1-2.6) and as much as 13.3 °C under high-emission scenario (SSP5-8.5) by end of the century (2080s) with respect to the baseline period (1986–2005). Figure 2.7 shows the average number of days of heat index falling under the category of "danger" during May to October (184 days) for the baseline and three future time periods under two different emission scenarios of SSP1-2.6 and SSP5-8.5. As can be seen in the figure, the number of days under "danger" category increases from less than 20 in the baseline to more than 75 in SSP5-8.5 by end of the century. Figure 2.8 shows 20-year average daily heat stress index values during May to October for the baseline and for the long-term period (2081–2100) corresponding to SSP1-2.6 and SSP5-8.5 emission scenarios. These daily values are averaged over Qatar. As can be seen in the figure, the average heat stress values in the baseline are within "extreme caution" category during the entire season. On the other hand, heat stress values of SSP1-2.6 enter "danger" category during the peak summer months. However, average heat stress values shift to "danger" category during the entire summer season in the future under SSP5-8.5 scenarios. One can also see temporal expansion of severe heat stress conditions beyond June to September on both sides besides the drastic increase in the intensity of the heat stress compared to the baseline. This can have serious implications for human comfort and on ecology in general.

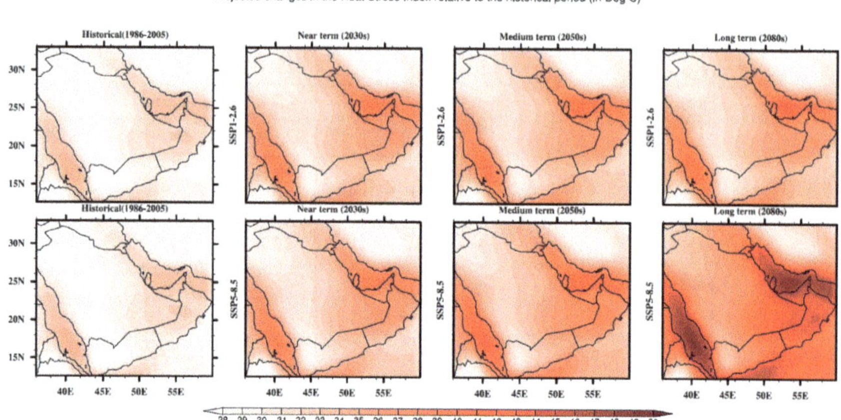

Fig. 2.6 Heat stress index (°C) estimated based on simulations from 20 different CMIP6 global climate models averaged for May to October period (considered here as summer season) over Qatar and adjoining region for the historical (1986–2005) as well as for three future time slices corresponding to 2030s, 2050s, and 2080s under two greenhouse gas emission scenarios: of SSP1-2.6 and SSP5-8.5

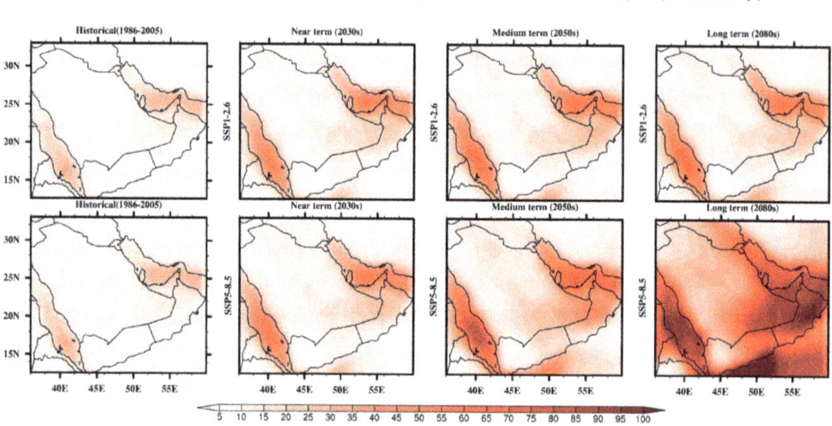

Fig. 2.7 Spatial maps showing the 20-year average number of days of heat stress under "danger" category in summer season (May to October) in the baseline and three future time periods under SSP1-2.6 and SSP5-8.5 emission scenarios over the Arabian Peninsula

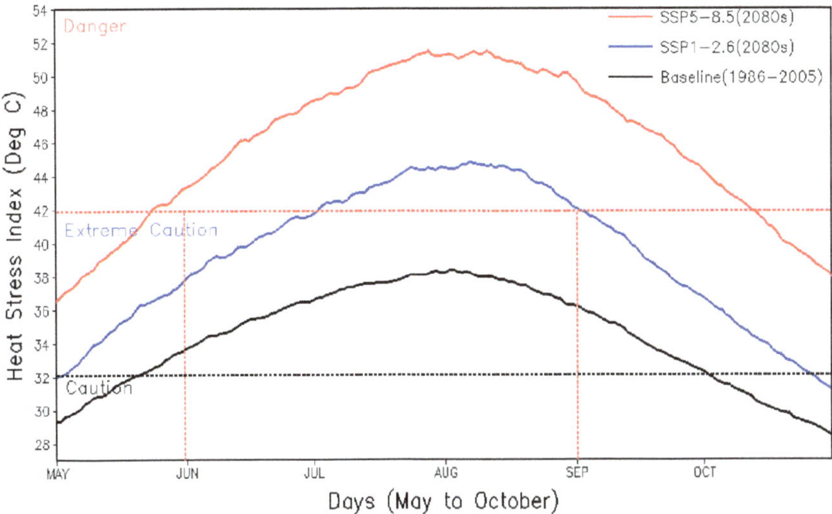

Fig. 2.8 20-year mean daily heat stress index averaged over Qatar during 1 May to 31 October for the baseline and long term (2081–2100) corresponding to SSP1-2.6 and SSP5-8.5 emission scenarios. Heat stress index levels of caution, extreme caution, and danger are marked in the graph to show the intensifying and expanding severe heat stress conditions in the future relative to the baseline period

2.3.3 Rainfall

Figure 2.9 shows expected changes (%) in the mean annual rainfall over the Arabian Peninsula corresponding to 2030s, 2050s, and 2080s with respect to the baseline period. As can be seen from these figures, rainfall is expected to increase in this region under all three emission scenarios considered. The increase is more from low-to-high emission scenarios and from near term to long term. For the high-emission scenario (SSP5-8.5), the rainfall is projected to increase by 14, 17, and 44% for the near-, mid-, and long-term time periods, respectively, when compared to the baseline. Table 2.4 provides rainfall amounts averaged over Qatar for different months and for different scenarios and future time periods including the baseline. It may be noted here that while the percentage change in rainfall is quite high, the region receives low rainfall. However, the temporal distribution of increased rainfall in the future will be of interest to know if the rainfall will occur in short spells leading to flooding or will be distributed uniformly across the rainy season. This is beyond the scope of the present work. However, Li et al. (2021) showed that at a global warming of 4°C relative to the preindustrial level, very rare heavy precipitation events would become more frequent and more intense than in the recent past in all continents and in all AR6 regions that includes ARP (Arabian Peninsula) region.

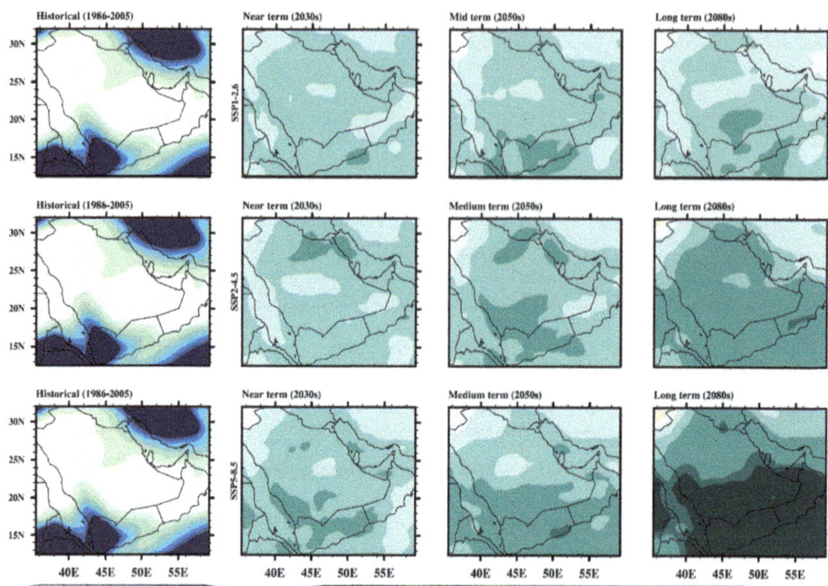

Fig. 2.9 Projected changes in total annual precipitation (%) for three future time slices (2030s, 2050s, and 2080s) under three emission scenarios of SSP1-2.6, SSP2-4.5, and SSP5-8.5. These changes (%) are relative to the baseline period of 1986–2005 (mm) and the mean of 19 model simulations

2.3.4 Sea Level Rise

In its sixth assessment report, IPCC explained that "coastal risks will increase over the twenty-first century due to committed sea level rise impacting ecosystems, people, livelihoods, infrastructure, food security, cultural and natural heritage at the coast" (IPCC 2022). According to the report, it is certain that global mean sea level will rise by 0.23m by 2050 and 0.77m by 2100 (with a medium confidence) relative to the period 1995 to 2014 under CMIP6 high-emission scenario (SSP5-8.5). Few studies show that, globally 570 coastal cities are vulnerable to a 0.5m sea level rise. In particular, most coastal cities across the Persian Gulf region which are located on low lying coastal zones are vulnerable to sea level rise, consequent upon global warming and Qatar is one of those coastal regions which is extremely vulnerable to sea level rise (ICDG report 2013; INDC report 2015; Laurent & Lambert, 2015). We have not assessed the magnitude of future sea level rise over the Arabian Gulf region (especially for the Qatar coast) in the present study.

Table 2.4 Monthly and annual total rainfall (mm) averaged over Qatar for the baseline and three different future time periods corresponding to two emission scenarios of SSP1-2.6 and SSP5-8.5

Monthly mean rainfall (mm)	1986–2005	Near term (2021–2040)		Medium term (2041–2060)		Long term (2081–2100)	
Month	Historical	SSP1-2.6	SSP5-8.5	SSP1-2.6	SSP5-8.5	SSP1-2.6	SSP5-8.5
January	6.3	6.5	8.0	6.8	7.3	6.9	9.9
February	7.4	7.5	7.4	6.7	6.5	7.1	9.0
March	8.1	9.8	8.9	9.7	9.6	9.7	10.3
April	10.1	13.0	11.8	11.4	12.6	13.1	14.4
May	6.2	7.0	6.0	7.5	6.7	6.5	9.4
June	1.0	0.9	1.0	1.0	1.0	1.0	1.2
July	1.1	1.6	1.4	1.9	1.6	1.4	2.2
August	1.9	2.8	2.4	2.2	2.5	2.0	4.2
September	1.2	1.8	1.6	1.7	1.8	1.3	2.9
October	4.2	4.4	5.1	3.8	4.7	3.4	6.4
November	11.3	14.2	14.9	13.4	12.3	11.0	16.7
December	9.2	9.3	9.8	9.7	12.9	9.1	11.2
Annual total rainfall	67.9	78.8	78.3	75.9	79.5	72.6	97.8

2.4 Summary

Given the focus of the report being the assessment of ecology for the State of Qatar, the important elements of climate like rainfall and temperature including heat stress are assessed here. There is a significant warming trend in the mean annual temperature in Doha during the past 60 years with a slight decreasing trend in annual total rainfall. During this period, night-time temperatures have warmed much more than day time leading to a decrease in the diurnal temperature range. Analysis of future temperature trends from a large suite of CMIP6 models indicate very pronounced warming in the future under high-emission scenario (SSP5-8.5) by the end of the century while the trends are relatively lower for low-to-medium emission scenarios (SSP1-2.6 and SSP2-4.5). Increasing heat stress could be a serious problem for Qatar and the entire region where it is going to shift from "extreme caution" to "danger" levels during the entire summer season under high-emission scenarios by the end of the century. This can also increase the chances for the occurrence of "extreme danger" category days in the future leading to severe human discomfort and potential implications for the fragile ecosystems in the region. There is a general consensus with majority of the models indicating increased rainfall in this region including Qatar in the future— more so with high-emission scenarios. Understanding of how the other aspects of the climate like the extremes in rainfall, temperature, heat index as well as strong wind events known as Shamals and regional sea level are going to change in the future is equally important but are beyond the scope of this report.

References

Almazroui M, Saeed S (2020) Contribution of extreme daily precipitation to total rainfall over the Arabian Peninsula. Atmos Res 231:104672. https://doi.org/10.1016/j.atmosres.2019.104672

Atif RM et al (2020) Extreme precipitation events over Saudi Arabia during the wet season and their associated teleconnections. Atmos Res 231:104655. https://doi.org/10.1016/j.atmosres.2019.104655

Chen C-A, Hsu H-H, Liang H-C (2021) Evaluation and comparison of CMIP6 and CMIP5 model performance in simulating the seasonal extreme precipitation in the Western North Pacific and East Asia. Weather Clim Extrem. 31:100303. https://doi.org/10.1016/j.wace.2021.100303

Hamed MM, Nashwan MS, Shahid S (2022) Inter-comparison of historical simulation and future projection of rainfall and temperature by CMIP5 and CMIP6 GCMs Over Egypt. Int J Climatol, 42, 4316–4332. https://doi.org/10.1002/joc.7468

ICDG report (2013) State of qatar, ministry of municipality and urban planning, interim coastal development guidelines, 11. http://www.mme.gov.qa/QatarMasterPlan/Downloads-qnmp/Regulation/English/Group3-Guideline/036_Interim%20Coastal%20Development%20Guidelines%20August%202017.pdf

INDC report (2015) State of qatar, ministry of environment, intended nationally determined contributions report. https://www4.unfccc.int/sites/submissions/INDC/Published%20Documents/Qatar/1/Qatar%20INDCs%20Report%20-English.pdf

IPCC (2022) Climate change 2022: impacts, adaptation and vulnerability. Contribution of working group II to the sixth assessment report of the intergovernmental panel on climate change [Pörtner H-O, Roberts DC, Tignor M, Poloczanska ES, Mintenbeck K, Alegría A, Craig M, Langsdorf S, Löschke S, Möller V, Okem A, Rama B (eds)]. Cambridge University Press. Cambridge University Press, Cambridge, UK and New York, NY, USA, pp 3056. https://doi.org/10.1017/9781009325844

Laurent A, Lambert (2015) Climate change risks for the state of qatar. Presentation made at the conference "climate change: facts, risks and solutions." Qatar University. Doha, Qatar. September, 27–28

Li C et al (2021) Changes in annual extremes of daily temperature and precipitation in CMIP6 models. J Clim 34(9):3441–3460. https://doi.org/10.1175/jcli-d-16-0213.1

Masson-Delmotte V, Zhai P, Pirani A, Connors SL, Péan C, Berger S, Caud N, Chen Y, Goldfarb L, Gomis MI et al (2021) IPCC summary for policymakers. In climate change 2021: the physical science basis. contribution of working group I to the sixth assessment report of the intergovernmental panel on climate change; cambridge university press: Cambridge, UK

Mohammed AM, Hazrat B, Rajesh G, Krishna Kumar K, Tareq AA (2021) Impact of the urban heat island effect on the climate of the State of Qatar. Int J Glob Warm. 24, 3–4

O'Neill BC, Tebaldi C, Van Vuuren DP, Eyring V, Friedlingstein P, Hurtt G, Knutti R, Kriegler E, Lamarque JF, Lowe J et al (2016) The scenario model intercomparison project (ScenarioMIP) for CMIP6. Geosci Model Dev. 9:3461–3482. https://doi.org/10.5194/gmd-9-3461-2016

O'Neill BC, Kriegler E, Ebi KL, et al (2017) The roads ahead: narratives for shared socioeconomic pathways describing world futures in the 21st century. Glob Environ Change 42:169–180.

Pal JS, Eltahir EAB (2016) Future temperature in southwest Asia projected to exceed a threshold for human adaptability. Nat Clim Chang 6:197–200. https://doi.org/10.1038/nclimate2833

Riahi K, Van Vuuren DP, Kriegler E, Edmonds J, O'neill BC, Fujimori S, Bauer N, Calvin K, Dellink R, Fricko O, Lutz W (2017) The shared socioeconomic pathways and their energy, land use, and greenhouse gas emissions implications: an overview. Global Environ Change 42:153–168

Riahi K, Van V, Detlef P, Kriegler E, Edmonds J, O'Neill BC (2017) Fujimori, Shinichiro; Bauer, Nico; Calvin, Katherine; Dellink, Rob; Fricko, Oliver; Lutz, Wolfgang. The Shared Socioeconomic Pathways and their energy, land use, and greenhouse gas emissions implications: An overview. Global Env Chan. 42:153–168

Steadman RG (1979) The assessment of sultriness. Part I: A temperature-humidity index based on human physiology and clothing science. J Appl Meteor 18:861–873

Tebaldi C et al (2021) Climate model projections from the scenario model intercomparison project (ScenarioMIP) of CMIP6, Earth Syst. Dynam 12:253–293. https://doi.org/10.5194/esd-12-253-2021

Zamani Y, Hashemi Monfared SA, Azhdari Moghaddam M, Hamidianpour MA (2020) Comparison of CMIP6 and CMIP5 projections for precipitation to observational data: The case of Northeastern Iran. Theor Appl Climatol 142:1613–1623. https://doi.org/10.1007/s00704-020-03406-x

Open Access This chapter is licensed under the terms of the Creative Commons Attribution 4.0 International License (http://creativecommons.org/licenses/by/4.0/), which permits use, sharing, adaptation, distribution and reproduction in any medium or format, as long as you give appropriate credit to the original author(s) and the source, provide a link to the Creative Commons license and indicate if changes were made.

The images or other third party material in this chapter are included in the chapter's Creative Commons license, unless indicated otherwise in a credit line to the material. If material is not included in the chapter's Creative Commons license and your intended use is not permitted by statutory regulation or exceeds the permitted use, you will need to obtain permission directly from the copyright holder.

Chapter 3
International and National Environmental Laws and Ethics

E. Athwal, G. Dimitropoulos, D. Olawuyi, A. S. Weber, R. Al-Sehlawi,
A. M. Dheen Mohamed, A. Amato, and A. D. Chatziefthimiou

Abstract This chapter offers a comprehensive legal and ethical analysis of Qatar's environmental governance system, examining the interplay between international, regional, and domestic frameworks. It highlights Qatar's progressive incorporation of global environmental treaties—including the UNFCCC, CBD, and Paris Agreement—into national strategies such as the Qatar National Vision 2030 and the National Environment and Climate Change Strategy. The chapter critically assesses Qatar's mixed approach to enforcement—ranging from command-and-control and market-based mechanisms to behavioral "green nudges"—and identifies key gaps in legal coordination, biodiversity protection, and regulatory enforcement. A central

E. Athwal (✉) · G. Dimitropoulos · D. Olawuyi
UNESCO Chair of Environmental Law and Sustainable Development, College of Law, Hamad
Bin Khalifa University, Doha, Qatar
e-mail: eiathwal@gmail.com

G. Dimitropoulos
e-mail: gdimitropoulos@hbku.edu.qa

D. Olawuyi
e-mail: dolawuyi@hbku.edu.qa

A. S. Weber
Weill Cornell Medicine Qatar, Doha, Qatar
e-mail: alw2010@qatar-med.cornell.edu

R. Al-Sehlawi
Co-founder Atlas Bookstore for the Natural and Built Environment, Doha, Qatar
e-mail: reem@atlasbookstore.org

A. M. Dheen Mohamed
Education Above All Foundation, Doha, Qatar
e-mail: ahmadmujthaba@yahoo.com

A. Amato · A. D. Chatziefthimiou
Earthna, Center for a Sustainable Future, Qatar Foundation, Doha, Qatar
e-mail: aamato1uk@mac.com

A. D. Chatziefthimiou
e-mail: a.d.chatziefthimiou@gmail.com

© The Author(s) 2026

A. D. Chatziefthimiou et al. (eds.), *Pathways to Nature Conservation and Resilience in Hot and Arid Lands*, Advances in Geographical and Environmental Sciences,
https://doi.org/10.1007/978-3-032-11046-6_3

contribution of the chapter is its integration of Islamic environmental ethics into legal discourse, exploring the potential of Quranic principles such as *tawhid* (unity), *khalifa* (stewardship), and *mīzān* (balance) to inform interspecies rights and ecocentric policy design. Drawing on the National Biodiversity Strategy and Action Plan, it outlines opportunities to strengthen nature conservation through the Islamic legal tradition, biodiversity offsetting, and improved EIA practices. The chapter concludes with forward-looking policy recommendations that bridge legal theory and implementation, emphasizing the need for institutional reform, inclusive governance, and ethical coherence to enable an effective, theologically grounded, and legally robust response to biodiversity loss and climate change in Qatar.

Keywords International environmental law · Environmental governance · Qatar environmental policy · Climate change law · Biodiversity protection · Islamic environmental ethics · Ecocentrism · Environmental impact assessment (EIA) · Enforcement mechanisms · Sustainable development

3.1 Introduction

In the context of global climate change, Qatar has crafted several broad legal frameworks to support sustainable social and economic development. Qatar has pledged to promote balanced and holistic social and environmental policies that work in tandem with economic growth across all key sectors. However, while the motivation and enthusiasm for sustainability is swiftly growing in Qatar, legal innovation is essential to provide greater coordination and coherence in the implementation of such efforts.

Qatar recognizes the need to balance economic growth, social development, and environmental protection, and created the Qatar National Vision (QNV) 2030, which expresses Qatar's aim to achieve "a diversified economy that gradually reduces its dependence on hydrocarbon industries" by the year 2030 (State of Qatar 2021a, p. 11). The QNV 2030 has four main pillars: human, social, economic, and environmental development. The environment pillar of the QNV 2030 specifically identifies the need for an "agile and comprehensive legal system that protects all elements of the environment, responding quickly to challenges as they arise" (General Secretariat for Development Planning 2008, p. 36).

As part of the fulfillment of this national vision, Qatar's Ministry of Environment and Climate Change was established by a ministerial decision No. 57 of 2021, attended by Sheikh Tamim bin Hamad Al Thani, Emir of the State (State of Qatar 2022a, b, c, d). Over the last years, the Ministry of Environment and Climate Change has issued a wide range of substantive laws and policies that cast a special spotlight on environmental protection in the country (Olawuyi 2022c). Additionally, institutional measures have also been put in place, such as the landmark achievement of Qatar's establishment of a National Climate Change Committee, an administrative body tasked with monitoring climate change (Olawuyi 2022c). Qatar has also launched a

National Environment and Climate Change Strategy which outlines practical steps to achieve the environment pillar of the QNV 2030 (State of Qatar 2021a, 2021b).

This document outlines the range of environmental laws in Qatar. The document is organized into five sections, this Introduction being the first. Section 3.2 is divided into three parts: Part 1 provides an overview of international, regional, and domestic legal frameworks in Qatar, pertaining to the environment and climate change. Section 3.2, Part 2 looks in detail at primary and secondary legislation, and the process of implementation of those laws, including various command-and-control, market-based, and green nudges approaches. Section 3.2, Part 3 examines the National Biodiversity Strategic Action Plan. Section 3.3 is an interrogation of various ethical approaches to the environment in Islam, to determine whether there is an avenue for closer coordination and interconnection between theological and legal environmental prescriptives. Section 3.4 examines several gaps and provides a brief review of the key recommendations, and Sect. 3.5 is the concluding section.

3.2 Laws

3.2.1 International Frameworks

From the time when the United Nations Framework Convention on Climate Change (UNFCCC) was adopted in 1992 as one of the Rio Conventions, Qatar has increasingly and proactively mainstreamed international law on climate change into its domestic strategies (Olawuyi 2022b). Qatar ratified the Convention on Biological Diversity (CBD) on the November 19, 1996, along with its two protocols, the Cartagena Protocol, on June 12, 2007, and the Nagoya Protocol, on April 25, 2017 (CBD). In addition, Qatar ratified the United Nations Convention to Combat Desertification (UNCCD) on March 15, 1999 (UNCCD). These important international frameworks safeguarding the environment helped to shape Qatar's National Biodiversity Strategy and Action Plan, which was finalized in October 2004. In 2011, the Ministry of Environment of Qatar submitted its *Initial National Communication* to the UNFCCC, which, in effect, acknowledged Qatar's pledge to support global efforts to tackle the problem of climate change (State of Qatar 2011a, b, c). In light of this initial communication, and in order to support global climate change negotiations, Qatar, in 2012, hosted the 18th Conference of the Parties (COP) to the UNFCCC (United Nations 2012).

Qatar took the initiative to be one of the first few countries in the Arab world to ratify the Paris Agreement, a treaty which sets a clear goal to reduce emissions and keep the rise in global temperature under 2 °C (3.6°F) and limit it to 1.5 °C above pre-industrial levels starting in the year 2020; Qatar ratified the Kyoto Protocol in January 2005 and the Paris Agreement in June 2017 (United Nations 2015a, 2015b, 2015c). Furthermore, Qatar recognized the Doha Amendment (the second commitment round of the Kyoto Protocol) in 2020 (United Nations 2012).

In the final quarter of 2022, Qatar submitted a National Voluntary Commitment to the UN at the Ocean Conference, to turn 30% of its territorial waters into Marine Protected Areas by 2030, a 28% jump from current levels (Personal Communication with MoECC). This development is in support of the QNV 2030. The newly synthesized document, the Qatar National Action Plan for Conservation and Management of Marine Resource, is the 30×30 pledge many countries in the world over are making to protect 30% of their territory by the year 2030, in line with SDG 14, on the conservation of oceans, seas, and marine resources (QNA). The 30×30 pledge is now part of the Kunming-Montreal Global Biodiversity Framework (GBFa, GBFb) adopted on December 19, 2022, during COP15 in Montreal, Canada, of which Qatar is a signatory. Of the four goals and 23 targets for 2030, one essential component is to "ensure and enable that by 2030 at least 30 percent of terrestrial, inland water, and coastal and marine areas, especially areas of particular importance for biodiversity and ecosystem functions and services, are effectively conserved and managed," while recognizing "indigenous and traditional territories, where applicable" (UN Press Release 2022).

In addition, Minister of Environment and Climate Change, His Excellency Sheikh Dr. Faleh bin Nasser bin Ahmed bin Ali Al Thani, inaugurated the environmental sustainability program, "Our Sustainable Environment," at Qatar's pavilion at the UN Climate Summit in Sharm El Sheikh, Egypt, for the UNFCCC 27th Conference of the Parties (COP27) (The Peninsula 2022).

3.2.2   *Regional Frameworks*

In terms of regional initiatives, Qatar endorses a diverse array of declarations and instruments across the Gulf and wider Arab region (see Table 3.1). First, Qatar has developed collaborative programs on environmental protection with some international environmental organizations that have regional offices and mandates in the region, such as the UNEP Office for West Asia (UNEP), the UNESCO Gulf States and Yemen Office (UNESCO), and the United Nations Economic and Social Commission for Western Asia (UNESCWA). The UNESCO regional office in Doha implements a wide range of environmental law education programs in partnership with higher education institutions in Qatar such as the UNESCO Chair on Environmental Law and Sustainable Development at the Hamad Bin Khalifa University (UNESCO). A joint initiative between HBKU, a Qatar Foundation (QF) member, and UNESCO, the Chair holder's multifaceted role will be to develop a program to foster UNESCO's emerging work on sustainability science, including research, with an emphasis on the legal dimension of sustainable development, with other institutions in Qatar, the Arab region, and across the world. Furthermore, in addition to its work with UNESCO, Qatar attaches great importance to the UNEP Regional Seas Programme, and is aligning its strategic environmental plans along these lines (QNA).

Second, regional intergovernmental bodies play an important role in developing the priorities of the Arab states as a whole. Some examples range from the League of

Table 3.1 Key declarations and instruments relating to climate change by regional bodies in the Middle East and North African (MENA) region

Instrument	Highlights
Arab declaration on environment and development and future perspectives, Cairo, 1991	The Declaration called on all Arab countries to limit the degradation of the environment and natural resources and manage both in a sustainable manner
Arab declaration to the world summit on sustainable development, 2001	It renewed the commitment of Arab countries to work together to advance environmental protection and sustainable development, especially by addressing the vulnerabilities of Arab countries to water scarcity, food insecurity, and climate change
Abu Dhabi declaration on the future of environmental action in the Arab world, 2001	It identifies the priority of environmental problems in the Arab world as: 'the acute shortage and deteriorating quality of sources of water; the paucity and deteriorating quality of exploitable land; the imprudent consumption of natural resources; urban sprawl and its associated problems; the degradation of marine, coastal and watered areas'
	It also identifies the need for clean production methods and technologies to reduce emissions
Arab charter on human rights, 2004	Article 38 states that 'every person has the right to an adequate standard of living for himself and his family, which ensures their well-being and a decent life, including food, clothing, housing, services and the right to a healthy environment. The States parties shall take the necessary measures commensurate with their resources to guarantee these rights'
The Arab ministerial declaration on climate change of December 6, Cairo, 2007	It contains regional aspirations by all Ministers responsible for environment and climate change to include 'policies to deal with climate change issues in all sectors within national and regional policies for sustainable development'
	The declaration also provides that: 'Adaptation to measures that address climate change shall be fully consistent with the economic and social development and in such a way so as to achieve sustainable economic growth and eradication of poverty'
Cairo Declaration on Development Challenges and population dynamics in a changing Arab world, 2013, (*paras 66, 85, and 87*)	It called on Arab countries to develop regional and local climate change response measure that take into account the distribution, vulnerability, and resilience of the targeted populations

(continued)

Table 3.1 (continued)

Instrument	Highlights
Rabat Islamic declaration on environment protection and achieving sustainable development goals, 2017	Called on Islamic countries to advance climate change mitigation and adaptation in all aspects of planning at national levels. It also affirms the importance of a green economy transition
Arab framework action plan on climate changes, 2010–2020	It creates a ten-year master plan for climate mitigation and adaptation in the Arab region
Pan-Arab renewable energy strategy, 2030 (League of Arab states)	It identifies the integration of renewable energy as a tool for addressing climate change vulnerablies in the Arab region. It sets a target of increasing installed renewable energy power generation capacity across the region by the year 2030

Source Law and Governance Innovations on Sustainability in Qatar: Current Approaches and Future Directions (Olawuyi and Athwal 2022)

Arab States' Resolution No. 4738 of 1986 establishing the Council of Arab Ministers Responsible for the Environment (CAMRE), and the GCC Ministers Responsible for Environmental Affairs (GCC), who held their last meeting in December 2022, regarding the post-2020 global biodiversity framework. The Regional Organization for the Protection of the Marine Environment (ROPME), founded in 1979, provides support for the Gulf States, and was an important mandate in ensuring that the pollution of oil and gas activities in one country do not pollute the marine environment of the region (Olawuyi 2022c, 69).

Qatar's legal framework on the environment also reflects a wide range of regional declarations and action plans on sustainable development such as the 1991 Arab Declaration on Environment and Development for Future Prospects (United Nations). While such soft law instruments are not legally binding, they lay a foundation for Qatar to agree on the balanced implementation of infrastructure and development programs in a manner that averts social and environmental challenges (Olawuyi and Athwal 2022). Furthermore, they support the need to incorporate low-carbon emission provisions into national policies, strategies, and planning along the United Nations Sustainable Development Goals (UN SDGs). One example in this regard is the *Pan-Arab Renewable Energy Strategy, 2030* of the League of Arab States, which establishes a goal of aggregating installed renewable energy power generation capacity across the region by the year 2030 (IRENA/League of Arab States 2014). Along these lines, Qatar has delineated strategies to generate 20% of its electricity from solar power by 2030 (Qatar General Secretariat for Development Planning 2008; State of Qatar 2011a, b, c).

3.2.3 Domestic Frameworks

3.2.3.1 Constitution

Qatar is one of the few countries in the world with a clear constitutional provision on environmental protection. Article 33 of Qatar's Constitution of 2003 specifies that the state "endeavors to protect the environment and its natural balance, to achieve comprehensive and sustainable development for all generations" (Permanent Constitution of the State of Qatar).

3.2.3.2 Primary Legislation

In terms of primary legislation, Qatar's Law No. 30 of 2002 on Environmental Protection outlines general provisions for the protection of the environment, including ensuring environmental quality, avoiding pollution and damage from development projects, protection of the public health in addition to the local flora and fauna, and lastly provides for environmental awareness under Article 7 (State of Qatar 2002). Article 6, in particular, states that all public and private bodies must include an environmental protection and pollution control clause in local and international agreements and contracts which may be detrimental to the environment, and these agreements and contracts shall include applicable penalties and the obligation to bear the costs of repairing the environmental degradation and harm (State of Qatar 2002). Article 8 provides for the Ministry's oversight of all approvals, control, and supervision of private and public development projects, including setting standards and measures related to environmental impact assessment (EIA) of projects, and the procedures and requirements for issuing authorizations (State of Qatar 2002). Additionally, legislation in part 2 establishes the basis for the governmental authorities to penalize and sanction all forms of air pollution, i.e., any chemical, physical, or biological change or modification of the natural characteristics of the atmosphere, in proportions that could be harmful to human life and nature, including contributions to climate change (State of Qatar 2002) (Table 3.2).

Ministry of Environment and Climate Change (MoECC) Process of Issuance of Permits and Fines

The Ministry of Environment and Climate Change (MoECC) directs regulations and restrictions preserving the environment, and executing the necessary mechanisms for the protection of the environment (Olawuyi and Athwal 2022).

- **Permits**

Any infrastructure or construction project in Qatar must obtain a permit after providing a scope of work report to the MoECC (State of Qatar, Hukoomi E-Government 2022). The process of issuing permits follow three main stages: first, at the design stage, an environmental impact assessment (EIA) related to the proposed

Table 3.2 Key domestic environmental regulation in Qatar

Instrument
Permanent constitution of the state of Qatar (2003), Article 33
Law No. 30 of 2002 on environmental protection
Decision No. (4) of 2005 of the president of the supreme council of environment and natural protection concerning the issuance of the executive regulations of the environmental protection
Law issued by Decree-Law No. (30) of 2002 (the "Executive By-Law")
Law No. (4) of 1977 on the conservation of petroleum resources and the conduct of petroleum operations within Qatar
Law No. 4 of 1983 on the exploitation and conservation of living aquatic resources in Qatar
Law No. (8) of 2004 concerning protection of the maritime facilities of petrol and gas
Law No. 5 of 2006 on the regulation of trade in endangered wildlife fauna and flora and their products 5/2006
Public hygiene Law No.18 of 2017
Law No. 20 of 2015 amending some provisions of Law No. 26 of 2008 on the rationalization of electricity and water consumption

Source Olawuyi (2022a), 43

project covering all the components must be undertaken before environmental clearance to the next step; second, at the construction stage, the temporary facilities must be approved, including site offices, laydown areas, worker facilities, fuel storage tanks, and so forth, as to the environmental impact and risk involved; third, at the operation stage, an operational environmental permit is required which covers the scope of the operation stage of the developed project and compliance with the operational requirements (State of Qatar, Environmental Assessment Department 2022). The EIA must include the impact of the construction and operations on the terrestrial, marine, air, and noise environment and potential mitigation measures, and include stakeholder guidelines and standards, such as those of Kahramaa Water and Design Guidelines, Ministry of Transport and Communication guidance, and Ashghal's Qatar Drainage design. The permit approval must be obtained from the MoECC office, Doha, Qatar; however, each municipality associated with the project can provide guidance and clarify any additional requirements (State of Qatar, Ministry of Municipality 2022).

- **Fines**

Penalties for violations of the law and/or damage in the State of Qatar are assessed by a dedicated department within the MoECC or a relevant municipal authority such as Wakrah Municipality, and so forth. Although the MoECC is the lead regulatory agency, municipalities do have the right to issue fines (State of Qatar, Ministry of Municipality 2022). The Qatar Coast Guard's role is limited to ports and harbors in Qatar, except under special circumstances in which help is needed. The Qatar Coast Guard uses the International Convention for the Prevention of Pollution from Ships (MARPOL) guidelines for the development related to the project (International Maritime Organization 2019).

3.2.3.3 Secondary Legislation

In terms of secondary legislation, the *Executive By-Law for the Environment Protection Law* comprises substantial provisions on reducing all sources of air pollution, especially the emission of several GHGs that cause climate change (State of Qatar 2002).

Qatar's Resolution of the Council of Ministers also inaugurates a Committee on Climate Change and Clean Development Mechanism (CDM) which outlines strategies, policies, and communications on climate change in Qatar (State of Qatar 2011b). Moreover, it provides that governmental and non-governmental organizations in Qatar adhere to the mandates of the UNFCCC and the Kyoto Protocol and create the basis for the preparation of databases, national assessments, reviews, and evaluations (State of Qatar 2011b).

3.2.4 Enforcement Approach

Qatar employs a mix of regulatory instruments and approaches to enforce and advance compliance with the wide range of primary and secondary legislation on the environment.

3.2.4.1 Command-and-Control Measures

Conventionally, in the world today, environmental as well as climate change laws and regulations are comprised of command-and-control (CAC) measures and market-based approaches, CAC being the oldest and most widely utilized instrument for climate change regulation (Hacker and Dimitropoulos 2017). In Qatar, the Supreme Council for the Environment and Natural Reserves (SCENR) is required to set out necessary basics, criteria, specifications, and restrictions for assessing the environmental impact of "Projects" and "Establishments" (Decree Law No. 11 of 2000, Articles 1 and 11). SCENR has now been incorporated into the new Ministry of Environment and Climate Change in 2022 (MoECC). This must be carried out by coordinating with competent Administrative Authorities (i.e., the Ministry, governmental bodies, public institutions, public organizations) (see Hacker and Dimitropoulos 2017). The divestment of power relating to EIA oversight which was previously vested in the Environment Department has overcome political problems that previously existed relating to other Ministries attempting to circumvent EIA procedures (Al-Marri 2001).

In general, the EIA process requires that any licenses issued for projects must only be granted after an EIA study has been presented and approved by the SCENR (Decree Law No. 11 of 2000, Article 13). This same process is applicable to any extensions or renovations of existing Projects (Decree Law No. 11 of 2000, Article 13). Projects are defined to mean utilities or facilities (onshore, offshore), which include any "building, structure, installation, project, establishment or any activity likely to be the source of environmental pollution or degradation," as referred to in the Executive Regulations (Decree Law No. 11 of 2000, Article 1, para 17).

3.2.4.2 Market-Based Approaches

Typically, market-based approaches (MBAs) are "regulations that encourage behavior through market signals rather than through explicit directives regarding pollution control levels or methods," and are generally classified under four categories: pollution charges, tradable permits, market-barrier reductions, and government subsidy reductions (Stavins and Whitehead 1997). The international climate change regime includes three MBAs in the Kyoto Protocol to the United Nations Framework Convention on Climate Change (UNFCCC): the Clean Development Mechanism (CDM), Emission Trading (ET), and Joint Implementation (JI); these three international market mechanisms give incentives to industrialized as well as developing countries to trade emission reduction credits or develop emission reduction projects (Olawuyi 2010, p. 21).

3.2.4.3 Green Nudges

Green nudges are relatively cheap and easy to administer regulatory tools informed by the insights of cognitive psychology and behavioral economics that use salience (Reisch and Sandrini 2015), default rules (Sunstein and Reisch 2013), and social norms (Sunstein 2011), to steer an individual's behavior toward more environmentally friendly actions. Green nudges can be divided into three categories, namely, pollution reducing nudges (e.g., green footprints), energy efficiency nudges (e.g., energy efficiency labels, fuel economy labels), and climate change nudges (Hacker and Dimitropoulos 2017).

Although environmental and climate change laws and regulations, in particular, command-and-control and market-based approaches, are measures that assume individuals are rational, green nudges, as an innovative regulatory approach, are measures based on cognitive psychology and behavioral economics (see Hacker and Dimitropoulos 2017; also Thaler 1996).

Hence, green nudges function as a type of behavioral adjustment mechanism. Thus, in order to raise environmental awareness in Qatar, under Article 7 of the Environmental Protection Law No. 30, using green nudges, all education authorities in Qatar are required to include environmental awareness subjects in all educational stages (State of Qatar 2002). Furthermore, the MENA Region's first behavioral insights unit was established in Qatar as the Qatar Behavioral Insights Unit (QBIU) under the Supreme Committee for Delivery and Legacy, now entitled the B4Development Foundation (B4D 2022). The B4D builds and designs programs and initiatives for capacity building, knowledge exchange, and research dissemination.

A specific example of a green nudge utilized in the Qatari context is the Kahramaa Tarsheed program, in which Kahramaa launched an Awareness Campaign for the Conservation Law No. 20 of 2015, to promote energy efficiency (Kahramaa 2021).

The Impact of Rapid Development on the Terrestrial and Marine Environment in Qatar

Qatar's extreme environmental conditions put many of the species living in and around the different ecosystems at their upper tolerance limits (Carpenter et al. 1997). The discovery of oil and gas has seen rapid industrialization and unparalleled growth in the last couple of decades (Furlan 2016). This development and increase in anthropogenic activity, such as discharges from the petrochemical industry, desalination plants, and wastewater treatment plants, have added immense pressures on these already stressed ecosystems (Burt et al. 2017; Khan 2007). As a result, the last few decades have seen a great loss of many of these precious ecosystems such as mangrove forests, seagrass beds, and coral reefs. Those that have prevailed these natural and anthropogenic stressors are not in the best state.

One of the major threats toward the coastal and marine ecosystems is coastal development. The reclamation of land and dredging processes result in the direct destruction of these habitats such as coral reefs and seagrass beds as well as result in increased siltation and water turbidity which in turn smother corals, seagrass beds, and algae mats, inhibiting their ability to photosynthesize (Al-Ghadban and Price 2002). It also prevents corals from recovering (Rezai et al. 2004), and affects fish populations by degrading their habitats (Jones et al. 2007). These anthropogenic activities coupled with the already existing natural stressors have also led to thermal impacts to coral communities leading to coral bleaching (Burt et al. 2016).

While these impacts have been affecting offshore regions more recently than coastal and nearshore environments, they are just as alarming (Burt et al. 2017). The creation of channels by dredging has severely degraded salt marshes, mangrove forests, seagrass and oyster beds, and coral reefs. (Smyth et al. 2016; Balakrishnan 2012; Sheppard et al. 2012; Erftemeijer and Shuail 2012; Al Kuwari and Kaiser 2011; Richer 2008). Moreover, increase in desalination and other power plants have inevitably led to an increase in various brine and thermal effluents, which are chronic stressors and further degrade these coastal ecosystems (Darwish et al. 2013; Richer 2008). In addition, there are numerous other pressures arising as a result of development and industrial growth, driven by an increasing population, including but not limited to pollution from various industrial sectors, overexploitation of natural resources, and invasive species being introduced either deliberately or through other means such as ballast water discharges (Richer 2008).

While steps have been taken in the past to protect and restore ecosystems that have been degraded (Al-Khayat and Balakrishnan 2014), as well as to look at how to reduce future impacts from upcoming projects and development (Darwish et al. 2013; Shomar et al. 2014), along with looking to relocate less tolerant but vital species from areas where human development cannot be prevented from taking place (Deb et al. 2014), the sheer scale and severity of the damage that ecosystems have already sustained seems to suggest that these ecosystems in their current state will still be vulnerable for quite some time.

These pressures combined with climate change suggest that Qatar's coastal ecosystems are very likely to be affected by rising sea levels. Computer models predict sea level rises between 1 and 5 m, potentially flooding Qatar's coasts from 3 to 13%, respectively (El Raey 2010). This rise in sea levels will greatly increase the phenomena known as coastal squeeze, further impacting valuable coastal ecosystems (Burt et al. 2017).

Qatar has taken steps over the last few decades to put various legislation in place to address environmental issues arising from various aspects such as those mentioned above and more. "Law No. 30 of 2002 Promulgating the Law of the Environment Protection" encompasses five parts addressing different areas of Environmental Protection. Chapter 1 deals with "Environment and Sustainable Development" including 10 Articles aimed at protecting the environment from pollution. Article 2 specifically addresses the need to preserve the environment and protect it from harm while simultaneously developing natural resources for human development in a sustainable way by maintaining the natural balance. Article's 3–10 state the need to ensure conservation of natural resources, giving priority to environmental considerations and rationalize any type of exploitation of biotic and abiotic resources. There are also requirements to include environmental protection in any agreements or contracts and integrate environmental awareness as part of the curriculum and the educational system. In addition, environmental protection measures need to be reviewed and updated while the council needs to ensure that all necessary measures are taken to either avoid, prevent, or mitigate any possible environmental harm. Article 9 clearly states the prohibition of capturing wildlife and deforestation while at the same time establishing nature reserves.

Chapter 2, Articles 11–20 of "Law No. 30 of 2002 Promulgating the Law of the Environment Protection," addresses the above-mentioned concerns related to the environmental impacts arising from the rapid development of projects. As per the laws, any development projects need to undergo strict assessments of the environmental impacts through environmental impact assessment (EIA) studies along with the development of contingency plans and mitigation measures to address any adverse impacts related to the project and submitted to the relevant administrative authorities, subject to approval.

Chapter 3, Articles 21–23 of Law No. 30 of 2002 Promulgating the Law of the Environment Protection, outline the framework of emergency contingency plans in response to environmental disasters that do occur as part of a coordinated effort between the Supreme Council for the environment and natural reserves, the permanent emergency committee and ministries, and government bodies and relevant organizations (administrative authority).

The issue of waste generation and disposal due to population increase and human development, including handling hazardous materials, is addressed in Chap. 4 of Law No. 30 of 2002 Promulgating the Law of the Environment Protection in Articles 24 to 27.

Part 2 of Law No. 30 of 2002 Promulgating the Law of the Environment Protection addressing the issue of air pollution states the requirement of industries emissions to be within accepted limits (Article 28), prohibited compounds such as pesticides (Article 29) and vehicular emission limits (Article 31). It also addresses water effluent discharges and waste treatment protocols to be carried out as per executive regulations (Article 32).

Part 3 states the Protection of the Aquatic Environment from Pollution of Law No. 30 of 2002 Promulgating the Law of the Environment Protection. Chapter 1, Articles 40 and 41, and Chap. 2, Articles 42 to 61, deal with the protection of surface water and groundwater and their sustainable use and distribution, protection of the marine environment from oil pollution, hazardous substances, sewage and garbage arising from different vessels as well as pollution from land-based sources.

The administrative and judicial procedures by the general secretariat, the council, administrative authorities as well as law enforcement officers are addressed in Part 4 (Articles 62 to 65) of "Law No. 30 of 2002 Promulgating the Law of the Environment Protection," followed by the applicable sanctions in Part 5 (Articles 66 to 75).

While there seems to be commitment toward environmental protection and although the country's leadership has been taking active steps toward ensuring proper environmental management, the fact remains that Qatar's coastal zone has been drastically changing over the last few decades and is still seeing major developments taking place. Therefore, despite the improvements in awareness and environmental management, significant degradation of valuable ecosystems has already taken place (Burt et al. 2017).

What is the Ethical Basis for the Sustainability Precept of "Equity" and Interspecies Rights Globally, and What is the Current Potential for Interspecies Rights in Qatar?

There have been numerous efforts in different parts of the world to address the concept of interspecies rights from an ecocentric point of view, particularly in New Zealand, with the protection of the Whanganui River and the Te Urewera Park (Te Awa Tupua Act 2017; Te Urewera Act 2014). In Bangladesh, in 2019, legal personhood was granted to all of its rivers (BDNews24.com 2019), and in Ecuador, even the Constitution of 2008, granted legally enforceable rights to trees, plants and non-human objects, called *Pacha Mama*, or Mother Earth (Constitution of Ecuador of 2008, Articles 10, 71–74). India, Colombia, and the United States of America have also followed suit, providing some legally recognizable rights to nature (Rivers 2018). Nevertheless, in the Arab world, and most particularly here in Qatar, a close reading of the legal texts of the Constitution and environmental laws demonstrates in the original Arabic text a leaning toward a more anthropocentric, rather than ecocentric, viewpoint, which would most likely create a series of legal hurdles and hindrances to achieving rights for nature. It appears that there is much headway to be made before this might be possible, as evidence toward the contrary is still sparse. Nevertheless, the world does appear to be more and more open toward an ecocentric viewpoint at the level of law and policy, giving more "moral and legal consideration commensurate with that of humans" (Olawuyi 2022c, 12).

Yet while the anthropocentric reading of the legal text presents difficulties, the designation of Islam as the state religion and Sharia as its main source of legislation in the Constitution of Qatar allows for a development of a theocentric legal viewpoint. Theocentrism in Islam develops from the principle of *Tawhīd* (Unity) where God is the only creator and Sustainer of the Earth and all created beings are equal, while humans are designated as guardians or stewards of the Earth, entrusted by God to act according to a pre-set code of ethics (see Sect. 3.3, Dr. Alan Weber) (Worrell and Appleby 2000).

All sources of Islamic law provide a foundation for the rights of nature. The Quran is inherently environmental as it speaks of *khalq* (Creation) in over 250 verses to describe what we see, feel, and sense in the world. *Ayah* or plural *Ayat,* a common term used in the Quran for the signs of the Creator on earth, is also a term applied in the Quran to the phenomena of nature as well as of what is within the souls of man: "There are signs on the Earth for people with certainty. And in your selves as well. Do you not then see"? (Quran, 51:20–21) (Khalid 2019, 151). It is noteworthy that here we find the human self and the natural world placed in equal status, designating inherent value on both equally. Beyond its inherent value, each created entity has practical or instrumental value as an integral component participating in the overall ecological balance (*mīzān*) of the ecosystem that supports life on earth (Haq 2003, 157). In another verse, the Quran states that "There is no animal on earth, or any bird that wings its flight, but is a community like you" (Quran, 6:38). It is again particularly noteworthy that the concept of community, a highly significant theme in Islamic tradition and literature, is used for animals (Özdemir 2003, 17).

These parallels drawn between humans and other created entities in the Quran bear the seeds for the recognition and legal application of interspecies rights that modern law still has difficulty establishing. Constitutions like that of Qatar that encompass the Islamic supremacy clause bear the potential for the enforcement of any legal provisions in the body of Islamic law, yet this possibility is not utilized in the framework of environmental protection. However, the Constitution of Qatar also encompasses environmental principles such as Article 33 on the preservation of the environment and its natural balance. One approach could be the combination of both the Islamic supremacy clause and the environmental protection clause to constitute the foundations of Islamic environmental law (Idllalène 2021, 125). Egypt offers an applied example of the combination of the Islamic Supremacy clause and the Environmental clause; pushed by activists it resulted in an inclusion of an Islamic conception of the principle of care toward animals in the new constitution. The Islamic formulation of the principle appears clearly in the Arabic text that reads *"Al-Rifq-Bi-AlHayawan"* (care toward animals), which includes the same formula adopted by Imam al Bukhari in his compilation of Hadiths (nineteenth century) (Idllalène 2021, 125). Using this combination approach activists in Egypt conveyed an Islamic principle which was integrated as such into the constitution instead of a generic formula following the Western model (Stilt 2017). The same approach could be adopted in Qatar for the incorporation of interspecies rights founded on the Islamic principles mentioned above.

Another way in which interspecies rights can be integrated into sustainability efforts is through the incorporation of social and cultural factors into environmental impact assessments (EIAs) (Olawuyi 2017), and other measures such as biodiversity offsetting. In the context of biodiversity offsetting, there is great risk that biodiversity is seen as too "utilitarian," and hence great harm can be exacerbated on the biological diversity that the mechanisms are meant to protect (Ives and Bekessy 2015). Thus, an approach that can help identify the potential wide-ranging effects of offsetting schemes, and facilitate the inclusion of local communities in decision-making processes, can be extremely beneficial. Biodiversity offsetting policies should be designed to include robust monitoring and evaluation mechanisms to ensure their effectiveness in achieving conservation goals, and can play an essential role in providing the necessary information to inform such monitoring and evaluation (Olawuyi 2017). A collaborative approach that includes a variety of stakeholders, such as regulators, project developers, and local communities; that focuses on the overall ecological value of biodiversity; and the ethical implications integral to its protection can help ensure that biodiversity offsetting policies are implemented successfully and appropriately (Ives and Bekessy 2015).

Islamic Values and Qatar's National Biodiversity Strategy and Action Plan

Qatar's national priorities align closely with the global action plan for biodiversity expressed in CBD's 2011–2020 Strategic Plan for Biodiversity and its 20 Aichi targets, and harmonizes them with recently developed national strategies including the Qatar Vision 2030 and the Qatar National Development Strategy 2011–2016. However, Qatar's National Biodiversity Strategy and Action Plan (NBSAP), finalized in 2004, is also deeply rooted in Islamic values and ethical principles that have guided the state's laws, policies, and national development. The NBSAP contains seven Islamic principles for the conservation of nature:

1. Conservation of the natural environment is a moral and ethical imperative.
2. Ethical teachings should be backed by legislation and effective enforcement of injunctions and prohibitions.
3. The development of the earth should be planned and implemented in accordance with natural constraints, ecological values, and sensitivities.
4. Ecologically sustainable economic development needs to integrate social and economic practices acceptable to local populations.
5. Scientific and technical knowledge of the natural environment and its conservation should continually be improved and developed.
6. Development projects undertaken in one country should not lead to damage or harm in the natural environment of another country.
7. The natural environment and natural resources should not be subjected to any irreparable damage resulting from military actions.

These principles show that the protection of the environment and its biological diversity is deeply rooted in Islam as a theological and ethical edict. This view is also more ecocentric than the anthropocentric norm under which development policies generally operate.

Furthermore, Qatar's Constitution clearly states that the principle source of law in Qatar is *shariah* law, and that Islam is the state religion (Permanent Constitution of the State of Qatar). Thus, Qatar as a state adheres to Islamic principles. One important principle is the precept of *khalifa*, which can be translated to mean "steward" or "trustee" (see also Sect. 3.3, Dr. Alan Weber,). The concept of *khalifa* has certain responsibilities attached to it, such as the duty to protect resources for future generations, as well as to govern and preserve the natural environment so no harm will come to it (IUCN 1983). It is stated in the NBSAP that.

> "As we cannot be aware of the beneficial functions of all things created by Allah, we cannot base our conservation efforts solely on their benefits to man because this would lead to a distortion of the dynamic equilibrium set by Allah. However when we base the conservation and protection of the environment on its value as the signs of the Creator, we cannot omit anything, for every element and species has its individual role to play. Man should not ignore his responsibility of stewardship on earth. It is only when our ethical horizons extend to embrace not only mankind, but all generations and created beings, that we can perform the noble role of stewardship on earth" (NBSAP 2004).

Some inventive ways of protecting the environment through legal means have been through this concept of trusteeship. In countries like Morocco, for example, the principle of *waqf*, which is the means of strengthening and supporting the principle of *khalifa*, through *waqf* trusts or assets (Idllalène 2021). Civil society has increasingly put forward the innovative idea that the climate system is in itself a *waqf* asset that should be safeguarded and preserved for future generations, and that governments, in order to fulfill their fiduciary duties, should reframe their environmental laws and legislation according to the principle of *khalifa* (Idllalène 2021, 56). Nevertheless, more institutional coordination is needed between the environmental ministries and Islamic affairs, as both the principles of *khalifa* and *waqf* imply a fiduciary duty similar to the public trust doctrine (PTD), and can be adopted into law through major reform in legal and institutional systems. The various stages of reform can be through raising awareness and capacity building—such as introducing ecology and environmental law to religious authorities; reinforcing cooperation—such as between *ulama*, academia, and non-governmental organizations; integrating and strengthening funding mechanisms—such as green financing; and fostering a broad conception of *waqf* as an overarching principle (Idllalène 2021, 54–57). Some scholars have researched the potential of Islamic finance in mobilizing and leveraging private climate finance in the Middle East and North Africa (MENA) region, as well as various ethical proposals for an effective mobilization of climate finance at the national level (see Aassouli 2021).

3.3 Ethical Approaches to the Environment in Islam

The Qatar National Biodiversity Strategy and Action Plan, 2015–2025, recognizes that "environmental problems cannot be solved through knowledge and technology alone. Only moral conviction and ethical consciousness, on both individual and social levels, can motivate people to forgo some of the short term profits of this life, and to make personal sacrifices for the common good" (NBSAP 2014). Therefore, it is important not only to examine what existing and future regulations based on both civil and *shariah* law can be harnessed in Muslim-majority countries to achieve sustainability goals, but also to explore how Islamic theology orients itself toward the environment, and humankind's relationship to nature as established in the Quran and ahadith.

Theologians in early Islam, Christianity, and Judaism during the Hellenistic period in an attempt to interpret ambiguous passages on nature and creation in their sacred texts turned in part to Greek philosophy (Plato, Aristotle, the Stoics, the atomic school, and later the Neoplatonists) to understand the natural world and matter. Explicit in Neoplatonic philosophy is the central idea that nature (earth) is fragmented, dirty, evil, and imperfect, and the idea that the divine soul is imprisoned in nature, and seeks liberation to join into union with a unitary God. Also, the monotheistic religions all developed within cultures who were worshiping multiple gods (al-Uzza, Manāt, etc. on the Arabian Peninsula), or animals, statues, and natural places like springs, rivers, and forests (Roman paganism). To make a clear distinction between monotheism and traditional religions, theologians thus reduced the status of the natural world or were openly hostile toward it. Whether God was part of nature, in nature, or transcendent of nature was also a serious area of debate, with transcendence becoming standard Islamic orthodoxy. Thus all the monotheistic religions have struggled with such negative views of nature as a place completely separate from the divine, which is created solely for human needs, which can be used in any manner deemed fit by humans, and which can be modified or destroyed according to human desires.

However, numerous passages in the Quran, ahadith, and opinions (*fatawa*) of Islamic scholars provide another, quite different, perspective on nature. For example, *Jannah* (heaven or paradise) is associated with divine pleasure, water, and gardens. Nature in Islam is designed by the Creator (Allah) and is purposeful. As a creation of the Almighty, it must be respected and cannot be abused or misused, since disrespect for Allah's creation is synonymous with disrespect for the Creator. Also, the study and contemplation of nature can serve as a means of understanding and loving the Creator since the environment is a system of visible signs (*ayat*) proving the benevolence and omnipotence of Allah: "And of His portents [ayat] is the creation of the heaven and the earth, and of whatever beasts He hath dispersed therein" (Quran, 42:29). Also, all of the created world demonstrates gratitude to the Creator: "Hast thou not seen that unto Allah payeth adoration whosoever is in the heavens and whosoever is in the earth, and the sun, and the moon, and the stars, and the hills, and the trees, and the beasts, and many of mankind, while there are many unto whom the doom is justly due" (Quran, 22:18).

Three fundamental environmental concepts can be identified in the classic Islamic conception of nature: *Tawhid* (the unity of everything), *Khalifa* (stewardship), and *Akhirah* (judgment in the hereafter, or accountability). Originally, the concept of *Tawhid* referred narrowly to the oneness of Allah, transcendent of the natural world, but has been applied analogically to the modern environmental concepts of the interconnectedness of living systems, i.e., ecological communities. S.H. Nasr has argued additionally that humankind must acknowledge that God is the real environment, or the All-Encompassing (*Muhit*), and that the Divine environment and the one that humans see around them are essentially the same. Thus, "the destruction of the environment is the result of modern man's attempt to view the natural environment as an ontologically independent order of reality, divorced from the Divine Environment without whose liberating grace it becomes stifled and dies" (Nasr 1990, p. 35). Since creation is a single entity, every human action must be examined for its impact on the environment, its ethical impact on the *umma* (Muslim society), with an avoidance of *fitna*, and whether an act is pleasing or displeasing to God.

Two Quranic passages clearly define man's duty as a steward, vice-regent, or manager of nature: "Hast thou not seen how Allah hath made all that is in the earth subservient unto you?" (Quran, 22:65) and "He it is Who hath placed you as viceroys of the earth and hath exalted some of you in rank above others, that He may try you by (the test of) that which He hath given you" (Quran, 6:165). These passages strongly indicate that man holds responsibilities toward nature (he will be "tested" to see that he has made good use of Allah's gifts); man has been entrusted to look after nature's well-being.

Akhirah, or the afterlife, implies accountability for one's life on earth (*dunya*)—how did one conduct one's life, and did one wisely and justly use the fruits of the earth (animals, fields, oceans, and forests) that were entrusted to man as a proper trustee or steward? Therefore, the concepts of stewardship, viceregency, and accountability are closely interrelated.

A moral imperative in Islam to respect and manage the environment properly clearly exists. In addition to a strong theological basis in environmental protection, Islam has also developed practical resource management strategies such as *Hima* (protected or forbidden area) in which water use, grazing, collecting of plants, and hunting were regulated under Islamic principles for the benefit of the community (Gari 2006).

3.4 Recommendations

Qatar has established a sophisticated foundation for environmental protection through its laws and legal frameworks. Nevertheless, in order to streamline and enhance conservation efforts, after the studies and research conducted, some recommendations that may be helpful would be the following:

1. Governance

Multi-sectoral committee: Establish a multi-sectoral committee that promotes environmental sustainability and nature conservation, to coordinate conservation efforts and avoid policy conflicts. This has been proposed recently by the Shura Council, in particular, to achieve cooperation between all relevant official authorities, enhance control over violations, ensure the preservation of the environment, dispose of polluting waste in appropriate ways, monitor violations at the highest levels, and follow up on what may result from industrial and medical violations (State of Qatar, Shura Council, May 29, 2023).

Review and revise legislation: In addition, the Shura Council has also reviewed a draft law amending some provisions of Law No. 29 of 2006 regarding building control, which was referred to the Council by the government (State of Qatar, Shura Council, May 29, 2023). The Council decided the draft law to be referred to the Legal and Legislative Affairs Committee to scrutinize and issue a report regarding it. The MoECC is making a robust effort to review and modernize the existing legislation, as well as response plans to environmental incidences and develop action mechanisms in all MoECC's structures and departments through bolstering the quality of environmental data and striving to disseminate and modernize them (QNA, June 05, 2023).

Engagement with the Private Sector: Consider involving representatives from the private sector in the multi-sectoral committee. Many businesses, particularly in the energy sector, play significant roles in environmental sustainability and could provide valuable insights.

Public Participation: Ensuring that citizens have a voice in decision-making can enhance the acceptance and success of environmental policies. Consider incorporating public consultation mechanisms, particularly for major projects or legislative changes.

2. Policy and Regulation

Command-and-control approaches: Collaborate with MoECC, MOM, and Civil society to impose fines on violations of nature conservation regulations, such as illegal hunting, using incentives and penalties as policy tools. One new proposal by the Shura Council when reviewing current conservation strategies is to link the inspectors' bonuses and financial incentives to the violations detected, after the Shura Council reviewed the report of the Public Services and Utilities Committee on violations committed by some individuals and companies and their impact on wildlife (State of Qatar, Shura Council, May 29, 2023).

Regular Review and Update of Regulations: Given the rapid advancements in technology and understanding of environmental impacts, it is beneficial to regularly review and update environmental regulations to reflect these changes.

Capacity Building for Law Enforcement: Regular training for environmental law enforcement officials, inspectors, and relevant personnel can improve the effectiveness of regulation enforcement.

3. National Accounting and Valuation

Natural capital: Incorporate natural capital in national budgeting and economic impact assessment measures. Commit a certain percentage of GDP to habitat protection and mandate environmental impact assessments to account for habitat loss costs.

Improve environmental data: Official environment statistics of the State of Qatar support the implementation of the fourth pillar of the Qatar Vision 2030 and provide a single source of trusted environmental information, which can be used for multiple purposes such as research, awareness raising, or national and international state of the environment reporting (State of Qatar, Planning and Statistics Authority, <u>n.d.</u>).

Green Economy Transition: Consider policies that encourage a shift toward a green economy, such as subsidies for renewable energy or tax incentives for businesses, that reduce their environmental impact.

4. **Promote Research**

Increase funding for research: This is particularly important in environmental sustainability and terrestrial ecology, collaborating with MoECC and QRDI to declare unified pillars for environmental research.

International Collaboration: Partner with international research institutions to benefit from global expertise and knowledge sharing in environmental sustainability and terrestrial ecology.

5. **Nature Conservation and Protection**

Monitor conservation efforts: Implement existing management plans for sensitive areas, assign protected status to important bird areas, restrict access to sensitive receptor zones, and develop a centralized repository of baseline environmental data from EIA reports. The MoECC envisions that all infrastructure projects and architectural and engineering constructions fulfill the country's vision in conserving the environment and reducing the pollution with its all forms (Gulf Times, June 06, 2023).

Species-Specific Conservation Efforts: Develop conservation programs that focus on protecting Qatar's unique and endangered species.

Climate Change Adaptation and Mitigation Plans: Develop comprehensive plans to protect the country's natural ecosystems from the impacts of climate change, outlined in Qatar's National Vision 2030 document, and further defined in the country's Nationally Determined Contributions (NDCs) submitted to the United Nations Framework Convention on Climate Change (UNFCCC).

6. **Behavioral Nudges**

Develop a culture of environmental protection: Develop ecotourism policies, encourage whistleblowing on environmental violations, and shift toward value-based approach to nature conservation, inclusive of theocentric approaches.

Public Awareness Campaigns: Leverage traditional media and social media to educate the public about the importance of environmental conservation and the specific actions they can take to help.

School Education Programs: Incorporate environmental education into school curriculums to foster a culture of environmental responsibility from a young age, and promote understanding of the natural environment and the impact of human behaviors through educational programs across sectors.

3.5 Future of Environmental Law

How environmental law will be interpreted in the future appears to rest upon three competing international discourses: ecocentric, anthropocentric, and theocentric views. The ecocentric perspective sees nature—not as mere utility—but rather as a holistic, nature-centered view based on ecological philosophy (Olawuyi 2022c, 10). Anthropocentric perspectives, on the other hand, sustain the human-dominating phase that some have called the Anthropocene (Crutzen and Stoermer 2000), in which there is a great deal of human-caused ecological change (The Guardian 2019), and a pressing need to tackle overconsumption, excessive waste, degradation, and pollution of our environment and biological diversity (Kotzé and French 2018, 5). The theocentric viewpoint is derived from a religious view on nature. In Islam, the concepts of *Tawhid* (unity), *Khalifa* (stewardship), and *Akhirah* (judgment) demonstrate that not only does one have duties toward that which God has made, but also one has responsibilities that one will be held accountable for in the hereafter (Weber, Sect. 3.3). Hence, these three diverging viewpoints on what nature is, and how one should care for and preserve it, provide, in many ways, both competing discourses, and yet also, converging points of commonality, in that nature must be preserved, no matter whether it is for itself, for humans, or for God.

Alongside the interpretation of environmental law, the implementation of environmental law depends on the relationships between international, regional, and national legal instruments. In order to tackle this wide array of issues, which reflect the need to mitigate environmental devastation, regional efforts must be cooperative and collaborative to establish pan-Arab environmental solutions to protect and conserve the environment for future generations (Olawuyi 2022c, 11, 54). The State of Qatar has been at the forefront of these efforts in many ways. Several gaps still remain; first, although Islamic ethical values have been incorporated into the National Biodiversity Strategic Action Plans, the incorporation of the link between theological into state policies and laws would be recommended. Second, it would be preferable to ensure that the latest scientific data is incorporated into Islamic ethical environmental knowledge through capacity building, as well as into state policies and laws. And third, the protection of biodiversity in states such as Qatar remains somewhat incoherent and inconsistent to a certain degree, particularly with regard to ongoing and robust anthropogenic activities, such as the sheer scale of construction and development, particularly terrestrial and coastal development, while including the environmental ramifications of the processing of desalinated water to meet the needs of the current population. Hence, the divide between environmental law and policy must be bridged if biodiversity of the terrestrial and marine environments is going to be safeguarded and preserved.

Most recently, the Shura Council in Qatar has proposed several environmental protection measures to the government, aiming to limit encroachments in wildlife areas. The proposal calls for a coordinating committee to collaborate with relevant authorities and enhance control over violations, and it also plans to tie inspectors' bonuses to the number of detected violations and suggests increasing recycling efforts and spreading environmental preservation culture. In a broader context, the council aims to align these measures with the Qatar National Vision 2030, emphasizing a balance between environmental, economic, social, and human development (Gulf Times, May 30, 2023). The Ministry of Environment and Climate Change (MoECC) in Qatar is working on the Climate Change and Environmental Sustainability Strategy within the third National Development Strategy 2023–2030, including 99 strategic projects across five work areas. The ministry is also implementing the National Climate Change Action Plan, which outlines 300 actions to mitigate the effects of climate change. During a Shura Council meeting, the Minister of Environment and Climate Change outlined the ministry's efforts in preserving biodiversity, digitizing services, enforcing environmental laws, and enhancing data quality (The Peninsula, June 06, 2023).

Green Growth

With regard to Islamic theological principles and perspectives, green growth and a low-carbon transition can be seen through the concepts of "green innovation" (*ijtihad*) and "green lifestyle" (*zohd*). If these concepts can be translated more closely into domestic and regional laws and initiatives, such as through the mechanisms within the GCC, the Regional Center for Renewable Energy and Energy Efficiency (RCREEE), and green building councils in Qatar and throughout the Arab region (Earthna, MENA GBCs), it would accelerate the transition to low carbon and green growth infrastructural development. Furthermore, to assist with these efforts from a theological basis, public and private partnerships (PPPs or P3s) may provide even clearer guidelines as to how to approach these efforts more efficiently and pragmatically, but in order to accomplish this, some legal barriers may have to be modified or removed (Olawuyi 2022c, 81).

3.6 Conclusion

In conclusion, in order to ensure there is a substantial, comprehensive, and long-lasting transition to green growth and a low-carbon economy in the Arab world, and most particularly in Qatar, there must be an openness to reimagining both the ongoing adherence to the *status quo* on the part of business and industry and environmental laws and policies. In order to overcome these legal and institutional hurdles, there will need to be a much more comprehensive implementation of the regional national visions and strategic action plans, if the protection of biological diversity, marine and terrestrial habitats, and the commitment to a transition to renewable energy and green growth is to be achieved.

Taking into account the many rich and diverse approaches to interpreting and implementing international, regional, and domestic environmental law, as well as the global efforts toward coordination and collaboration, through the sharing of knowledge and best practices, it appears that there is much hope for headway in this regard, particularly when the levels of education and awareness are rising, giving way to demands for change. Lastly, clear and comprehensive frameworks, such as the post-2020 global biodiversity framework, in which 30 percent of terrestrial and marine environments are to be safeguarded, give a level of optimism that much can be and will be achieved, as long as the diverse array of efforts from many different international, regional, and domestic organizations and institutions are mobilized and focused toward the goal of saving our planet.

References

Aassouli D (2021) Mobilizing Islamic climate finance. In: Olawuyi D (ed) Climate change law and policy in the Middle East and North Africa region. Routledge, pp 204–230. https://www.taylorfrancis.com/chapters/edit/10.4324/9781003044109-14/mobilizing-leveraging-islamic-climate-finance-mena-region-dalal-aassouli?context=ubx&refId=816e9d67-4f56-44ea-bb77-e5f028da72d2

Al Kuwari NY, Kaiser MF (2011) Impact of North Gas Field development on landuse/landcover changes at Al Khor, North Qatar, using remote sensing and GIS. Appl Geogr 31(3):1144–1153

Al-Ghadban AN, Price ARG (2002) Dredging and infilling. In: Khan NY, Munawar M, Price ARG (eds) The Gulf eco-system: health and sustainability. Backhuys

Al-Khayat J, Balakrishnan P (2014) Avicennia marina around Qatar: tree, seedling and pneumatophore densities in natural and planted mangroves using remote sensing. Int J Sci.

Al-Marri A (2001) Environmental impact assessment in Qatar (12 April 2001) Beirut. http://digitallibrary.un.org/record/466244/files/E_ESCWA_ENR_2001_WG.1_CP.3-EN.pdf

Al Jazeera (2022) Qatar takes part in coordinative meeting of ministers responsible for environment affairs in GCC states (11 December 2022). https://www.qna.org.qa/en/News-Area/News/2022-12/11/0049-qatar-takes-part-in-coordinative-meeting-of-ministers-responsible-for-environment-affairs-in-gcc-states

B4Development (2023) B4Development foundation (B4D) is the MENA region's first behavioral insights and nudge unit. http://b4development.org/about/

Balakrishnan P (2012) Application of remote sensing for mangrove mapping: a case study of Al-Dhakira, the State of Qatar. J Earth Sci Eng 2(10):602

BDNews24.com (2019) Bangladesh court gives Turag, other rivers status of 'legal person' to save them from encroachment (30 January 2019)

Biodiversity for Development (B4D) (2022) Qatar Biodiversity Data Bank Report

Burt JA, Smith EG, Warren C, Dupont J (2016) An assessment of Qatar's coral communities in a regional context. Mar Pollut Bull 105(2):473–479

Burt JA, Ben-Hamadou R, Abdel-Moati MA, Fanning L, Kaitibie S, Al-Jamali F, Warren CS (2017) Improving management of future coastal development in Qatar through ecosystem-based management approaches. Ocean Coast Manag 148:171–181

Carpenter KE, Krupp F, Jones DA (1997) Living marine resources of Kuwait, Eastern Saudi Arabia, Bahrain, Qatar, and the United Arab Emirates. Food & Agriculture Org

Crutzen PJ, Stoermer EF (2013) The 'Anthropocene' (2000) The future of nature: documents of global change. In Libby R, Sverker S, Paul W, New Haven: Yale University Press, pp 479–490. https://doi.org/10.12987/9780300188479-041

Darwish M, Hassabou AH, Shomar B (2013) Using seawater reverse osmosis (SWRO) desalting system for less environmental impacts in Qatar. Desalination 309:113–124

Deb K, Al-Shaikh I, Al-Hitmi H, Al-Khayat J (2014) Coral relocation as habitat mitigation for impacts from the Barzan gas project pipeline construction offshore eastern Qatar. In: Proceedings of the international petroleum technology conference (IPTC), Doha, Qatar

Ecuador Constitutional Assembly (2008) Constitution of 2008. https://web.archive.org/web/200903 20002938/. https://www.asambleaconstituyente.gov.ec/documentos/constitucion_de_bolsillo. pdf

El Raey M (2010) Impact of sea level rise on the Arab Region. University of Alexandria, Arab Academy of Science, Technology, and Maritime

Erftemeijer PL, Shuail DA (2012) Seagrass habitats in the Arabian Gulf: distribution, tolerance thresholds and threats. Aquat Ecosyst Health Manage 15(sup1):73–83

Furlan R (2016) Urban design and livability: the regeneration of the corniche in Doha

Gari L (2006) A history of the Hima conservation system. Environment and History 12(2):213–228

General Secretariat for Development Planning (2008) Qatar national vision 2030. https://www.psa. gov.qa/en/qnv1/Documents/QNV2030_English_v2.pdf

Gulf Times (2023) Shura Council proposes steps to protect Qatar's environment (30 May 2023). https://www.gulf-times.com/article/661991/qatar/shura-council-proposes-steps-to-protect-qatars-environment

Hacker P, Dimitropoulos G (2017) Behavioural law & economics and sustainable regulation. In Mathis K, Huber B (eds) Environmental law and economics. Economic analysis of law in European legal scholarship, vol 4. Springer, Cham. https://doi.org/10.1007/978-3-319-50932-7_ 7

Haq SN (2001) Islam and ecology: toward retrieval and reconstruction. Daedalus 130(4):141–177. http://www.jstor.org/stable/20027722

Idllalène S (2021) Rediscovery and revival in Islamic environmental law: back to the future of nature's trust. Cambridge University Press

Idllalène S (2022) The role of environmental Waqf. In: Olawuyi D (ed) Climate change law and policy in the Middle East and North Africa region. Routledge, pp 43–62. https://www.taylorfrancis.com/chapters/edit/10.4324/9781003044109-4/role-enviro nmental-waqf-addressing-climate-change-mena-region-samira-idllalène

International Convention for the Prevention of Pollution from Ships (MARPOL) (1973) Adoption: 1973 (Convention), 1978 (1978 Protocol), 1997 (Protocol - Annex VI); Entry into force: 2 October 1983 (Annexes I and II). https://www.imo.org/en/About/Conventions/Pages/Internati onal-Convention-for-the-Prevention-of-Pollution-from-Ships-(MARPOL).aspx

International Maritime Organization (2019) MARPOL Consolidated Edition

International Renewable Energy Agency (IRENA) and League of Arab States (LAS) (2014) https:// www.irena.org/-/media/Files/IRENA/Agency/Publication/2014/IRENA_Pan-Arab_Strategy_J une-2014.pdf

International Union for Conservation of Nature (IUCN) (1983) Islamic principles for the conservation of the natural environment, IUCN Environmental Policy and Law Paper 20. https://por tals.iucn.org/library/node/6307

Ives CD, Bekessy SA (2015) The ethics of offsetting in nature. Front Ecol Environ. https://esajou rnals.onlinelibrary.wiley.com/doi/abs/10.1890/150021

Jones DA, Ealey T, Baca B, Livesey S, Al-Jamali F (2007) Gulf desert developments encompassing a marine environment, a compensatory solution to the loss of coastal habitats by infill and reclamation: the case of the pearl city Al-Khiran, Kuwait. Aquat Ecosyst Health Manage 10(3):268–276

Kahramaa (2021) Tarsheed. https://www.km.qa/Tarsheed/Pages/TarsheedIntro.aspx

Khalid FM (2019) Signs on the earth: Islam. Kube Publishing, Modernity and the Climate Crisis

Khan NY (2007) Multiple stressors and ecosystem-based management in the Gulf. Aquat Ecosyst Health Manag 10(3):259–267

Kotzé L, French D (2018) The anthropocentric ontology of international environmental law and the sustainable development goals: towards an ecocentric rule of law in the anthropocene. Glob J Comp Law 7:5–36. https://doi.org/10.1163/2211906X-00701002. https://www.resear chgate.net/publication/322943742_The_Anthropocentric_Ontology_of_International_Environ mental_Law_and_the_Sustainable_Development_Goals_Towards_an_Ecocentric_Rule_of_ Law_in_the_Anthropocene

Ministry of Education (MOE) (2014) Qatar national biodiversity strategy and action plan 2015–2025. State of Qatar. Retrieved March 28, 2023 from https://www.cbd.int/doc/world/qa/qa-nbsap-v2-en.pdf

Nasr SH (1990) Islam and the environmental crisis. MAAS J Islam Sci 6(2):31–51

New Zealand Parliamentary Council Office (2014) Te Urewera Act 2014. https://www.legislation.govt.nz/act/public/2014/0051/latest/whole.html

New Zealand Parliamentary Council Office (2017) Te Awa Tupua (Whanganui River Claims Settlement) Act 2017. https://www.legislation.govt.nz/act/public/2017/0007/latest/whole.html

Olawuyi D (2010) From Kyoto to Copenhagen: rethinking the place of flexible mechanisms in the Kyoto protocol's post 2012 commitment period. 6/1 Law, Environ Dev J 21. http://www.lead-journal.org/content/10021.pdf

Olawuyi D (2022a) Nature and sources of climate change law and policy in the MENA region. In: Olawuyi D (ed) Climate change law and policy in the Middle East and North Africa region. Routledge, pp 3–20

Olawuyi D (2022b) Climate change law and policy in the Middle East and North Africa region. Routledge

Olawuyi D (2022c) Environmental law in Arab States. Oxford University Press

Olawuyi D, Ako R (2017) Environmental justice in Nigeria: divergent tales and future prospects. In: The Routledge Handbook of environmental justice

Olawuyi DS, Athwal EI (2023) Law and governance innovations on sustainability in Qatar: current approaches and future directions. In: Cochrane L, Al-Hababi R (eds) Sustainable Qatar. Gulf Studies, vol 9. Springer, Singapore. https://doi.org/10.1007/978-981-19-7398-7_3

Özdemir I (2003) Towards an understanding of environmental ethics from a *Qur'anic* perspective. In: Foltz RC, Denny F, Baharuddin A (eds) Islam and ecology: a bestowed trust. Harvard University Press, pp 3–37. https://www.researchgate.net/publication/350726446_Towards_An_Understanding_of_Environmental_Ethics_from_a_Qur'anic_Perspective

Qatar General Secretariat for Development Planning (2008) Qatar national vision 2030. https://www.gco.gov.qa/wp-content/uploads/2016/09/GCO-QNV-English.pdf

Qatar News Authority (QNA) (2022) Minister of environment: Qatar attaches great importance to seas, integration between environment, social, economic governance (28 June 2022). https://www.qna.org.qa/en/News-Area/News/2022-06/28/0037-minister-of-environment-qatar-att aches-great-importance-to-seas,-integration-between-environment,-social,-economic-govern ance

Qatar News Authority (QNA) (2022) Qatar takes part in coordinative meeting of ministers responsible for environment affairs in GCC states (11 December 2022). https://www.qna.org.qa/en/News-Area/News/2022-12/11/0049-qatar-takes-part-in-coordinative-meeting-of-minist ers-responsible-for-environment-affairs-in-gcc-states

Regional Organization for the Protection of the Marine Environment (ROPME). https://ropme.org

Rezai H, Wilson S, Claerebolidt M, Riegl B (2004) Coral reef status in the ROPME sea area: Arabian/Persian Gulf, Gulf of Oman and Arabian Sea in status of the coral reefs of the world

Richer R (2009) Conservation in Qatar: impacts of increasing industrialization. SSRN Electron J. https://doi.org/10.2139/ssrn.2825865

Richer R (2008) Conservation in Qatar: impacts of increasing industrialization. CIRS Occas Pap (2)

Reisch L, Sandrini J (2015) Nudging in der Verbraucherpolitik: Ansätze verhaltensbasierter Regulierung. Nomos Verlagsgesellschaft

Rivers J (2018) Granting rivers legal rights: a strategy for environmental protection

Sheppard C, Al-Husiani M, Al-Jamali F, Al-Yamani F, Baldwin R, Bishop J, Zainal K (2012) Environmental concerns for the future of Gulf coral reefs. In: Coral reefs of the gulf. Springer, Dordrecht, pp 349–373

Shomar B, Darwish M, Rowell C (2014) What does integrated water resources management from local to global perspective mean? Qatar as a case study, the very rich country with no water. Water Resour Manage 28(10):2781–2791

Smyth D, Al-Maslamani I, Chatting M, Giraldes B (2016) Benthic surveys of the historic pearl oyster beds of Qatar reveal a dramatic ecological change. Mar Pollut Bull 113(1–2):147–155

State of Qatar (2002) Law No. 30 of 2002 promulgating the law of the environment protection. https://almeezan.qa/LawPage.aspx?id=4114&language=en

State of Qatar (2004) National biodiversity and strategic action plan. https://faolex.fao.org/docs/pdf/qat180906.pdf

State of Qatar (2005) Resolution No. 4 of 2005 by the chairperson of the supreme council of the environment and natural reserves (SCENR) issuing executive bylaw for law No. 30 of 2002 on environment protection ['Executive by-law']. https://www.almeezan.qa/LawPage.aspx?id=2108&language=en

State of Qatar (2011) Qatar national development strategy 2011–2016. https://www.psa.gov.qa/en/nds1/Documents/NDS_ENGLISH_SUMMARY.pdf

State of Qatar (2011b) Qatar's resolution of the council of ministers No. 15 of 2011 establishing the committee on climate change and clean development at the Ministry of Environment. https://www.almeezan.qa/LawView.aspx?opt&LawID=3172&TYPE=PRINT&language=en

State of Qatar (2011) Ministry of environment. In: Initial national communication to the United Nations framework convention on climate change. https://unfccc.int/sites/default/files/resource/final%20climate%20change.pdf

State of Qatar (2021a) Ministry of municipality and environment, nationally determined contribution (NDC), August 2021. https://www.mme.gov.qa/pdocs/cview?siteID=2&docID=23348&year=2021

State of Qatar (2021b) PM launches Qatar's national environment. Clim Strat (31 October, 2021b). https://hukoomi.gov.qa/en/news/pm-launches-qatar-s-national-environment-climate-strategy

State of Qatar (2022a) Hukoomi E-Government. https://hukoomi.gov.qa/en/service/request-new-environmental-permit-for-development-projects

State of Qatar (2022b) Ministry of environment and climate change (MoECC). https://www.mecc.gov.qa

State of Qatar (2022c) Ministry of municipality. Environmental assessment department. https://www.mme.gov.qa/cui/view.dox?id=1451&contentID=3836&siteID=2

State of Qatar (2022d) Official website of ministry of environment and climate change inaugurated (11 April 2022). https://hukoomi.gov.qa/en/news/official-website-of-ministry-of-environment-and-climate-change-inaugurated

Stavins NR, Whitehead B (1997) Market-based environmental policies. In: Chertow MR, Esty DC (eds) Thinking ecologically: the next generation of environmental policy. New Haven: Yale University Press, pp 105–117

Stilt KA (2017) Constitutional innovation and animal protection in Egypt. Law Soc 43(3):1364–1390. https://onlinelibrary.wiley.com/doi/abs/10.1111/lsi.12312

Sunstein CR, Reisch LA (2013) Green by Default. Kyklos 66(3):398–402

Sunstein CR (2011) Empirically Informed Regulation. Univ Chic Law Rev 78(4):1349

Te Awa Tupua (2017) Whanganui River Claims Settlement Act. NZ

The Guardian (2019) The anthropocene epoch: have we entered a new phase of planetary history? (30 May 2019) https://www.theguardian.com/environment/2019/may/30/anthropocene-epoch-have-we-entered-a-new-phase-of-planetary-history

The Peninsula (2022) Environment Minister launches sustainability programme at COP27 (17 November 2022). https://thepeninsulaqatar.com/article/17/11/2022/environment-minister-launches-sustainability-programme-at-cop27

The Peninsula (2023) Shura council discusses environment issues, reviews conservation strategies (06 June 2023). https://thepeninsulaqatar.com/article/06/06/2023/shura-council-discusses-environment-issues-reviews-conservation-strategies

Thaler R (1996) Doing economics without homo economicus. In: Medema SG, Samuels WJ (eds) Foundations of research in economics: how do economists do economics? Edward Elgar, Cheltenham/ Northampton, pp 227–237

The Noble Quran (2022) Trans. Muhammad Pickthal. Retrieved January 15, 2023 from https://www.searchtruth.com

United Nations Convention on Biological Diversity (CBD) (2022) Kunming-Montreal global biodiversity framework (19 December 2022). https://www.cbd.int/doc/decisions/cop-15/cop-15-dec-04-en.pdf

United Nations Convention on Biological Diversity (CBD) (n.d.) Qatar: main details. Status and trends of biodiversity overview. https://www.cbd.int/countries/profile/?country=qa

United Nations Convention to Combat Desertification (UNCCD) (n.d.) Qatar. https://www.unccd.int/our-work-impact/country-profiles/qatar

United Nations Department of Economic and Social Affairs (UN DESA) Sustainable Development (n.d.) Council of Arab ministers responsible for the environment (CAMRE), Cairo, Egypt. https://sdgs.un.org/events/council-arab-ministers-responsible-environment-camre-6727

United Nations Department of Economic and Social Affairs (UN DESA). Sustainable Development. (n.d.) The 17 goals. https://sdgs.un.org/goals

United Nations Economic and Social Commission for Western Asia (UNESCWA). (n.d.). https://www.unescwa.org/publications/regional-initiative-promote-small-scale-renewable-energy-applications-rural-areas-arab

United Nations Environment Programme (UNEP) (2014) Press release: UNEP signs landmark agreement with league of arab states to strengthen partnership on environment and sustainable development (10 November 2014). https://www.unep.org/news-and-stories/press-release/unep-signs-landmark-agreement-league-arab-states-strengthen

United Nations (1992) The 1991 Arab declaration on environment and development and future prospects, E/ESCWA/ENVHS/1992/1. https://digitallibrary.un.org/record/1292205?ln=en

United Nations (2012) Doha amendment to the Kyoto protocol. U.N. Doc. FCCC/KP/CMP/2012/13/Add.1, Decision 1/CMP.8. https://unfccc.int/process/the-kyoto-protocol/the-doha-amendment

United Nations (2015a) Conference of the parties, adoption of the Paris agreement, Dec. 12, 2015a, UN Doc FCCC/CP/2015a/L.9/Rev/1. https://unfccc.int/sites/default/files/resource/parisagreement_publication.pdf

United Nations (2015b) Islamic declaration on global climate change, August 18, 2015b. https://unfccc.int/news/islamic-declaration-on-climate-change

United Nations (2015c) Transforming our world: the 2030 agenda for sustainable development. UN Doc. A/RES/70/1. https://sustainabledevelopment.un.org/content/documents/21252030%20Agenda%20for%20Sustainable%20Development%20web.pdf

United Nations (2022) Press release: nations adopt four goals, 23 targets for 2030 in landmark UN biodiversity agreement (19 December 2022). https://www.un.org/sustainabledevelopment/blog/2022/12/press-release-nations-adopt-four-goals-23-targets-for-2030-in-landmark-un-biodiversity-agreement/

United Nations Education, Scientific and Cultural Organization (UNESCO) Gulf States & Yemen (2022) HBKU establishes new UNESCO chair on environmental law and sustainability (22 May 2022). https://en.unesco.org/news/hbku-establishes-new-unesco-chair-environmental-law-and-sustainability

United Nations Education, Scientific and Cultural Organization (UNESCO) Gulf States & Yemen (n.d.) About UNESCO Gulf States & Yemen. https://www.unesco.org/en/fieldoffice/doha

Worrell R, Appleby MC (2000) Stewardship of natural resources: definition, ethical and practical aspects. J Agric Environ Ethics 12:263–277. https://link.springer.com/article/10.1023/A:1009534214698

Yale Environment 360 (2018) Should rivers have rights? a growing movement says it's about time https://e360.yale.edu/features/should-rivers-have-rights-a-growing-movement-says-its-about-time

Open Access This chapter is licensed under the terms of the Creative Commons Attribution 4.0 International License (http://creativecommons.org/licenses/by/4.0/), which permits use, sharing, adaptation, distribution and reproduction in any medium or format, as long as you give appropriate credit to the original author(s) and the source, provide a link to the Creative Commons license and indicate if changes were made.

The images or other third party material in this chapter are included in the chapter's Creative Commons license, unless indicated otherwise in a credit line to the material. If material is not included in the chapter's Creative Commons license and your intended use is not permitted by statutory regulation or exceeds the permitted use, you will need to obtain permission directly from the copyright holder.

Chapter 4
Terrestrial Ecology, Threats, and Prioritization of Conservation Needs and Actions

A. D. Chatziefthimiou, A. T. Conkey, M. A. Nawaz, R. Richer, S. Hameed, S. AlHajri, J. Lari, S. Tull, R. Yousif, and A. Amato

Abstract In this chapter, we review and present Qatar's terrestrial natural capital, the threats that negatively affect it, and the positive conservation measures that are in place. Additionally, we present original work conducted especially for this publication: a preliminary analysis of habitat loss due to land use change over the last 70 years in selected sites; identification, analysis and ranking of and mitigation measures for global and local-based threats; as well as an assessment of risk of extinction for reported species that formed the basis for the development of species priorities-based conservation strategy also presented here. Based on these, we have identified that the essential elements for conservation and management of the natural capital that are missing, include: the current status of habitats and biodiversity; the definitive number of species per taxonomic group; an assessment of species conservation status;

A. D. Chatziefthimiou (✉) · A. Amato
Earthna, Center for a Sustainable Future, Qatar Foundation , Doha, Qatar
e-mail: a.d.chatziefthimiou@gmail.com

A. Amato
e-mail: aamato1uk@mac.com

A. T. Conkey
Independent Researcher, Texas, USA
e-mail: aprilanntc@gmail.com

M. A. Nawaz
Department of Biological and Environmental Sciences, Qatar University, Doha, Qatar
e-mail: anawaz@qu.edu.qa

R. Richer
Duke Kunshan University, Kunshan, China
e-mail: rricher@richerenvironments.com

S. Hameed
Snow Leopard Foundation, Islamabad, Pakistan
e-mail: sshoaibh@gmail.com

S. AlHajri · J. Lari
Ministry of Environment and Climate Change, Doha, Qatar
e-mail: SHALHAJRI@mecc.gov.qa

© The Author(s) 2026

A. D. Chatziefthimiou et al. (eds.), *Pathways to Nature Conservation and Resilience in Hot and Arid Lands*, Advances in Geographical and Environmental Sciences, https://doi.org/10.1007/978-3-032-11046-6_4

a centralized species database; as well as an assessment of desert specific ecosystem functions. In the concluding section, we offer recommendations and action planning on how to address these and other identified gaps, organized by topic and stakeholder.

4.1 Environmental Profile

4.1.1 Geography, Geology and Geomorphology, Water Resources

Geographically situated in the Middle East, Qatar is a peninsular country, an offshoot from the eastern Arabian Peninsula. It covers an area of 11,581 km^2, or 0.36% of the total land area of the Arabian Peninsula, with a single, 60 km long, land border to Saudi Arabia in the town of Abu Samra (Fig. 4.1). Qatar runs 160 km north to south, and 90 km at its widest from east to west. Except for urban centers and built/man-made habitats, distinct, yet limited in their productivity desert ecosystems are carved in the land area. Namely, rocky desert, rocky hills, sand dunes, wadis, rawdas, runnels, sinkholes, and caves. The relief of the land area is slight, with a hypsometric maximum of 103 m in the rocky hills in Tuwayyir al Hamir in the south.

A very interesting fact and maybe something that is not considered often, is that unlike other drylands, Qatar's hyper-arid mainland phases into very productive, albeit extreme, coastal marine ecosystems of the Perso-Arabian Gulf sea (Fig. 4.1). Three of these ecosystems hugging the 562 km of the peninsula's coastline, namely mangroves, seagrass beds and coral reefs, are the most productive in the coastal marine environment creating an antithesis to the low-productivity desert ecosystems. Additional coastal marine ecosystems include sabkha (salt flats), intertidal, subtidal, oyster beds, and open sea. Qatar takes up 15% of the 240,000 km^2 Perso-Arabian Gulf sea area, and on a bathymetric scale, Qatar's sea level ranges from 0 m to about a depth of 60 m, by the sea border with Iran (Butler et al 2020).

Although in geological times, freshwater was running in rivers and streams originating from Saudi Arabia in the west and Mesopotamia in the north, and was sipping up from wells in these wadis, currently Qatar is characterized by the lack of renewable

J. Lari
e-mail: JLari@mecc.gov.qa

S. Tull
Qatar Bird Records Committee, Doha, Qatar
e-mail: simonjtull@gmail.com

R. Yousif
Department of Geological Sciences & Geological Engineering, Queen's University Kingston, Kingston, Canada
e-mail: 19ray@queensu.ca

Fig. 4.1 Map of Qatar in a regional perspective (*Credit* Google Earth Pro)

freshwater resources, and by aquifers containing groundwater which is not replenished to the same extent to which it is abstracted, with corresponding deteriorating quality mainly seen as increasing salinity (Parker et al. 2006; Heggy et al. 2022).

For an overview of Qatar's coastal marine environment, please refer to the Chapter on Marine Ecology, and for an overview of Qatar's geology and water resources, please refer to the Chapter on Geology and Hydrology.

4.1.2   Climate

Qatar is a hot, hyper-arid desert, with an Aridity Index of <0.03, minimal rainfall (average annual 76.3 mm) during the winter months, high evaporation rates (average 6 mm per day), in addition to high temperatures which can reach to more than 50 °C in summertime (Zomer and Trabuco 2022). Seasonality is also exhibited in the dust (sand) storm events that affect Qatar. During the spring and dry summer months, the northwesterly Al Shamal wind, raises dust from the alluvial plains in the Tigris-Euphrates basin of Iraq, depositing considerable dust loads in the Arabian Region (Goudie 1983). Also during the spring months, dust and sand particles originating from the Sahara and Sahel deserts of eastern Africa, transverse to the Arabian Peninsula and reach Qatar via Saudi Arabia (Al-Dousari et al 2013). These storm events are especially felt in the central and southern parts of Qatar.

Another extreme that characterizes this desert country is the high Global Horizontal Irradiance (GHI) that can reach a yearly total of 2,020 KWh/m^2, comparable only to the Atacama (Chile and Peru), the driest hot desert in the world (Knight 2016; SSE 2017). Average annual rainfall and GHI create a dichotomy between north and south, being higher in the north, while the rocky hill range along the Qatar anticline splits the country into east and west.

For an overview of Qatar's climate, historical weather record, and future climate predictions, please refer to the Chapter on Climate Modeling and Forecasting.

4.2 Biodiversity

Official reports and scientific publications state that in Qatar, there are about 2,000 species of organisms, almost equally distributed, half in the terrestrial and half in the coastal marine environment (MOE 2014). When zeroing in on the exact numbers of each taxonomic order or species group, it becomes fast apparent that no definitive numbers exist. There are many reasons for this, primary of which is the lack of systematic and exhaustive baseline surveys and monitoring campaigns. These are necessary to assess and measure the species in the environment, to validate species records in published scientific reports, and to determine how these change over time and across space. Based on our understanding of the unexplored biodiversity in Qatar, if these efforts were to be conducted, many new species would be discovered.

In lieu of this accounting, species lists are often compiled with most likely outdated observations from the field, or with extrapolated distribution of species that have been found elsewhere in the Arabian Peninsula and are projected to be present in Qatar as well. And in this same manner, the number of threatened species in the country is also declared. Between the years of 2009 and 2024, based on the IUCN, the number of threatened species jumped from 18 to 98, which is striking both because of the steep increase and lack of place-based information (IUCN 2024).

For the purposes of this study, and to help develop a priorities-based conservation strategy, we devised a methodology and ran an analysis to determine the Species-at-Risk of extinction, and the degree of risk for each. To do this, we drew species information from disparate sources, to curate a centralized list of species that are terrestrial and/or use the terrestrial environment for any part of their life-cycle, i.e., Hawksbill turtles that birth on land but otherwise lead an exclusively marine existence. The methodology and results of this analysis are presented in Section: Threatened species/Species-at-Risk: methodology and results.

In the following sub-sections, species in this curated list are presented in a number of ways. As a general overview of individuals under their taxonomic groupings and in the communities they assemble, in terms of their unique adaptations to life in the hyper-arid hot desert, the specific habitats they colonize, and the ecosystems that develop in that physical space.

4.2.1 Microbes

The microbial kingdom of life may be the least explored. Moubasher and Al-Subai (1987) reported 142 fungal species, most likely a vast underestimate and certainly outdated. No definitive species lists for bacteria or archaea exist, even if many research teams study and report on microbes in all terrestrial and coastal marine ecosystems in Qatar, including biocrust and marine mat communities (Moubasher and Al-Subai 1987; Al-Thani and Bassam 2021; Shariah et al 2021; Chatziefthimiou et al 2024 and references therein).

4.2.2 Flora

Flora of Qatar is a unique assemblage of species that reflects African and Asian centers of plant diversity with the influence of European species as well (Miller and Cope 1996; Richer et al. 2022). Species richness exceeds 500 with new species records both likely and expected to emerge with appropriate field studies. Richer et al. (2022) document 470 species with new species records for Qatar. An additional 74 species having been recorded by other authors (Abulfatih et al. 2001). The status of species populations and how those populations are changing through time and space is currently unknown.

Endemic species have not yet been recorded in Qatar. However, this may change with appropriate taxonomic work. Recently, new species identifications have been made of species in Qatar. *Bienertia sinuspersici* (Akhani et al. 2005) is a succulent species with unique leaf anatomy. It is only the third terrestrial C4 species to be identified without the characteristic Kranz anatomy. *Suaeda iranshahrii* (Freitag et al. 2013) is a newly identified coastal succulent. Both species were identified with taxonomic work done on Iranian samples as this is reflected in the scientific names. Thus, as more work on the flora is encouraged, new species records for Qatar and species new to science are likely to continue to arise.

Even if there have been a plethora of publications on the floral species found in Qatar, to the best of our knowledge, the communities of plants in Qatar have only been surveyed, recorded, and described by Batanouny in his 1981 seminal book: Ecology and Flora of Qatar (Batanouny 1981). As such, segments of the presented information are likely outdated, since major developments (industrial, infrastructural, residential, diplomatic crises, etc.) have taken place in Qatar between that time and the present, incurring stresses that may have influenced not just the assembly of these communities but their distribution as well.

Another interesting yet positive development that occurred post publication of Batanouny's book is the conservation effort led by the ministry of environment (then SCERN) starting in the 1980s, to afforest the coastline of Qatar with the mangrove species *Avicenna marina* (Abdel-Razik 1994). This has led to a broader distribution of mangroves across the country than what Batanouny described. Also, through time the forests have phased into secondary successional development of communities that include more than what Batanouny described mangrove to be: "its own dense pure population with no other associates". Farther below, we describe the present day community of this species that has been surveyed and recorded for the purposes of this publication.

To begin with, a total of 20 plant communities are described, 11 in the terrestrial environment and 9 in the coastal marine environment, otherwise mentioned as halophytic. There are 4 tree species making up 3 communities, *Acacia tortilis* and *Lycium shawii* being co-dominant keystones of the same community, and the rest of the keystones all being perennials, either herbs/shrubs or grasses. In the terrestrial environment, communities have more numerous associates, with the tree species *Ziziphus nummularia*, locally Sidr, enumerating 31 annual and perennial associates, followed by the *Acacia tortilis* and *Lycium shawii* community of 26 associates, and the herb/shrub species *Zygophyllum qatarense*, the most common species with the most extensive distribution in the country, and 24 combined associates. Unlike *Z. nummularia*, the other two communities have distinct associates in their northern distribution versus the southern distribution range. In the coastal marine environment, the perennial shrubs *Suaeda verniculata* and *Limonium axillare* (sea lavender) have the most associates, at 5 each.

The terrestrial habitats that are strewn with these plant communities include runnels and rawdas of the wadi system, sand sheets and sand dunes, and in less abundance the rocky desert habitat. For a description of the flora communities in each of these terrestrial habitats, please refer to Sect. 2.7. The halophytic communities take root in habitats inundated with seawater to different degrees and frequency, namely the sabkha, sandy beaches, and intertidal habitats.

Finally, and specific to the plant community associated with the mangrove *Avicennia marina*. During surveys conducted between October 2022 to April 2023, in the Purple Island, in Al Thakira, and in Umm Tais island, we observed that the mangrove community is made up of all the halophytic keystones as described by Batanouny (1981), except for the perennial grasses *Aeluropus lagopoides, Halopyrum mucronatum* and *Sporobolus ioclados*. Other common associates include *Arthrocnemum macrostachyum, Haloxylon salicornicum,* and in some cases the seagrass *Halodule universis* when its meadows hug the same intertidal habitat with the mangroves.

4.2.3 Fauna

Faunal species are distributed across the terrestrial habitats, as well as the sabkha and mangrove habitats in the coastal marine environment. In approximate numbers, there are 21 mammal species, 32 species of reptiles (21 species of lizards and possibly 11 species of snakes), 3 species of amphibians, 3 species of bats, 2 species of scorpions. (Harrison 1991; Egan 2008; Castilla et al. 2014, 2017; Abdulrahman et al. 2021).

The official list of birds occurring in the State of Qatar amounts to some 378 species and subspecies, of which 364 occur regularly, with 77 known or suspected to have bred (https://qatarbirds.org). The remainder occur on autumn/spring passage or are resident during the winter months, reflecting the location of the country on the African-East Eurasian and Central Asian Flyways. The peak period for avian diversity is thus from September to April. According to Birdlife International, 13 bird species that occur in Qatar are globally threatened (https://datazone.birdlife.org/country/fac tsheet/qatar). These include coastal/marine species such as Grey Plover (*Pluvialis squatarola*), and Socotra Cormorant (*Phalacrocorax nigrogularis*), farmland species such as Sociable Lapwing (*Vanellus gregarius*), and waterbirds such as Common Pochard (*Aythya ferina*) and Ferruginous Duck (*Aythya nyroca*).

The documentation of insect diversity in Qatar began in the 1980s, with a total of 441 insect species having been recorded in Qatar by 2025, with many more expected to be discovered with additional field studies (SM Table 4.1). Pittaway (1980) identified 14 butterfly species, while Abdu and Shaumar (1985) published a list of 170 insect species from various habitats across the country. Walker and Pittaway (1987) also documented insect species from the Eastern Arabian Peninsula, including records from Qatar. Abushama (1997, 1999) studied the arthropod fauna of Qatar and noted that most desert arthropods belonged to the class Insecta. In a later study, Abushama (2006) showed that ants (Formicidae) and rove beetles (Staphylinidae) were among the most frequently encountered groups.

Abdel-Dayem (2007) investigated insect species associated with the Ghaf tree (*Prosopis cineraria*) and reported 65 species. Háva (2007) recorded various Dermestidae (Coleoptera) from the Arabian Peninsula, including one species from Qatar, and later listed 12 Dermestidae species from Qatar specifically (Háva 2008). Harten et al. (2008, 2009) documented arthropods of UAE, including several records from Qatar. Moreover, Soldati (2009) documented darkling beetles (Tenebrionidae), one of the most dominant insect groups in the region, listing 51 species from Qatar. In a regional context, Lush (2009) recorded several ant species from the Middle East, including one species from Qatar.

Grunwell (2010) listed nine species of dragonflies and two of damselflies from various sites in Qatar. AlHajri (2013) documented 110 insect species from different habitat types in northern Qatar using pitfall traps. Garcia-Paris et al. (2013) reported the first record of the blister beetle *Meloe (Mesomeloe) coelatus* Reiche, 1857 in Qatar, while Mas-Peinado et al. (2013) reported the first record of the darkling beetle *Scaurus puncticollis* Solier, 1838. Wetterer (2013) studied the geographic distribution of the sword ant *Brachyponera sennaarensis* Mayr, 1862, including its presence in Qatar.

Kardousha (2015, 2016a) documented 16 mosquito species from the northeastern part of the country. Further, Kardousha (2016b) recorded 11 aquatic insect species from man-made wetlands. Sharaf et al. (2015, 2016) reported the first records of three ant species in Qatar: *Strumigenys membranifera* Emery, 1869, *Trichomyrmex destructor* Jerdon, 1851, and *T. mayri* Forel, 1902. Later, AlHajri and Agosti (2019) reported nine additional ant species for the first time, including *Tapinoma melanocephalum, T. simrothi, Camponotus thoracicus, Cataglyphis livida, Cardio-condyla emeryi, Crematogaster antaris, Pheidole indica, Trichomyrmex lameeri*, and *Tetramorium sp.*

More recently, Sharaf et al. (2020) reported a list of 23 ant species, including nine recorded for the first time in Qatar *Camponotus oasium, Cataglyphis arenaria, Lepisiota bipartita, L. gracilicornis, Monomorium abeillei, M. areniphilum, M. subopacum, M. venustum,* and *Pheidole sinaitica.*

4.2.4 Adaptations of Organisms to the Desert Extremes

Desert plants have evolved a suite of adaptations to withstand and thrive in climatic conditions that push the limits of biological existence. In general, the desert plants are separated in terms of their life history strategies into annuals (r strategy) and perennials (K strategy) (Friedman 2020). The annuals go through their whole life cycle within just a few short months following the rain season. They invest in fast germination, growth, flowering and seeding, and come the hotter temperature fronts of the later Spring, they go into dormancy.

The perennials on the other hand, follow the K life history strategy, and are highly efficient in extracting limiting nutrients from their surrounding environment, as they are physiologically adapted to conserve energy and prevent water loss. Since leaves and flowers are the organs where most water is lost during transpiration, and it follows that larger leaves and flowers will lead to greater water loss, desert perennials have small or absent leaves relying on photosynthesis in the stem, and small or succulent or absent petals on flowers. Perennial plants are also likely to use alternative photosynthetic processes such as C4 and CAM that reduce water loss (Zahedi et al 2021).

In saline ecosystems, such as coastal and inland sabkhas, many plant species will have associations with microbial communities that ameliorate salt conditions in the soil. In addition, many plants in high saline conditions will actively excrete salt via salt glands in the leaves. Sea Lavender (*Limonium axillare*) and mangroves (*Avicennia marina*) are equipped with this adaptation, in addition to the active salt filtration that occurs in the pneumatophore structures in the mangroves (Yasseen and Abu-Al-Basal 2008).

Qatar's animals possess unique physiological and behavioral adaptations for dealing with a lack of water and the temperature extremes of hot deserts. Small animals (less than 20 g) tend to gain and lose heat quickly because they have more exposed body surface area compared to their small volume of body size. They evade the extreme summer heat, a behavioral adaptation, by living under shaded vegetation, rock crevices, and emerge when temperatures are more favorable, at dawn and dusk or at night in the case of nocturnals (Ward 2016). Many desert animals, among others the Arabian Red Fox (*Vulpes vulpes arabica*), the Spiny-tailed Agama (*Uromastyx aegyptia microlepis*), and the Desert Hare (*Lepus capensis*) form burrows in the desert floor because there is a significant temperature differential between the surface and just a few centimeters deep into the soil, which helps minimize water loss from evaporation. Furthermore, humidity exuded from the side walls of the burrow, allows animals, especially reptiles, to maintain their skin moistened (Wilmer et al. 2000).

Desert animals that are active during the day (diurnal) have body features and exhibit behaviors that help them reduce excessive heat. Many desert animals have a high tolerance for environmental heat and a reduced metabolic rate that may help with energy conservation and reduce water loss. Some animals, including camels (*Camelus dromedarius*), will minimize exposure to direct sun by standing in line with the sun's rays to shade the middle of the body. To reduce heat conduction from the ground to the body, they often have long legs to keep the body raised, which allows for greater airflow under the body. Desert insects, such as the Darkling Beetles (of the Tenebrionidae family), have long legs for running between patches of shade and have waxy filaments across the body that increase reflectance of sunlight and help reduce water loss. Some lizards run on their tip-toes and alternately raise their feet off the ground to minimize contact with hot soil. Likewise, desert snakes like the Horned Viper (*Cerastes gasperettii*) often use sidewinding locomotion to reduce the amount of their body in contact with the sand.

Mammals lose water and salts when they pant or sweat to cool their bodies. But many of the large desert ungulates, including the Dromedary Camel (*Camelus dromedarius*) and Arabian Oryx (*Oryx leucoryx*), have high heat tolerances. They use adaptive hyperthermia as a way of conserving energy and water. Their body temperatures rise several degrees with the heat of the day, and then they radiate that extra heat off their bodies at night. This allows them to minimize evaporation loss due to sweating. In addition, they have a counter-current cooling system where blood is cooled as it passes through a network of veins in the nasal cavity and helps cool the blood traveling to the brain (this system also works to reclaim water as moisture condenses along the elongated nasal tissues and is reabsorbed). They are also able to reclaim water through their intestines and kidneys and produce dry fecal matter and concentrated urine (Fuller et al 2014).

4.3 Habitats, Communities, and Ecosystems

The distinct yet limited in their productivity habitats that are carved in the terrestrial environment include rocky desert (*hazm, hazoon, hamada*), Barchan sand dunes (*tu's, kheyt, nigyan*), and the watershed system (*wadi*). The individual habitats within the wadi are the rocky hills (*jebel, jabal*) at their highest elevation reaching 102 m, the water channels (runnels) that are created after the rains, carrying water into the fertile depressions (*rawda, rodah, riyad*), the sinkholes and the caves (dahl) (Batanouny 1981); These are described in more detail below.

Fig. 4.2 Panorama of the rocky desert habitat (*Credit* Cohen R)

4.3.1   Rocky Desert Ecosystem

The rocky desert is the habitat most abundantly seen in Qatar. The rocks are formed by weathering of local bedrock, while finer particles may also be transported on site by sand- or rainstorms. Rocky desert is found in undulating higher ground, and usually surrounds rawdas, sand dunes, as well as rocky hills, and borders with inland and coastal sabkha (Fig. 4.2).

Even if they seem inhospitable, many desert dwellers make their home in this ecosystem while they forage and hunt in the surrounding ecosystems where vegetation abounds. For this reason, the animal community can be thought of as shared among them (See Sections below for animal community on an ecosystem basis). For example, Spiny-tailed agamas, dig burrows in the soil among and below rocks, where they reside in times of rest, and to avoid predators and extreme environmental conditions alike, and forage on plants in the rocky desert, runnels and rawda.

The plant community takes root in areas between rocks where soil is exposed, and is dominated, indicatively, by *Ziziphus nummularia, Acacia tortilis,* and *Lycium shawii* among annuals of *Zygophillum quatarense, Savignya parviflora,* and *Helianthemum lipii.*

4.3.2   Sand Dune Ecosystem

A striking feature of the Qatari terrestrial environment is the crescent-shaped sand dune, or Barchan. Barchan morphology is a byproduct of the unidirectional impact of Al Shamal winds, blowing sand onto the windward face. Sand passes through the top thin line of the crest, falling off to the leeward face, and on either side to the horns of the dune (Fig. 4.3).

Sand dunes may appear stationary, yet they actively travel, merge with one another and break off again. On average, Barchans may move about 10 m per year. It is thought that Barchan sand originated from the floodplains of Mesopotamia in Iraq, migrating in geological times across the Gulf basin when it was still dry to Qatar

Fig. 4.3 Panorama of the sand dune habitat (*Credit* Cohen R)

and all the countries to the south (Edgell 2006). Flooding of the Gulf after the last ice age, has inadvertently cut off the Barchans from their source of sand, and thus new ones have not been formed since about 6,000 years ago. In Qatar the sand dune fields commence at the center of the country running southeasterly. Eventually, and in geological times, the slow migration of the sand dunes will lead them to their "sinking" into the Gulf and this habitat will be forever lost.

Temperature and UV light on the surface of the sand dune are significantly higher than inside the dune (Louge et al 2013). Skink lizards (*Scincus mitranus*), the only organism living *exclusively* inside sand dunes, avoid the scorching heat by swimming inside dunes. Sand cats may form burrows at the base of sand dunes, yet unlike skinks, they may reside in any terrestrial ecosystem as long as the desert floor is made out of soft or sandy soil. Other animals that live around dunes include fox, owls, snakes, scorpions, geckos, camel spiders, multiple bird species, and camels.

The base of sand dunes located on top of rocky desert in central Qatar may be dominated by *Cyperus conglomeratus*, *Helianthemum* species, *Cornulaca aucheri*, and multiple grass species. The plant community of sand dunes on top of inland and coastal sabkha in the south is enriched with halotolerant plants like *Seidlitzia rosmarinus* and *Suaeda vermiculata*, growing on all surfaces of the dune.

4.3.3 Wadi Ecosystem

Wadi, meaning valley in Arabic, is an ephemeral watercourse, flash flooded during the rain season. Topographically, the wadi ecosystem lies enclosed within rocky hills and low-lying areas of undulating rocky desert around Qatar (Fig. 4.4).

Rainwater carrying soil particles, rocks, and nutrients drains from the tops of rocky hills and rocky desert into the valley in runnels, and if its course is blocked, water and soil end up and get deposited into the rawda, or into the sabkha in the case of the coastal marine environment. Runnels may be long or short, deep or shallow, line- or oval-shaped and sometimes they are collectively called wadi (Fig. 4.4). Nutrient deposition and water passage make runnels very fertile ground where diverse

Fig. 4.4 Panorama depicting the wadi ecosystem. Higher relief rocky desert is strewn with runnels in the foreground, leading to the rawda habitat (*Credit* Cohen R)

communities of plants take root, consequently attracting diverse communities of animals.

The plant community includes the brown lily *Dipcadi erythraeum*, *Aizoon canariense*, *Blepharis ciliaris*, *Anchusa hispida*, and *Glossonema varians*. *Helianthemum kahiricum* and *lippii* may be present and are indicator species for the desert truffles that crack the surface in runnels and rawdas after the first rains.

The animal community is often transient, using runnels as feeding grounds, while making their burrows in other parts of the wadi or in the nearby rawda and rocky desert. These may include birds, migrant and local, spiny-tailed agamas, geckos, monitors, snakes, jerboas, camels, hedgehogs, foxes, and others.

4.3.4 Rawda and Cave Ecosystems

Rawda in Arabic means garden, meadow, or fertile depression (Edgell 2006). Geologically, rawda along with sinkholes, caves and solution hollows, are karst features, and provide evidence of a wetter Arabia. Karst forms from the dissolution of carbonate rock by circulating groundwater, which in Qatar, left behind caves in the limestone. With time, cave openings are created, then caves get filled in with aeolian sand and alluvial soils, and when the arch opening collapses, a rawda or a sinkhole is created (Chatziefthimiou 2019). In the literature, the terms cave and sinkhole are oftentimes used interchangeably.

Although the number of reported depressions in Qatar may vary from 870 to 9,736 depending on the source, caves are not that numerous, probably because they have already been filled in with sand and soil naturally being moved by the force of air and water (Sadig and Nasir 2002; Orndorff et al. 2018). The investigated and reported sinkholes and caves in Qatar are found in the northern and central parts of the country, and only 2 of those have a big enough opening and cavern space for animal movement and activity, namely the Hamam or Dhal al Hamam sinkhole in Doha, and Musfer or Dahl al Musfer in Mukaines (Fig. 4.5). Dhal al Hamam is

Fig. 4.5 Panorama depicting the cave habitat, specifically the Dahl al Musfer cave in Mukaines (*Credit* Cohen R)

reported as the only cave that has/had saline water, and is connected to the Arabian Gulf waters subterraneously. Dahl al Musfer on the other hand, as all other known caves in Qatar, is connected to ancient groundwater sources.

Moving on to the surface expression of sinkholes, the rawda is lusciously-green and most productive after the winter rains when the understory annuals emerge, while during the summer it is seemingly inactive, brown, with a cracked desert floor and dusted perennials (Fig. 4.6).

Rawda receives only ephemeral surface water, contrasting oases, receiving a constant supply of groundwater throughout the year. It should be mentioned here, that to the best of our knowledge, there are no oases present at this time in Qatar. Flows from direct rainfall, runoff from uplands, and through runnels transport soil particles, rocks, and nutrients that accumulate in the depression providing for the establishment of denser plant and faunal assemblages. Biocrusts, often covering inter-plant spaces, turn green from brown after the rains, stabilizing and making soil more fertile, setting the stage for higher plants to take root.

Native annuals include *Tetraena qatarensis*, *Fagonia bruguieri*, *Plantago* spp., and *Zygophyllum simplex* among others. Rawda in secondary succession, additionally

Fig. 4.6 Panorama depicting the rawda habitat (*Credit* Cohen R)

supports *Rhanterium epapposum* shrublets and *Lycium shawii* shrubs, a relative of tomato, and the trees *Acacia tortilis, Acacia ehrenbergiana, Ziziphus nummularia,* and more rarely *Phoenix dactylifera* and *Prosopis cineraria. Ziziphus nummularia* and *Lycium shawii* are unique since phytogenic mounds (nebkhas) form at their base, where rainwater is retained and wildlife create burrows to reach deeper and wetter ground, a common behavioral adaptation to avoid extreme conditions and predators. Fauna include Spiny-tailed Agamas, Jerboas, Arabian fat-tailed Scorpions, birds, foxes, hedgehogs, and others.

The community of organisms inhabiting caves is not well described. Often flora is absent, some birds occur feeding on insects, co-inhabiting with the bats: Trident Leaf-nosed Bat, Naked-rumped Tomb Bat, and the Desert Long-eared Bat (Abdulrahman et al 2021).

4.3.5   Rocky Hill (Jebel) Ecosystem

Rocky hills, or jebels in Arabic, are surrounded by sloping erosional plains and aeolian sands. The top surfaces of the jebels are usually covered with weathered rocks, and vegetation is sparse (Fig. 4.7).

The texture and composition of the hillsides depends on the bedrock material. Sometimes it is just exposed white limestone, void of vegetation, while other times the hillsides are covered with soils that nourish a healthy plant community. Jebels are mostly positioned along the Qatar arch, which separates the country approximately in the middle and runs from the northern most Al Ruwais to the southern most inland waters of Khor Al Adaid, or along the Dukhan anticline and the Al Huriyeh syncline proximal to the west coast, running from Dukhan all the way to Abu Samra. The highest reported elevation is 103 m in Tuwayyir al Hamir in the south.

The plant community resembles that of rocky desert, comprising the annuals *Aizoon canariense, Blepharis ciliaris, Anchusa hispida, Helianthemum lippii* and others. Sometimes, although very sparingly, *Lycium shawii* is observed and *Acacia*

Fig. 4.7   Panorama depicting the rocky hill habitat (*Credit* Cohen R)

tortilis, Acacia ehrenbergiana, Ziziphus nummularia, and more rarely *Prosopis cineraria.*

Inside crevices of jebels, many an animal make their transient or permanent home. The Pharaoh Owl Eagle, the Arabian Red Fox, and the Spiny-tailed Agama are quite often observed in these hollows, either through direct sightings or signs of their activity (prints, pellets, scat, etc.). Young of both the Pharaoh Owl Eagle and the Arabian Red Fox have also been observed inside these crevices during the reproductive season of early to mid-winter.

4.3.6 Built Habitats

Built, artificial, or man-made habitats may also host a variety of organisms. Urban parks, of which a total of 83 exist, such as Al Bidda in Doha, farms such as Irkaya Farm in Ar Rakiyah, and Treated Sewage Effluent Storage Lagoons like the one in Al Karaana, are just a few examples of built habitats, that due to an abundance of water and food sources have become hotspots of biodiversity in this hyper-arid country. If birds can be used as a bio-indicator of the overall biodiversity benefit, approximately 75% of Qatar's bird species have been recorded at Irkaya Farm, while the 9 eBird top hotspot lists include Irkaya Farm and 3 more built habitats.

4.4 Threats to Biodiversity

In general, the five threats to biodiversity in order of their severeness are: habitat loss (land/sea use change or conversion, habitat degradation, etc.); species over-exploitation (over-grazing, over-exploitation of fisheries, etc.); pollution (plastics, noise, air, chemical, etc.); invasive species and disease; and climate change (drought, desertification, unpredictable weather patterns, etc.) (Fahrig 2003, Brondizio et al. 2019, WWF 2020).

4.4.1 Habitat Loss, Fragmentation, Degradation, and Land Use Change

Unequivocally, of the five universally recognized threats to biodiversity, the most grave is that of habitat loss, since organisms not only lose settlement and homing grounds but natural resources as well, and of course plants are decimated in the process rendering the land infertile, a leading driver to desertification. Data for the whole of Qatar on habitat loss or land use change over time are not readily available, yet this change within the confines of Doha may be indicative. Namely, between the years 1997 and 2010, there was a 288% increase in land use due to development (Qatar Atlas 2020).

For the purposes of this work, we ran a preliminary analysis of habitat loss due to land use change *in selected rural sites in Qatar*, using aerial imagery retrieved in 1963, 1971, and 1995 superimposed against current (2023) satellite imagery in the same locations (Zhang et al 2020). In the terrestrial environment in the last 60 years, we observe a 73% loss of rawda and sinkhole fertile depressions, i.e., the most productive habitat in the terrestrial environment, due to development (Fig. 4.8).

The habitat loss in the depicted sites is due to the demands of an ever-growing population, in terms of housing and infrastructure amenities (all panels), as well as energy production/supply (Ras Laffan Industrial City in northeast panel A), international transportation of human population (Hamad International Airport in panel B), and trade of goods (Hamad International Port in southeast panel C).

Sand dunes in specific areas of the country have also been lost as their sand is extracted to be used in construction and in agriculture (Fig. 4.9).

As is the case for the loss of rawdas and sinkholes shown above, loss of the sand dune habitat directly translates into a lost community of organisms, especially the skink (*Scincus mitranus*) population, since these lizards are exclusively associated with the sand dunes and thus this habitat is essential for their survival. If ever the

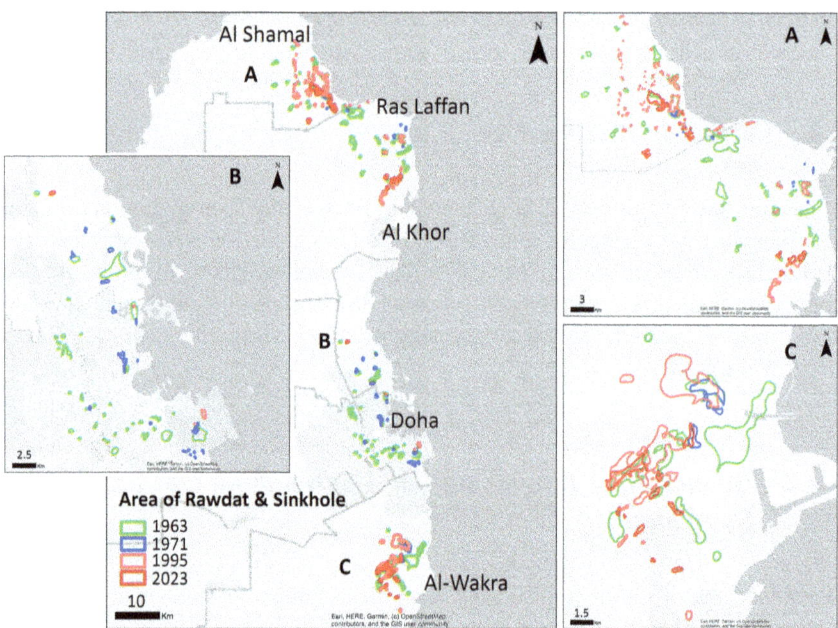

Fig. 4.8 Map of loss of rawda and sinkhole habitats in selected rural sites in the northeast of Qatar **a**, Doha metropolitan area **b**, and in the southeast. The extend of these areas is shown for the following years: 1963 (green), 1971 (blue), 1995 (pink), and 2023 (red) (*Credit* Yousif R)

Fig. 4.9 Maps depicting a Barchan sand dune field (black parallelogram) in the center of the country prior to extraction in 2016, during sand extraction in 2017, and post extraction works in 2022 (*Credit* Google Earth Pro)

extend of loss of this habitat becomes pervasive, then this may lead to a habitat-species type of extinction for the skinks. In addition, in this specific geographic location depicted in Fig. 4.9, the extend, and magnitude of the pressure on the habitat was so severe, that the threat has now become irreversible, considering of course that sand dunes are geological formations that cannot be re-constructed no matter the willingness and intensity of conservation efforts that could be applied to mitigate the lost habitat.

4.4.2 Species Over-Exploitation

Species over-exploitation of terrestrial wildlife in Qatar can include over-harvesting, wildlife trafficking, illegal hunting and poaching (killing). The last two are often carried out for food for humans or kept falcons, sport, or trade. Snakes and lizards (Cox et al. 2012), owls, eagles, desert foxes, scorpions, and bats may be killed out of fear, superstition, or presumed competition with or predation on domestic animals. Brochet et al. (2019) estimated that 61% of bird species regularly occurring in Qatar are among those illegally poached, most of which were taken for sport.

Over-harvesting (including trapping and capture) may be done for sport, consumption, or the pet trade. Falcons and other raptors may be taken from the wild to be kept and trained for falconry. The Houbara Bustard had been over-hunted due to falconry. The Desert Cape Hare may be threatened by sport and consumption hunting with dogs. Reptiles face over-harvesting from the pet market and pet trade. In particular, the Spiny-tailed Agama (also called Dhub) may be over-harvested for consumption in traditional meals and pet markets (Conkey et al. 2023).

Wildlife trafficking includes biological items poached in Qatar or in other locations and illegally brought into or through the country. For example, falcons may be illegally trapped in another country and imported into Qatar (Brochet et al. 2019). Some global wildlife trafficking has been linked to organized crime and terrorist organizations.

Fig. 4.10 Picture depicting a site in Umm Bab, on the west coast of the country, in its pristine state in 2017 (inlet picture with Starr) and in 2023 after all vegetation was lost to overgrazing (broader panorama) (*Credit* Chatziefthimiou AD, Cohen R)

Of course, species over-exploitation can be applied to plants as well. Usually, the cause is unmanaged grazing by goats, sheep, and camels, which at its most severe forms leads to the denuding of the desert floor, loss of fertile land and desertification (Fig. 4.10). In the past, mangrove forests around the country suffered losses from overgrazing, yet this has since been fully prohibited.

Figure 4.10 showcases an example of how a diplomatic conflict can lead to habitat loss and species over-exploitation not out of neglect or inefficient management of natural resources, but out of necessity and unforeseen emergent circumstances. The point in case refers to the Qatar diplomatic crisis between 2017 and 2021. During the time while the international relations were severed, Qatari owned camels kept all around the GCC region were returned to Qatar. The area from Umm Bab to Al Nakhsh was designated for camel farm development to accommodate this influx. The extend and magnitude of pressure on the flora was so high, that the threat has now become irreversible, unless directed conservation measures and efforts are applied to bring the barren soil to health, to reconstruct the natural floral communities, which will lead to recolonization of this area by the wildlife community.

4.4.3 Invasive Species and Disease

Invasive species are those that have been introduced to a new area and find the conditions favorable for survival, reproduction, and expansion of their distribution. The Global Biodiversity Information Facility lists 140 species of introduced and invasive species (including 112 plants) in Qatar from data compiled in 2019 (Elazazi et al. 2020).

Urban areas, ports, and animal markets are likely places where non-native species are first introduced. Invasive species often cause changes in the native ecosystems once they have become naturalized and can displace native species. The extremes of desert living may prevent many exotic species from becoming established in Qatar, however, access to water and food in urban and residential areas, parks and built habitats, and farms can provide suitable conditions for exotic species to take hold.

For example, Qatar has several invasive and introduced bird species that may become invasive, such as the Common Myna and parakeets (Conkey et al. 2023).

Invasive plant species spread quickly and can alter native ecosystem functions. One example, Fountain Grass (*Penniseum setaceum*), increases fire risks in areas where it has grown. In addition to out-competing native species, exotic and invasive species can introduce new diseases to Qatar. The international pet and animal parts trade is a well-known avenue for zoonoses to spread (Barradas et al. 2019; Seimenis 2008). Introduced arthropods, such as mosquitos, flies, lice, fleas, and ticks, can carry vector-borne diseases that can spread to domestic animals, wildlife, pets, and humans (Schaffner et al. 2021). Live animals brought into the country for meat and imported meat products can carry microbial pathogens like *Salmonella*, *Campylobacter*, *Toxoplasma gondii*, and toxic strains of *Escherichia coli* (Schaffner et al. 2021). Humans can be exposed to these microbes during slaughter or meat processing or contamination of food serving surfaces.

4.4.4 Pollution

Air Pollution

Air pollution, from natural sources and human activities in Qatar, frequently exceeds recommended air quality standards for human exposure (Lanouar et al. 2016). Energy producers and fossil fuel combustion are the greatest contributors to greenhouse gasses in Qatar (Lanouar et al. 2016). Industrial manufacturing emissions, construction activities, and vehicle emissions also contribute to air pollution (Lanouar et al. 2016).

Air pollution from dust, fumes, gas, vapor or mist, odor, and smoke can seriously impact our environment and people's health and well-being. Sources of outdoor air pollution include vehicle emissions, industrial emissions, dust storms, and airborne particulate matter from agriculture, manufacturing, and construction activities. Chemical reactions of noxious gasses in the air can produce ground-level ozone and volatile organic compounds. The contributions of natural sand storms versus the human induced deterioration of air quality from particulate matter loading is not well understood, yet in the Middle East as a whole, natural sources of dust account for about 50% of the urban pollution from PM2.5 (Metcalf et al 2021 and references therein.

Breathing in air pollutants can damage the airways and lungs leading to respiratory illnesses, such as asthma, bronchitis, chronic obstructive pulmonary disease, lung cancer, and a decrease in life expectancy (Cunningham et al. 2023). Small particulate matter can even enter the bloodstream and increase the risk of stroke, heart disease, neurological diseases, diabetes, and babies born with low-birth weight (WHO 2023). Breathing particulate matter nucleated by cyanobacteria and cyanotoxins, widely distributed in natural habitats in Qatar, has also been identified as a causative agent of neurological diseases (Richer et al 2012; Metcalf et al 2021 and references therein).

Although less is known about the relationship of air pollutants on vertebrate animals in the wild, negative effects of air pollution components have been documented in mammals (Gupta and Bakre 2013; Veltman et al. 2009), birds (Barton et al. 2023), and in lab experiments in reptiles (Mautz et al. 2004) and amphibians (Dohm et al. 2001, 2008). A review of the literature by Sanderfoot and Holloway (2017), revealed that exposure to air pollution in birds was related to "respiratory distress and illness, increased detoxification effort, elevated stress levels, immunosuppression, behavioral changes, and impaired reproductive success".

Air pollution can have negative impacts on plants as well. Plants take in air pollution through their stomata. Particulate matter can accumulate on the surface of vegetation and smog and haze reduce visibility; both factors can block out the sunlight necessary for photosynthesis. Acid rain can damage plant leaves, make the soil more acidic, and erode stone, metals, and paints on buildings. Airborne pollutants settle on the soil and contribute to soil pollution. Plants then absorb and incorporate these pollutants in their tissues as they take in water through their roots. All these impacts contribute to decreased plant production and growth (He et al. 2012).

Water Pollution

Water pollution includes a change in water quality that negatively affects the usability of water or makes the water unsuitable for living organisms. The main sources of water pollution are industrial wastewater, sewage, agricultural runoff, urban runoff, and trash. Pollution can contaminate surface-water and groundwater and harm aquatic life (Mushtaq et al 2020). Sewage and runoff of excess nutrients from fertilizers and animal feces can lead to eutrophication and Harmful Algal and Cyanobacterial Blooms. This uncontrolled microbial growth depletes the oxygen in the water that in conjuction with toxin accumulation and consequent exposure causes pet and wild animals including fish to die as has been reported numerous times in aquatic and near shore-habitats in the Arabian Gulf and elsewhere (Chatziefthimiou et al 2021 and references therein). The first intoxication report of a pet dog in a terrestrial setting was published in Qatar (Chatziefthimiou et al 2014). The incident occured as a result of consumption of rainwater from inside a natural depression lined with actively growing biocrusts and accumulated cyanotoxins. Finally, sewage and animal feces can contaminate water with disease causing parasites, bacteria, and viruses.

Soil and silt from land erosion can cover aquatic vegetation and fill ponds and wadis with sediment. Organic chemical pollutants, such as pesticides, plastics, detergents, oil, and petroleum, are common in agricultural, urban, and household runoff. Plastics in water bodies fragment into smaller pieces over time and are incorporated into the food web when they are consumed by terrestrial and aquatic organisms alike (Susanti et al 2020). Unmanaged and untreated brine discharge from desalination plants also has deleterious effects to the aquatic biota and to environmental health, and this is of significant importance in countries of the Arabian Gulf since the vast majority of drinking water is produced through desalination (Panagopoulos and Haralambous 2020).

Soil Pollution

Soil pollution can reduce plant productivity and growth and make edible crops unsafe to eat, as well as contaminate surface water and groundwater. Soil pollution is caused by the accumulation of harmful chemicals in the soil and is also related to air pollution particulates settling on and building up in the soil. Industrial emissions, runoff, and solid waste disposal are the main causes of soil pollution. Plastics break down into microplastics and can contaminate the soil.

Microplastics were documented in beach sediments at 8 coastal sampling sites in Qatar (Abayomi et al. 2017). A more recent study concluded that soil pollutants in Qatar were within allowable limits except for Strontium (Sr) at some sites (Shomar et al. 2023). Heavy metals (arsenic, chromium, mercury, and uranium) were more likely to be found in soils near industrial and urban areas and near major roads in Qatar (Bou Kheir et al. 2019).

Land Surface Pollution

Land-surface pollution is caused by the accumulation of solid waste on land. The main sources of land-surface pollution in Qatar are industrial waste, construction waste, and municipal solid waste (garbage and litter). Plastics (including bottles and shopping bags) are among the most common municipal waste products that litter the country. The adoption of recycling of plastics and other materials in Qatar households has been slow (Al-Maaded et al. 2012; Alsalama et al. 2021; Bello 2018). Some domestic waste is disposed of improperly (dumping) due to a lack of effort by individuals, an avoidance of unattractive and bad smelling garbage bins, and difficulties accessing garbage bins (Alsalama et al. 2021). In some rural areas of Qatar, open waste dump sites that lack a liner to prevent leaching into the soil have been used. These dump sites can attract pests, be sources of disease, and be sources of soil and water contamination.

Medical and pharmaceutical waste, batteries, and broken glass require special handling and disposal but are often components of household solid waste (Alsalama et al. 2021). Electronic waste (e-waste) contains plastics, metals, and other materials and is often disposed of in municipal solid waste as well. Of particular concern are the heavy metals, polycyclic aromatic hydrocarbons, and volatile organic compounds from e-waste (Kuang et al. 2022).

Thermal Pollution

Thermal pollution often refers to hot effluent discharged into natural waterbodies from cooling systems of electrical power plants, industrial facilities, and desalination plants. Thermal water pollution is an environmental problem because aquatic organisms are poorly adapted to rapid water temperature changes and there is less dissolved oxygen in warmer waters (Cunningham et al. 2023).

The urban heat island effect could be considered a form of thermal air pollution. Urban areas have large amounts of impermeable surfaces, such as concrete and brick that absorb and hold large amounts of solar radiation. Air conditioning systems also release waste heat into the air contributing to the heat island effect (Kamal et al.

2023). Tall buildings can create updrafts that circulate air pollutants, which can become concentrated in a heat island dome over the city (Cunningham et al. 2023). Air temperatures in cities are often 3–5 °C warmer than surrounding rural areas due to the heat island effect (Cunningham et al. 2023).

Light Pollution

The main sources of light pollution, excess artificial light at night, in Qatar are roadway lights, building lights, billboard lights, and industry facility lights. Light pollution can be disorienting and detrimental to plants (Signhal et al. 2019) and nocturnal animals including moths, bats, migrating birds, and nesting and hatching sea turtles (Azman et al. 2019); Kamrowski et al. 2012; La Sorte et al. 2022).

Light pollution shining in bedroom windows can even disrupt sleeping patterns and affect health in humans (Cupertino et al. 2023). Qatar was ranked among those countries with the highest exposure to light pollution, where 97% of the human population in Qatar lives in areas (16% of the country's area) with light pollution (Falchi et al. 2016; Singhal et al. 2019).

Noise Pollution

Noise pollution is excessive noise that can disrupt behavior, physiology, and communication in terrestrial animals (Sordello et al. 2020). Sources of noise pollution include transportation (roadways, road traffic, and airplanes), industrial facilities, construction and demolition sites, and large stadium events. Many studies of vehicle traffic noise on roadways have been conducted in Qatar in reference to human impacts (for example, Abdur-Rouf et al. 2022; Shaaban and Abouzaid 2021), but only the abstract of one study about noise pollution's effect on wildlife in Qatar appeared in a Google Scholar search. That abstract mentions the impact of noise on the pearl oyster spawning in a laboratory setting (Khalifa 2021), however, the full thesis was not available in an online search.

4.4.5 Climate Change

The effects of climate change (discussed in Chap. 2) will impact terrestrial biodiversity in several ways. Physiological stress may occur when increased temperatures and heat stress push organisms beyond their physiological tolerance limits. Higher atmospheric and soil temperatures increase evapotranspiration of plants and results in greater water loss. Trophic cascades may occur. Additional plant stress may lead to reducing plant cover over the landscape, which means there is less plant matter for herbivores to feed on. Animals may alter their behavior to avoid the hottest times of the day, which can impact their ability to forage and disrupt predator–prey dynamics. Rising temperatures and increased periods of drought associated with climate change bring the risk of habitat degradation affecting ground-nesting, desert-adapted birds. Birds reliant on man-made lagoons are threatened by increasing periods of desiccation and habitat shrinkage. An additional factor is that climate change shifts the timing and availability of food resources along migratory routes. Lastly, species that cannot adapt to the changing conditions may become locally extinct (Ward 2016; Willmer et al. 2000).

Qatar is extremely vulnerable to sea level rise (mainly from thermal expansion) that model projections predict will flood current shorelines and negatively impact mangrove forests, sea grass communities, fish and shellfish nurseries, and cause beach loss for nesting sea turtles and shorebirds (Ministry of Municipality and Environment 2017).

4.5 Threatened Species/Species-At-Risk: Methodology and Results

To help develop a priorities-based conservation strategy, we curated a centralized list of species and devised a weighted method, ranking criteria related to conservation status so we can assess the risk of extinction for each species. The four-step process for the sorting and grading of species that was used is as follows:

Step-1

A centralized list was curated with a total 855 species identified as occurring in Qatar. This list includes 28 mammals, 392 birds, 17 snakes, 4 turtles, 32 lizards, 382 plants, and no arthropods.

Step-2

Species were sorted out based on their IUCN Red List Status, their relevance (medicinal, cultural or religious), and whether they are native to Qatar. Selected species fall under these categories only: "Near Threatened" (NT), "Vulnerable" (VU), "Endangered" (EN) and "Critically Endangered" (CR), unless they have some cultural/religious importance or are native, in which case "Least Concern" (LC), and "Data Deficient" (DD) categories were also included. During this sorting process, only 280 out of 855 species of mammals, birds, snakes, turtles, lizards, and plants were selected for the next step.

Step-3

Nine risk assessment criteria were used to assess each species' Risk-of-Extinction in a comprehensive manner, and to rank them based on conservation priority needs. Each risk criterion was ranked 0 to 5 according to its critical level for conservation. Every "unknown" category, in other words when data were missing, this category was ranked as 1. The weight (% contribution in the final scoring and ranking) given to each criterion reflects the level of its importance in determining a species' conservation status and need. The 9 criteria are as follows:

1. **IUCN Red List Status (30%)**: The IUCN Red List is a global standard for assessing the extinction risk of species. The Red List criteria evaluate a species'

population size, population trend, and distribution, as well as threats to the species. The Red List Status is a vital tool in determining a species' conservation status, and it is based on a quantitative assessment of the likelihood of extinction. Therefore, the Red List Status carried the most weight in determining a species' risk assessment, with a weight of 30%. Categories were marked as: critically endangered (5), endangered (4), vulnerable (3), near threatened (1), not threatened (0), and not assessed/unknown (1).

2. **CITES (15%)**: CITES is an international agreement that regulates trade in wild animals and plants. Species listed under CITES receive varying levels of protection, depending on the degree of threat they face. The inclusion of a species on the CITES list is an important indicator of its conservation status and the level of protection it requires. Hence, the CITES status was be given a weight of 15%. Categories were marked as: Appendix I (5), Appendix II (2), Appendix III (1), Quotas (1), and not on CITES (0).

3. **Population Size (15%)**: Population size is a crucial factor in determining a species' risk of extinction. A small population size increases the risk of inbreeding depression, genetic drift, and demographic stochasticity, which can lead to extinction. Therefore, a species' population size was given a weight of 15%. Categories were marked as: rare (5), vagrant (5), unknown (1), and common (0).

4. **Population Trends (7%)**: Population trends indicate whether a species' population is increasing, decreasing, or remaining stable. Declining population trends are a critical indicator of a species' risk of extinction. Therefore, population trends was given a weight of 7%. Categories were marked as: Decreasing (5), unknown (1), stable (0) and increasing (0).

5. **Distribution (5%)**: The distribution of a species is essential in determining its risk of extinction. A species with a limited distribution is more vulnerable to habitat loss, fragmentation, and other threats. Therefore, the distribution of a species was given a weight of 5%. Categories were marked as: endemic (5), regional (3), global (1), and unknown (1).

6. **Relevance/Importance (10%)**: The relevance or importance of a species can be determined based on its ecological role, cultural significance, religious significance, economic value, or scientific interest. A species' importance can affect its conservation status and the allocation of conservation resources. Therefore, the relevance/importance of a species was given a weight of 10%. Categories were marked as: cultural (5), religious (5), and none (0).

7. **Harvesting (8%)**: Overharvesting can threaten the survival of a species, particularly if it has a slow reproductive rate or a limited distribution. Therefore, the harvesting of a species was given a weight of 8%. Categories were marked as: game/trophy (5), food/medicine/wood (4), byproduct (2) and none (0).

8. **Breeding Status (for Birds Only) (5%)**: Breeding status is a critical factor in determining a bird species' risk of extinction. A species that breeds only in a limited geographic area or during a particular season is more vulnerable to habitat loss, hunting, or climate change. Therefore, the breeding status of a bird species was given a weight of 5%. Categories were marked as: resident/breeding in Qatar (5), non-breeding (1), migratory/passage (1), vagrant (0).

9. **Captive/Cultivated/Introduced (5%)**: Captive breeding, cultivation, and introduction programs can help save species from extinction. These programs can increase the population size, genetic diversity, and distribution of a species. Therefore, captive/cultivated/introduced status was given a weight of 5%. Categories were marked as: wild (5), introduced/captive (4), cultivated (4), vagrant (1).

Step-4

In the final step of the analysis, species were clustered into 3 broader risk categories based on their weighted sum ranking. "Critical Species" were those with a weighted sum above 3, "High-risk Species" were those with a weighted sum below 3 and above 2, and "Moderate-risk Species" were those species which ranked below 2. The risk categories offer a prioritization in the conservation of each of these species at risk.

An example of our methodology is given in Table 4.1 below for the Eastern Imperial Eagle (*Aquila heliaca*).

Based on the results from our analysis, the Eastern Imperial Eagle was one of 8 species that are ranked as "Critical Species", the other 7 species being the marine turtles Hawksbill, Green, Loggerhead, and Leatherback, the Arabian Oryx, the Saker Falcon, and the Egyptian Vulture. Twenty seven species were ranked as "High-risk Species", and 238 as "Moderate-risk Species" (SM Table 4.2).

It should be mentioned here that there are some inherent limitations in the evaluation and ranking of species to create this list. Namely, the list was prepared using available species information that may have not been validated in recent times, e.g., species reported as being present in Qatar may not be present anymore, etc. Furthermore, the majority of data such as those from the IUCN Red List, and CITES are based on global species trends. Thus, it is possible that if the evaluation is performed

Table 4.1 Risk of extinction ranking for the Eastern Imperial Eagle (Aquila heliaca)

Species risk assessment criteria	Category	Rank
IUCN red list status	Vulnerable (VU)	3
Population trend	Decreasing	5
CITES appendix	I	5
Population size	Rare	5
Distribution	Global	1
Relevance/Importance	None	0
Harvesting	Game	5
Breeding status (for birds only)	Resident	5
Captive/Cultivated/Introduced	Wild	5
Total score		34
Weighted sum		3.7
Category assigned		Critical species

singularly for species in Qatar, that their status may be designated as being different than what we observe for the global populations.

Finally, this Species-at-Risk/conservation priority list is ready for use to inform baseline studies and be amended by their findings. Habitat loss/ change-of-land-use mapping and the above list can be worked at in tandem to synthesize and develop an IUCN Red List of Ecosystems and of Threatened Species Reports for Qatar, and a conservation strategy and action program. These IUCN reports can be used as headline indicators to monitor progress towards achieving 2030 Targets and 2050 Goals of the Kunming-Montreal Global Biodiversity Framework of the Convention on Biological Diversity (CBD).

4.6 Ranking of Threats: Methodology and Results

In this section we present the ranking of threats to the environment and to biodiversity, grouped in 10 distinct categories (Table 4.2). These include the universal 5 threats to biodiversity, as well as local-based threats due to gaps that we have identified in research, monitoring, regulation, and community engagement when it comes to conservation, protection, and management.

The threats have been ranked based on the degree of each of four criteria: geographic extent, magnitude, irreversibility and urgency of interventions. The total ranking has been assigned a level of severity from low (yellow) to moderate (orange) to severe (red). In more detail, the 4 criteria and their scoring are as follows:

1. Geographic Extent: Refers to the size of the geographic area affected by a given threat. Scoring here is highest (5) to lowest (1).
2. Magnitude: Refers to the scale or size of the impact on biodiversity, ecosystems, or genetic variation. Special attention is paid to threats that are rapidly escalating or causing severe damage right now. Scoring here is highest (5) to lowest (1).
3. Irreversibility: Refers to the degree of irreversible damage that a given threat may cause and whether there are opportunities for short or long-term recovery and restoration. Scoring here is: Effects of the threat are not reversible (5), effects are technically reversible, but not practically affordable, and/or would take >50 years to achieve (4), effects are reversible with a reasonable high commitment of resources and/or it would take 26–50 years to achieve (3), effects are reversible with a medium commitment of resources and/or within 6–25 years (2), effects are easily reversible at relatively low cost and/or within 0–5 years (1).
4. Urgency of Intervention: Refers to the speed at which a threat needs to be addressed and neutralized or mitigated through recovery and restoration. Scoring here is highest (5) to lowest (1).

Table 4.2 Ranking of categories of threats to the environment and to biodiversity

Threat	Extent	Magnitude	Irreversibility	Urgency	Total Ranking
Category 1: Land use change (= Habitat loss)					
1.1. Development/ construction: Roads, fences, and other man-made barriers	4	4	5	4	17
1.2. Lack of land use planning	4	3	4	4	15
1.3. Unmanaged recreational activities/ Tourism	4	4	3	4	15
1.4. Unmanaged harvesting of certain species (such as spiny-tailed lizards, *Uromastyx aegyptia*)	3	4	4	4	15
1.5. Unmanaged grazing	4	4	3	4	15
1.6. Direct and indirect impacts due to mineral exploration and mining	3	3	4	4	13
1.7. Species over-exploitation due to urbanization/ Bird-window collision	2	2	5	3	12
Category 2: Pollution					
2.1. Air pollution - ground-level ozone (urban heat island) and other contaminants	4	4	4	4	16
2.2. Water pollution -runoff	4	4	4	4	16
2.3. Plastics and microplastics - both on land and sea/ landfills leading to habitat loss	3	3	4	4	14
2.4. Light pollution - particularly disorienting to bats, migrating birds, and nesting /hatching sea turtles	2	2	2	2	8
2.5. Noise pollution - disrupts animal communication	2	2	2	2	8
Category 3: Climate Change					
3.1 Drought	5	5	4	4	18
3.2. Unpredictable rainfall patterns	5	4	3	4	16
3.3. Extreme temperature events (hot or cold) can affect plants and animals, seed set, germination & co-evolved species	4	4	4	4	16
3.4. Urban heat island effect	5	4	3	4	16
3.5. Flash floods/ storm surges/ highest intensity cyclones likely to cause damage to the built infrastructure and increased motor accidents	3	4	2	2	11
Category 4: Invasive species and diseases					
4.1. Introduction of invasive tree species/ importation of exotic species that displaces indigenous ones, i.e., parakeets, squirrels, toads, etc. invasive microbes, Harmful Algal and Cyanobacterial Bloom formations	3	3	3	3	12
4.2. Wildlife reduction due to diseases of wildlife	2	2	2	2	8

(continued)

Table 4.2 (continued)

Category 5: Lack of baseline on terrestrial species and ecosystems					
5.1. Species inventories are anecdotal and dated, no information on species distributions, and prime habitats, density pattern and major stronghold of key species not known, human-wildlife interactions and threats poorly known	5	4	2	4	15
Category 6: Limited capacity in ecological research and inter-agency coordination					
6.1. Latest field research methods are not tested and calibrated for Qatar/ modern analytical methods not adopted for biodiversity investigation	5	4	2	4	15
6.2. Human power not adequately trained in monitoring terrestrial species/ lack of institutional capacity	4	4	3	4	15
6.3. Lack of coordination between management agencies and academic institutions	4	4	1	4	13
Category 7: Representation of important habitats and species in Protected Areas (PA) network in view of 30x30					
7.1. Current PAs coverage in Qatar (24%) is impressive, however given knowledge gap on habitat suitability of key species and potential range shifts under climate change, there is risk of underrepresentation of certain species/habitats	5	4	1	4	14
Category 8: Insufficient involvement of local community in conservation initiatives –Urban & rural population					
8.1. Terrestrial habitats are intensively used for grazing, recreation, and farming. Public understanding of and involvement in conservation issues is limited	4	4	1	4	13
Category 9: Regulatory and monitory mechanisms and law enforcement					
9.1. Gaps in wildlife laws	4	4	1	4	13
9.2. Gaps in law enforcement and monitoring of wildlife crimes	4	4	1	4	13
Category 10: Hunting, poaching and Illegal Wildlife Trading (IWT)					
10.1. Wildlife reduction due to illegal hunting/ poaching	5	4	2	5	16
10.2. Wildlife reduction due to legal hunting	3	2	2	3	10
10.3. Wildlife reduction due to IWT of animals (live or dead), their parts, products and/ or derivatives.	3	2	2	3	10

4.6.1 Severe Threats

Qatar confronts a range of severe threats to its biodiversity, imperiling its delicate ecological balance and wildlife populations. The most prominent of these threats include significant alterations in land use, arising from factors like rapid urbanization, housing development, and infrastructure construction. This has led to habitat degradation, loss, and fragmentation, resulting in the decline of native species and the encroachment of non-native ones. Unmanaged tourism and recreational activities compound this issue, contributing to landscape degradation, disturbance of wildlife, and competition for limited resources. Despite initial intentions for ecotourism, existing facilities remain inadequate, recreational activities in natural settings lack proper management and oversight, leading to a lack of awareness regarding nature protection and proper code of conduct. For an overview on the progress and challenges facing Qatar's tourism development, as well as the transitional steps towards ecotourism and building up a regulatory framework for sustainability in this sector, please refer to the chapter on Tourism in Qatar.

Unsustainable harvesting of species like Spiny-tailed Agamas and medicinal plants, along with the persecution of wildlife, pose another critical threat. The problem of unmanaged grazing, primarily due to livestock, exacerbates these issues by causing the depletion of native vegetation, desertification, and soil erosion. Furthermore, traditional rangeland management methods have waned, leading to decreased vegetation density and overall biodiversity loss. The absence of comprehensive land use planning further aggravates these challenges, as it disregards considerations of land productivity and carrying capacity during development processes.

Moreover, the energy-intensive process of water desalination, a major source of drinking water and single source of water to meet household, industry and agricultural needs alike, relies heavily on fossil fuels. This pollution extends to water bodies, as wastewater from domestic and industrial sources contaminates areas around Doha, leading to odor complaints, insect issues, and health concerns. Efforts are underway to treat wastewater and restore groundwater quality.

Climate change poses a formidable challenge, marked by intensifying heat extremes, droughts, sand storms, and floods. Qatar's hyper arid climate, already under high water stress, is vulnerable to these shifts. Adaptive measures already in effect to different degrees include desalination, reservoir installations, managed aquifer recharge (MAR), and water conservation and re-use initiatives to address water scarcity. Erratic rainfall patterns further compound this challenge, with occasional heavy rainfall causing flooding, while Qatar relies on rainfall as its single freshwater source to recharge its brackish aquifers. The country is taking steps to manage extreme temperature events, increase vegetation cover, and transition to renewable energy sources, e.g., installation of solar energy collectors, to mitigate climate change.

However, fundamental gaps persist. Inadequate baseline data on terrestrial species and ecosystems hinder effective conservation planning. This deficiency in species distribution, density, and interactions with humans hampers efforts to devise targeted conservation strategies. Additionally, a limited capacity in ecological research and inter-agency coordination, coupled with inadequate expertise in utilizing modern research methods, restricts progress in understanding and mitigating these threats. To add to these concerns, illegal hunting, poaching, and the illicit trade in flora exacerbate the challenges faced by Qatar's biodiversity.

4.6.2 Moderate Threats

The introduction of invasive tree species and the importation of exotic species that displace indigenous ones, such as parakeets, squirrels, and toads, along with invasive microbes and Harmful Algal and Cyanobacterial Blooms, pose moderate threats to Qatar's wildlife. The nation's sparse natural vegetation includes important species like *Prosopis cineraria* (Ghaf), an endangered species, and *Acacia ehrenbergiana* (Salam), among others. While Qatar has made commendable progress in expanding Protected Areas (PAs) from 11 to 29.9% between 2005 and 2014, there's a risk of underrepresentation of certain species and habitats due to knowledge gaps in species' habitat suitability and potential shifts under climate change. Protected Areas, mostly terrestrial, aim to safeguard delicate desert ecosystems from development, over-grazing, and desertification, yet their effectiveness lacks monitoring and management measures, as well as law enforcing.

Insufficiently involving the local community in conservation initiatives, in both urban and rural settings, contributes to challenges. Terrestrial habitats, heavily used for grazing, farming, and recreation, suffer from limited public understanding and involvement in conservation efforts. Insufficient awareness campaigns and educational plans hinder the public's appreciation of the Protected Areas' purpose and the significance of encompassed species and habitats.

Moreover, law enforcement and monitoring of wildlife crimes require improvement, particularly concerning illegal hunting and trade. Wildlife reduction due to illegal trade (IWT) poses another concern. Qatar remains a significant wildlife trafficking hub, with local airlines being exploited by traffickers to transport items like ivory, rhino horn, reptiles, and birds. Demand for rare falcons, falcon eggs, and other exotic animals further exacerbates this issue, making effective enforcement and monitoring of IWT critical.

Addressing these multifaceted threats is of paramount importance to preserve Qatar's unique ecosystems, ensure the survival of its wildlife, and foster a sustainable future for the nation's natural heritage.

4.7 Conservation Efforts: Positive Actions and Identified Gaps

4.7.1 Protected Areas

At the time of this writing, and to in part satisfy national and international environmental conservation commitments, 11 areas have been declared protected, amassing 25.55% or 3,463.7 Km^2 of the country's total footprint. Of these, 2,743.2 Km^2 cover the land and coastal area, while 720.5 Km^2 cover the country's territorial sea waters, otherwise called Exclusive Economic Zone (EEZ). The names and geographic locations, as well as the coverage of the Protected Areas in the terrestrial and marine environment in Qatar, are showcased in Fig. 4.11.

In addition to the current Protected Areas' footprint, there are security sensitive areas where military activities and oil & gas operations take place, and these have become *de facto* Protected Areas. Their total footprint is not officially published, yet they include the Al Shaheen oil field in the northeast territorial waters, all Qatar Energy concession areas including in Ras Laffan, Dukhan and Messaid, and the topographic area wherein lies the Al Udeid Air Base for example.

Circa 2018, the land area of about 737.5 km^2 hugging the Al Ureiq and Al Maszhabiya Protected Areas, called Al Janoub, became itself a *de facto* protected area, reserved for military operations and traditional falconry hunting during the Marmi falconry festival in the month of February, organized by Al Ghannas Society. With this addition, the collective land and coastal footprint of Protected Areas totals 3,480.7 km^2 or 29.9%.

In 2007, the Al Reem Protected Area, covering 10% of the land area of Qatar, was enlisted under UNESCO's Man and Biosphere Programme (MAB). An Integrated Management Plan (2018–2022) was created in 2017, a collaboration between UNESCO and MoECC's Natural Reserves Department. The Plan is an exemplar in conservation, because of the detailed surveys and planning that went into determining zonation within the Protected Area, sensitivity areas identification for fauna, flora and landforms, etc. This methodology sets a precedent to be followed for other Protected Areas in Qatar. Indeed, there are anecdotal reports that the Al Thakira Protected Area is considered as a potential new UNESCO MAB, and Khor Al Adaid Protected Area has been nominated for a UNESCO Global Geopark or a World Heritage Site since the combination of natural and cultural heritage features it contains have been recognized as globally unique and important to conserve (Krupp et al 2014). For the Al Reem Protected Area the online interactive version of this Integrated Management Plan, showcasing the history of the area, its biodiversity, etc., was developed in collaboration with the GIS department of the ministry, and is available at https://gis.mme.gov.qa/alreem/#map.

Finally, there are a total of 70 rawdas that the ministry has declared protected, and accordingly manages.

It is worth noting here, that impressive as this feat is, i.e., for almost 30% of the land area to be declared as protected, there is risk of underrepresentation of

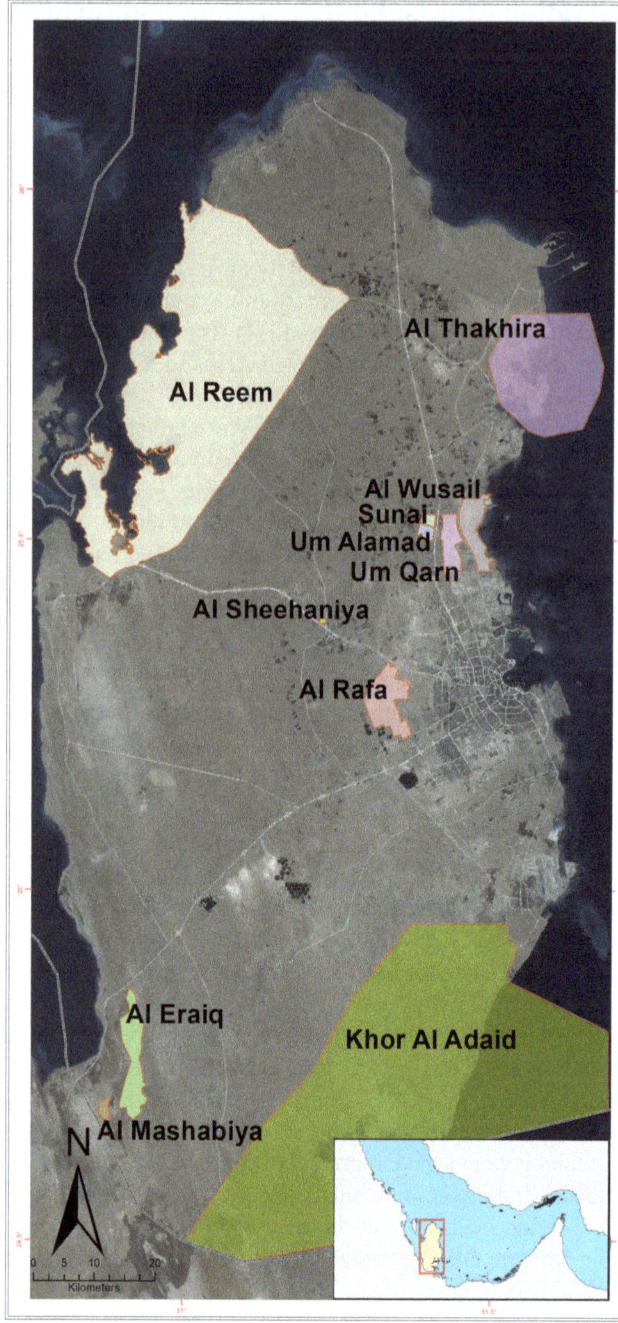

Fig. 4.11 Map showcasing the Protected Areas in Qatar (*Credit* MoECC—GIS Department)

certain species/habitats given the knowledge gap on habitat suitability of key species and potential range shifts under climate change. Furthermore, the lack of adequate enforcement of law, is a gap that requires attention if these areas are to be fully protected.

4.7.2 Conservation of Threatened Species

The Hawksbill turtle (*Eretmochelys imbricata*) is one of the 5 species of sea turtle found in the seawaters of Qatar, and the only one that births in its shores. Hawksbill turtles are Critically Endangered (CR) according to the IUCN (https://www.iucnre dlist.org/species/8005/12881238). Since the late 2000s, the different ministries of environment have implemented various conservation actions, culminating in the last years into protecting Fuwairit beach for the exclusive use by the turtles, rescuing laid eggs and transporting them in a safe hatchery nearby, as well as deploying scouting teams to search and collect laid eggs in multiple other shores of the country, returning them to the same hatchery (Pilcher et al. 2015). The associated research work to the safeguarding of the reproduction success of the Hawksbill turtle is carried out by Qatar University, under the patronage of MoECC, with Qatar Energy providing funding as well, making this a fruitful collaboration between government and academia. The ministry has also worked hand in hand with the children's museum Dadu, and the NGO Qatar Natural History Group, to create educational material, to raise awareness and increase community engagement in the conservation efforts. Community engagement and inclusion is essential and equally important for the success of ecological research and ecotourism as well.

Another conservation action is the captive breeding programs in Qatar and in other Gulf countries. These successfully revived populations of the Arabian Oryx (*Oryx leucoryx*), which had disappeared from the wild in some MENA countries and gone extinct in others in the early 1960s (Mallon et al. 2001). Its status in the IUCN Red List was switched from Endangered to Vulnerable in 2011, making it a great conservation success story, not only for the Region, but for the world over (https://www.iucnredlist.org/species/15569/50191626). Along with the Arabian Oryx, MoECC is carrying out conservation efforts for these endangered and/or rare species: Sand Gazelle, Idmi Gazelle, Addax, South African oryx, African tortoise, and North African ostrich. And more recently, two additional breeding programs have commenced: one for the Desert Cape Hare, and one for the Houbara Bustard which has been over-hunted due to falconry. The ultimate goal of course being to reintroduce these threatened species back to the wild.

MoECC and the peripheral municipalities, together assess penalties for violation of the environmental law and/or damage to the environment in Qatar and have authority to issue fines. MoECC has a dedicated hotline or unified call center, 16066, where citizens can call in complaints related to environmental violations, such as illegal hunting, illegal fishing, trashing in nature, driving through rawdas, violations that are subsequently investigated by the authorities and reported annually and made available on the National Planning Council website under Environmental Statistics (https://www.npc.qa/en/statistics/pages/subjectdetails.aspx). It is within this report that the number of Environmental Impact Assessments issued for new development projects are presented as well.

4.7.3 Identified Gaps

In addition to the threats identified, ranked, and addressed along with the species priorities-based conservation strategy presented previous sections, we have identified that the essential elements for conservation and management of the natural capital that are missing, include: the current status of habitats and biodiversity; the definitive number of species per taxonomic group, an assessment of species conservation status; a centralized species database; as well as an assessment of desert specific ecosystem functions and ecosystem services valuation mechanism.

Habitat and land use change maps that would reveal their current status are not continually updated, and baselines of distribution and abundance of biodiversity are grossly outdated, especially in the terrestrial environment. Furthermore, most of the species found in Qatar lack Regional IUCN Red List conservation status classifications, especially plant species, a fact that impedes our ability to determine the exact species that are most at risk and prioritize conservation actions accordingly. Lacking an assessment of ecosystem functioning and services provisioning in the regional and local scale, makes their economic valuation difficult. Inability to accurately assign a price tag to environmental impacts of large projects and national strategies, hinders proper assignment of compensation schemes to finance conservation restoration projects downstream.

4.8 Recommendations to Improve Nature and Biodiversity Conservation Efforts

Recommendations for ways to address gaps in conservation are presented in list form below, organized by topic. An important *take home message* is that to conserve and protect nature, as well as to adapt to and mitigate the climate crisis, we need plurality of voices and all hands on deck. In other words, irrespective of sector, educational background, etc., we all have a stake on Mother Earth's health, thus we all have to act in concert.

Governance

- Establishment of a centralized Committee made up of all relevant stakeholders whose mission in to promote environmental sustainability and nature conservation.

Fines and Penalties

- Improvement on enforcement of Legislation and Policy;
- Improvement on crackdown of illegal hunting of threatened species, especially birds.

Research

- Increased provision of funding for basic research and terrestrial ecology studies;
- Support for studies to value ecosystem services in the arid context;
- Declaration of unified Pillars for environmental research by funding bodies;
- Development of policy institutes or think tanks on environmental topics.

Economy and Commerce

- Commitment of a certain percentage of GDP for plant, animal and habitat protection;
- Incorporation in EIAs of a calculation for the cost associated with the habitat loss as a result of development.

Public Works-Ecology-Conservation Nexus

- Harnessing the urban ecology to provide ecosystem services within cities, such as storm water management, pollination etc.;
- Development of urban agricultural operations to prevent further land use conversion from pristine to developed. This step will ensure species protection as well as habitat conservation;
- Improvement of storm water management, designed so that the rainwater accumulated by the side of highways is injected back into the water horizon to replenish groundwater supplies.

Conservation, Protection

- Research and publication of IUCN Red List of Ecosystems and of the Threatened species in Qatar to be used as headline indicators towards achieving 2030 Targets and 2050 Goals of the Kunming-Montreal Global Biodiversity Framework;
- Improvement on monitor programs to measure success of conservation efforts, with reports for at least 5 years after completion of the project;
- Protected status should be assigned to important Bird Areas recognized by BirdLife International (See Irkaya Farm, and Al Karaana Lagoons), as well as the threatened species that winter in these. These are the artificial habitats, such as TSE storage lagoons and urban farms that due to an abundance of water and food sources have become hotspots of biodiversity in this arid country;
- Protect/restrict access to sensitive receptor zones within Protected Areas;
- Development of Integrated Management Plan for the terrestrial environment;
- Conservation/Preservation Zoning—Require habitat conservation buffer zones and wildlife movement corridor zones in development plans and biodiversity/species conservation plans;

- Mitigation—Developers create or restore lands of the same habitat type (with similar quality and size) to that which they impacted or developed. Created sites are rarely equivalent to natural sites, thus reducing negative impacts and restoring degraded areas should be prioritized over created mitigation sites (such as created wetlands).

Data Transparency, Accessibility, and Transparency

- Development of an English version of Ministry of Municipalities' Qatar Plant Gene Bank Information System;
- Development of a centralized repository of baseline environmental data from EIA reports; and
- Improvement on data-sharing and accessibility among public, private and academic sectors.

Tourism

- Development of Ecotourism and Slow Tourism Policy and Action Plan, as well as training programs for safari tour guides. As it stands right now, any activity taking place in nature is defined as ecotourism, without meeting any of the criteria set by IUCN or the UN World Tourism Organization.

Animal Welfare

- Drafting and passing into law the protection and welfare of animals, including clarifications for wildlife and kept animal treatment, sentiency rights, animal movement liberty, among others;
- Establishment of Wildlife Rescue Center in conjunction with stray animal sanctuaries for more efficient sharing of resources;
- Establishment of TNR programs/animal control programs and facilities within centers and sancturies, and as mobile units that can be deployed in geographic areas facing unmanaged stray populations and no ready access to vet facilities;
- Drawing regulations on importation of exotic species/invasive species.

Disease Control

- Establish One Health zoonotic disease management and monitoring program

Environmental Awareness

- Improvement on data-sharing and accessibility;
- Campaigns to raise environmental awareness and systemic understanding in the general public; and
- Improvement on the dissemination of conservation and nature protection efforts, as well as climate adaptation strategies.

Behavioral Change

- Strive for self-education, become better informed customers, know the full life-cycle of the products we buy so that we may better conserve resources and lead a sustainable lifestyle;
- Reduction of food waste and water use;
- Take ownership of nature, not in the capitalistic sense, but as a means to become her stewards without waiting for permission to act;
- Adopt the mode of action: "you see something you say something", calling in when witnessing environmental violations, including illegal hunting. This is crucial in helping the ministry improve enforcement of the law; and
- Change mindset—protect nature with the same intensity and care that we protect our properties or the country's infrastructure.

References

Abayomi OA, Range P, Al-Ghouti MA, Obbard JP, Almeer SH, Ben-Hamadou R (2017) Microplastics in coastal environments of the Arabian Gulf. Mar Pollut Bull 124(1):181–188

Abd Farag EAB, Bansal D, Mardini K, Sultan AA, Al-Thani MHJ, Al-Marri SA, Al-Hajri M, Al-Romaihi H, Schaffner F (2021) Identification and characterisation of mosquitoes from different locations in Qatar in 2017–2019. Parasites 28. https://doi.org/10.1051/parasite/2021079

Abdel-Razik MS (1994) Avicennia marina Al-Qurm: General study and propagation experiments in Qatar. Qatar university, Doha, Qatar. https://www.qu.edu.qa/file_source/qu/research/ESC/Books/99921-21-32-7.pdf

Abdel-Dayem M (2007) Plant-Insect Relationship in Prosopis cineraria. In: Abdel Bary EM (eds) The Ghaf Tree *Prosopis cineraria* in Qatar. National Council for Culture, Arts and Heritage; Doha, Qatar. 1st. ed., pp. 69–126

Abdu RM, Shaumar NF (1985) A preliminary list of the insect fauna of Qatar. Qatar Univ Sci J 5:215–232

Abdulrahman M, Gardner A, Yamaguchi N (2021) The occurrence and distribution of bats in Qatar. J Arid Environ 185:104379. https://doi.org/10.1016/j.jaridenv.2020.104379

Abdur-Rouf K, Shaaban K (2022) Measuring, mapping, and evaluating daytime traffic noise levels at urban road intersections in Doha, Qatar. Futur Transp 2(3):625–643

Abulfatih HA, Abdel Bari, EM, Alsubaey A, Ibrahim YM (2001) Vegetation of Qatar. Scientific and Applied Research Centre, University of Qatar, Doha, Qatar

Abushama F (1997) Density and diversity of the desert arthropoda of Qatar. Qatar Univ Sci J 17(2):359–375. http://hdl.handle.net/10576/9825

Abushama F (1999) Dominance value and community production of desert arthropoda in Qatar. Qatar Univ Sci J 18:137–153. http://hdl.handle.net/10576/10047

Abushama F (2006) Night and early morning flying insects in a residence backyard in Doha City, Qatar. Qatar Univ Sci J 26:83–90. http://hdl.handle.net/10576/9713

Ahmed RI (2015) The most common mosquitoes at Al-Rayyan municipality (Qatar state) and their potential for transmitting malaria (Master's thesis). http://hdl.handle.net/10576/3906

Al-Dousari AM, Al-Awadhi J, Ahmed M (2013) Dust fallout characteristics within global dust storm major trajectories. Arab J Geosci 6:3877–3884

AlHajri SH (2013) Seasonal changes in biodiversity and abundance of invertebrates in different ecological environments (Master's thesis). Qatar University

AlHajri S, Agosti D (2019) The ant fauna (Hymenoptera: Formicidae) of Qatar. In: Presented at the Asian myrmecological conference, Bangkok, Thailand: Zenodo. https://doi.org/10.5281/zenodo.3527970

Al-Thani RF, Bassam YT (2021) Microbial ecology of Qatar, the Arabian gulf: possible roles of microorganisms. Frontiers in Marine Science 8. https://doi.org/10.3389/fmars.2021.697269

Alkhayat FA, Ahmad AH, Rahim J, Dieng H, Ismail BA, Imran M, Sheikh UAA, Shahzad MS, Abid AD, Munawar K (2020) Charaterization of mosquito larval habitats in Qatar. Saudi J Biol Sci 27(9):2358–2365. https://doi.org/10.1016/j.sjbs.2020.07.006

Al-Maaded M, Madi NK, Kahraman R, Hodzic A, Ozerkan NG (2012) An overview of solid waste management and plastic recycling in Qatar. J Polym Environ 20:186–194

Akhani H, Barroca J, Koteeva N, Voznesenskaya E, Franceschi V, Edwards G, Ziegler H (2005) Bienertia sinuspersici (Chenopodiaceae): a new species from Southwest Asia and discovery of a third terrestrial C4 plant without Kranz anatomy. Systematic Botany, 30(2):290–301

Alsalama T, Koç M, Isaifan RJ (2021) Investigations on the domestic waste and its impact on human health-case study for Qatar to understand the needs and recommendations. Int J Glob Warm 25(3–4):516–526

Azman MI, Dalimin MN, Mohamed M, Bakar MA (2019) A brief overview on light pollution. In: IOP conference series: earth and environmental science. IOP Publishing, vol 269, no 1, p 012014

Barradas PF, Mesquita JR, Lima C, Cardoso L, Alho AM, Ferreira P, Amorim I, de Sousa R, Gärtner F (2020) Pathogenic *Rickettsia* in ticks of spur-thighed tortoise (*Testudo graeca*) sold in a Qatar live animal market. Transbound Emerg Dis 67(1):461–465

Barton MG, Henderson I, Border JA, Siriwardena G (2023) A review of the impacts of air pollution on terrestrial birds. Sci Total Environ 873:162136

Batanouny KH (1981) Ecology and Flora in Qatar, Qatar University, Doha, Qatar

Bello H (2018) Impact of changing lifestyle on municipal solid waste generation in residential areas: case study of Qatar. Int J Waste Resour 8(2)

Bou Kheir R, Greve M, Greve M, Peng Y, Shomar B (2019) A Comparative GIS tree-pollution analysis between arsenic, chromium, mercury, and uranium contents in soils of urban and industrial regions in Qatar. Euro-MediterrEan J Environ Integr 4:1–11

Brochet AL, Jbour S, Sheldon RD, Porter R, Jones VR, Al Fazari W, Al Saghier O, Alkhuzai S, Al-Obeidi LA, Angwin R, Ararat K, Pope M, Shobrak MY, Willson MS, Zadegan SS, Butchart SHM (2019) A preliminary assessment of the scope and scale of illegal killing and taking of wild birds in the Arabian Peninsula, Iran and Iraq. Sandgrouse 41:154–175

Brown G, Böer B (2005) Terrestrial habitats. In: Hellyer P, Aspinall S (eds), The emirates a Natural history. Environ Agency, Abu Dhabi, UAE

Brondízio ES, Aumeeruddy-Thomas Y, Bates P, Carino J, Fernández-Llamazares Á, Farhan Ferrari M, Galvin K, Reyes-García V, McElwee P, Molnár Z, Samakov A, Babu Shrestha U (2021) Locally based, regionally manifested, and globally relevant: indigenous and local knowledge, values, and practices for nature. Annual Rev Environ and Resources. 46:481-509. https://doi.org/10.1146/annurev-environ-012220-012127

Butler JD, Purkis SJ, Yousif R, Al-Shaikh I, Warren C (2020) A high-resolution remotely sensed benthic habitat map of the Qatari coastal zone. Marine Pollution Bulletin. 160, 111634. https://doi.org/10.1016/j.marpolbul.2020.111634

Castilla A et al (2014) The lizards Living in Qatar. Ministry of Environment, Doha, Qatar

Castilla AM, Riera R, Humaid MA, Garland T, Alkuwari A, Muzaffar S, Naser HA, Salman Al-Mohannadi A, Al-Ajmi D, Chikhi A, Wessels J, Al-Thani MAF, Takacs Z, Valdeón A (2017) Contribution of citizen science to improve knowledge on marine biodiversity in the Gulf Region. J Assoc Arab Univ Basic Appl Sci 24(1):126–135. https://doi.org/10.1016/j.jaubas.2017.06.002

Chatziefthimiou AD, Banack SA, Metcalf JS (2021) Harmful algal and cyanobacterial blooms in the Arabian Seas: current status, implications and future directions. In: Jawad L (ed) The Arabian Seas: biodiversity, environmental challenges and conservation measures. Springer, Switzerland. https://doi.org/10.1007/978-3-030-51506-5

Chatziefthimiou AD, Metcalf JS, Glover, WB, Powell JT, Banack SA, Cox PA, Ladjimi M, Sultan AA, Chemaitelly H, Richer R (2024) Cyanotoxin accumulation and growth patterns of biocrust communities under variable environmental conditions. ToxiconX. https://doi.org/10.1016/j.toxcx.2024.100199

Conkey AT, Purchase C, Richer R, Yamaguchi N (2023) Terrestrial biodiversity in arid environments: one global component of climate crisis resilience. In: Cochrane L, Al-Hababi R (eds), Sustainable Qatar: social, political and environmental perspectives, Springer, Singapore. https://doi.org/10.1007/978-981-19-7398-7_13

Cox NA, Mallon D, Bowles P, Els J, Tognelli MF (2012) The conservation status and distribution of reptiles of the Arabian Peninsula. UK and Gland, Switzerland, Cambridge

IUCN, Sharjah, UAE: Environment and protected areas authority

Chatziefthimiou AD (2019) Terrestrial ecosystems in Qatar: ecology, geology and climate. e-Nature app. https://www.enature.qa/ecosystem/terrestrial/

Chatziefthimiou AD, Richer R, Rowles H, Powell JT, Metcalf JS (2014) Cyanotoxins as a potential cause of dog poisonings in desert environments. Veterinary Record. 174: 19 484-485. https://doi.org/10.1136/vr.g3176

Cunningham WP, Cunningham MA, O'Reilly CM (2023) Principles of environmental science: inquiry & applications, 10th edition. McGraw-Hill, LLC

Cupertino MDC, Guimarães BT, Pimenta JFG, Almeida LVLD, Santana LN, Ribeiro TA, Santana YN (2023) Light pollution: a systematic review about the impacts of artificial light on human health. Biol Rhythm Res 54(3):263–275. https://doi.org/10.1080/09291016.2022.2151763

Dohm MR, Mautz WJ, Looby PG, Gellert KS, Andrade JA (2001) Effects of ozone on evaporative water loss and thermoregulatory behavior of marine toads (*Bufo marinus*). Environ Res 86(3):274–286

Dohm MR, Mautz WJ, Doratt RE, Stevens JR (2008) Ozone exposure affects feeding and locomotor behavior of adult *Bufo marinus*. Environ Toxicol Chem: Int J 27(5):1209–1216

Dudley N, Stolton S (1996) Air pollution and biodiversity: a review. Technical report of the world wildlife fund (WWF). http://www.equilibriumresearch.com/upload/document/airpollutionandbiodi4f9.pdf

Edgell HS (2006) Arabian deserts: nature, origin and evolution. Springer, Dordrecht, The Netherlands. https://doi.org/10.1007/1-4020-3970-0

Egan D (2008) Snakes of Arabia: a field guide to the snakes of the Arabian Peninsula and its shores. Motivate Publishing Ltd., Dubai, UAE

Elazazi EM, Wong LJ, Pagad S (2020) GRIIS checklist of introduced and invasive species—Qatar. Version 1.2. Invasive species specialist group ISSG. Checklist dataset https://doi.org/10.15468/cw5z8x. Accessed via GBIF.org on 27 May 2025

Fahrig L (2003) Effects of habitat fragmentation on biodiversity. Annu Rev Ecol Evol Syst 34(1):487–515. https://doi.org/10.1146/annurev.ecolsys.34.011802.132419

Falchi F, Cinzano P, Duriscoe D, Kyba CC, Elvidge CD, Baugh K, Portnov BK, Rybnikova NA, Furgoni R (2016) The new world atlas of artificial night sky brightness. Sci Adv 2(6):e1600377. https://doi.org/10.1126/sciadv.1600377

Friedman J (2020) The evolution of annual and perennial plant life histories: ecological correlates and genetic mechanisms. Annual Review Ecology, Evolution, and Systematics. 51:461–481. https://doi.org/10.1146/annurev-ecolsys-110218-024638

Freitag H, Brandt R, Chatrenoor T, Akhani H (2013) Suaeda iranshahrii, a new species of Suaeda subgenus Brezia (Chenopodiaceae) from the Persian Gulf coasts. Rostaniha 23;14(1):68–80

Fuller A, Hetem RS, Maloney SK, Mitchell D (2014) Adaptation to Heat and Water Shortage in Large, Arid-Zone Mammals. Physiology 29(3):159–167

García-París M, Ruiz JL, Jiménez-Ruiz Y, Castilla AM, Saifelnasr EO (2013) *Meloe (Mesomeloe) coelatus* Reiche, 1857 (Coleoptera: Meloidae): first record for the fauna of Qatar and COI mtDNA data. Qscience Connect 2013(1):20. https://doi.org/10.5339/connect.2013.20

Goudie AS (1983) Dust storms in space and time. Prog Phys Geogr Earth Environ 2:502–530

Grunwell MJ (2010) Dragonflies and damselflies in the state of Qatar. J QNHG (3)

Gupta V, Bakre P (2013) Mammalian faeces as bioindicator of urban air pollution in captive mammals of Jaipur Zoo. World Environ 3(2):60–65

Harrison DL (1991) Mammals of Arabia. Bates PJJ (ed). Sevenoaks, Harrison Zoological Museum, Kent, UK

Harten AV (2008) Arthropod fauna of the UAE, Dar Al Ummah, Printing, Abu Dhabi, UAE, vol 1, p 754

Harten AV (2009) Arthropod Fauna of the UAE, Dar Al Ummah, Printing, Abu Dhabi, UAE, vol 2, p 786

Háva J (2007) New species and new records of Dermestidae (Insecta: Coleoptera) from the Arabian Peninsula including Socotra Island. Fauna Arab 23:309–317

Háva J, Pierre E (2008) Contribution to the family Dermestidae (Coleoptera) from Qatar. J Entomol Res Sociey 10(2):37–41

He C, Clifton O, Felker-Quinn E, Fulgham SR, Juncosa Calahorrano JF, Lombardozzi D, Purser G, Riches M, Schwantes R, Tang W, Poulter B, Steiner AL (2021) Interactions between air pollution and terrestrial ecosystems: perspectives on challenges and future directions. Bull Am Meteor Soc 102(3):E525–E538

Heggy E, Normand J, Palmer EM, Abotalib AZ (2022) Exploring the nature of buried linear features in the Qatar peninsula: Archaeological and paleoclimatic implications. ISPRS J Photogrammetry and Remote Sensing, 183: 210–227

IUCN (2014) Total, threatened, and EX & EW endemic species in each country (totals by taxonomic group): VERTEBRATES (Mammals, Birds, Reptiles, Amphibians https://nc.iucnredlist.org/red list/content/attachment_files/2020_1_RL_Stats_Table_5.pdf

Kamal A, Mahfouz A, Sezer N, Hassan IG, Wang LL, Rahman MA (2023) Investigation of urban heat island and climate change and their combined impact on building cooling demand in the hot and humid climate of Qatar. Urban Clim 52:101704

Kamrowski RL, Limpus C, Moloney J, Hamann M (2012) Coastal light pollution and marine turtles: assessing the magnitude of the problem. Endanger Species Res 19(1):85–98

Kardousha MM (2015) Additional records of vector mosquito diversity collected from AlKhor district of North-eastern Qatar. Asian Pac J Trop Dis 5(10):804–807. https://doi.org/10.1016/ S2222-1808(15)60944-6

Kardousha MM (2016a) First report of some adult mosquito captured by CDC gravid traps from North-Eastern Qatar. Asian Pac J Trop Dis 6(2):100–105. https://doi.org/10.1016/S2222-180 8(15)60993-8

Kardousha MM (2016b) First record of some aquatic fauna collected from Qatari Inland waters with reference to Arabian Peninsula. Int J Curr Res 8(06):32963–3269. http://hdl.handle.net/ 10576/5188

Khalifa RTH (2021) Impact of environmental and anthropogenic conditions on the reproductive cycle of the pearl oyster *Pinctada Imbricata* in Qatar. (Master's thesis abstract) Qatar University

Knight O (2016) Assessing and mapping renewable energy resources. ESMAP Knowledge Series;025/16. © World Bank. http://hdl.handle.net/10986/24913

Krupp F, Böer B, Sloane BA (2014) Khor Al-Adaid nature reserve: Qatar's globally unique inland sea. World Heritage Review 72 https://unesdoc.unesco.org/ark:/48223/pf0000227959

Kuang H, Li Y, Li L, Ma S, An T, Fan R (2022) Four-year population exposure study: implications for the effectiveness of e-waste control and biomarkers of e-waste pollution. Sci Total Environ 842:156595

Lanouar C, Al-Malk AY, Al Karbi K (2016) Air pollution in Qatar: Causes and challenges. Qatar University, White Paper, vol 1, issue 3, pp 1–7

La Sorte FA, Aronson MF, Lepczyk CA, Horton KG (2022) Assessing the combined threats of artificial light at night and air pollution for the world's nocturnally migrating birds. Glob Ecol Biogeogr 31(5):912–924

Louge MY, Valance A, el-Moctar O, Xu J, Hay AG, Richer R (2013) Temperature and humidity within a mobile barchan sand dune, implications for microbial survival, J. Geophys. Res. Earth Surf., 118, 2392–2405. https://doi.org/10.1002/2013JF002839

Lush MJ (2009) Some ant records (Hymenoptera: Formicidae) from the Middle East. Zool Middle East 47(1):114–116. https://doi.org/10.1080/09397140.2009.10638357

Mallon DP, Kingswood SC (compilers) (2001) Antelopes. Part 4: North Africa, the Middle East, and Asia. Global Survey and Regional Action Plans. SSC Antelope Specialist Group. IUCN, Gland, Switzerland and Cambridge, UK

Mas-Peinado P, Ruiz JL, García-París M, Castilla AM, Valdeón A, Saifelnasr EO (2013) On the presence of *Scaurus puncticollis* Solier, 1838 (Coleoptera: Tenebrionidae) in Qatar. Qscience Connect 2013(1):25. https://doi.org/10.5339/connect.2013.25

Mautz WJ, Dohm MR (2004) Respiratory and behavioral effects of ozone on a lizard and a frog. Comp Biochem Physiol a: Mol Integr Physiol 139(3):371–377

Metcalf JS, Chatziefthimiou AD, Souza NR, Cox PA (2021) Desert dust as a vector for cyanobacterial toxins. In: Jawad, L.A. (eds), The Arabian Seas: Biodiversity, Environmental Challenges and Conservation Measures. Springer, Cham. https://doi.org/10.1007/978-3-030-51506-5_8

Miller AG, Cope TA (1996) Flora of the arabian peninsula and socotra, vol 1. Edinburgh University Press, Scotland

Ministry of Municipality and Environment (2017) Climate change strategy for urban planning and urban development sector in the state of Qatar: strategy report

Mirab-Balou M, Yang SL, Tong XL (2014) First record of nine species of thrips (Insecta: Thysanoptera) in Qatar. Arab J Plant Prot 32(3):278–282

Moubasher AH, Al-Subai AAT (1987) Soil fungi in the State of Qatar. The Scientific and Applied Research Council, Qatar University, Doha, Qatar

MOE (Ministry of Environment) (2014) National biodiversity strategy and action plan 2015–2025. https://www.cbd.int/doc/world/qa/qa-nbsap-v2-en.pdf

Mushtaq N, Singh DV, Bhat RA, Dervash MA, Hameed Ob (2020) Freshwater contamination: sources and hazards to aquatic biota. In: Qadri H, Bhat R, Mehmood M, Dar G (eds), Fresh Water Pollution Dynamics and Remediation. Springer, Singapore. https://doi.org/10.1007/978-981-13-8277-2_3

Orndorff RC, Knight MA, Krupansky JT, Alakhras K, Stamm RG, Samad U, Elalim A (2018) Linking geology and geotechnical engineering in the karst: the Qatar geologic mapping project. Geology. https://doi.org/10.5038/9780991000982.1015

Panagopoulos A, Haralambous K-J (2020) Environmental impacts of desalination and brine treatment - Challenges and mitigation measures, Marine Pollution Bulletin,161, Part B. https://doi.org/10.1016/j.marpolbul.2020.111773

Parker AG, Goudie AS, Stokes S, White K, Hodson MJ, Manning M, Kennet D (2006) A record of holocene climate change from lake geochemical analyses in southeastern Arabia. Quaternary Research 66(3):465–476. https://doi.org/10.1016/j.yqres.2006.07.0012006

Pittaway AR (1980) Butterflies (Lepidoptera) of Qatar. Entomologist's Gaz 31(2):103–111

Pilcher NJ, Al-Maslamani I, Williams J, Gasang R, Chikhi A (2015) Population structure of marine turtles in coastal waters of Qatar. Endang Species Res 28, 163–174. https://doi.org/10.3354/esr00688

Qatar Atlas (2020). https://gis.psa.gov.qa/QatarAtlas/

Richer R, Knees S, Norton J, and Sergeev, A (2022) Hidden beauty: an exploration of Qatar's native and naturalised flora. 522 pages Akkadia Press Edinburgh, UK

Richer R, Anchassi D, El-Assaad I, El-Matbouly M, Ali F, Makki I, Metcalf JS (2012) Variation in the coverage of biological soil crusts in the State of Qatar. J Arid Environ 78:187–190

Sanderfoot OV, Holloway T (2017) Air pollution impacts on avian species via inhalation exposure and associated outcomes. Environ Res Lett 12(8):083002

Sadiq AM, Nasir SJ (2002) Middle pleistocene karst evolution in the state of Qatar, Arabian Gulf. J Cave and Karst Studies. 64(2):132–139 https://legacy.caves.org/pub/journal/PDF/V64/v64n2-Sadiq.pdf

Schaffner F, Bansal D, Mardini K, Al-Marri SA, Al-Thani MHJ, Al-Romaihi H, Sultan AA, Al-Hajri M, Farag EABA (2021) Vectors and vector-borne diseases in Qatar: current status, key challenges and future prospects. J Eur Mosq Control Assoc 39(1):3–13

Sharaf M, Al-Hajri SH, Aldawood A (2015) First record of the ant genus *Strumigenys* S. Smith, 1860 (Hymenoptera: Formicidae) from Qatar by the invasive species *S. membranifera* Emery, 1869. Zool Middle East 61(4):362–367. https://doi.org/10.1080/09397140.2015.1095514

Sharaf M, Salman S, Al Dhafer H, Akbar S, Abdel-Dayem M, Aldawood A (2016) Taxonomy and distribution of the genus *Trichomyrmex Mayr*, 1865 (Hymenoptera: Formicidae) in the Arabian Peninsula, with the description of two new species. Eur J Taxon 246:1–36. https://doi.org/10.5852/ejt.2016.246

Sharaf M, Abdel-Dayem M, Mohamed A, Fisher B, Aldawood A (2020) A preliminary synopsis of the ant fauna (Hymenoptera: Formicidae) of Qatar with remarks on the zoogeography. In: *Annales Zoologici*. Museum and Institute of Zoology, Polish Academy of Sciences, vol 70, no 4, pp 533–560. https://doi.org/10.3161/00034541ANZ2020.70.4.005

Skariah S, Abdul-Majid S, Hay AG, Acharya A, Kano N, Al-Ishaq RK, de Figueiredo P,Han A,Guzman A,Dargham SR, Sameer S, Kim GE, Khan S, Pillai P, Sultan AA (2023) Soil properties correlate with microbial community structure in Qatari arid soils. Microbiol Spectr11:e03462-22. https://doi.org/10.1128/spectrum.03462-22

Shomar B, Sankaran R, Solano JR (2023) Mapping of trace elements in topsoil of arid areas and assessment of ecological and human health risks in Qatar. Environ Res 225:115456

Seimenis AM (2008) The spread of zoonoses and other infectious diseases through the international trade of animals and animal products. Veterinaria Italiana 44(4):591–599

Singhal RK, Kumar M, Bose B (2019) Eco-physiological responses of artificial night light pollution in plants. Russ J Plant Physiol 66:190–202

Shaaban K, Abouzaid A (2021) Assessment of traffic noise near schools in a developing country. Transp Res Procedia 55:1202–1207

Soldati L (2009) The darkling beetles (Coleoptera: Tenebrionidae) of Qatar. Natura optima dux Foundation, Warszawa, Poland, p 101

Sordello R, Ratel O, Flamerie De Lachapelle F, Leger C, Dambry A, Vanpeene S (2020) Evidence of the impact of noise pollution on biodiversity: a systematic map. Environ Evid 9:1–27

SSE (Surface meteorology and Solar Energy – NASA). Atmospheric Science Data Center (2017). https://eosweb.larc.nasa.gov/sse/

Veltman K, McKone TE, Huijbregts MA, Hendriks AJ (2009) Bioaccumulation potential of air contaminants: combining biological allometry, chemical equilibrium and mass-balances to predict accumulation of air pollutants in various mammals. Toxicol Appl Pharmacol 238(1):47–55

Walker DH, Pittaway AR (1987) Insects of Eastern Arabia. Macmillan, London, UK, pp 16 + 175

Ward D (2016) The biology of deserts, second edition. Oxford University Press

Wetterer JK (2013) Geographic spread of the samsum or sword ant, *Pachycondyla (Brachyponera) sennaarensis* (Hymenoptera: Formicidae). Myrmecol News 18:13–18

Willmer P, Stone G, Johnston I (2000) Environmental physiology of animals. Blackwell Science LTD

World Health Organization. Air quality and health: health impacts. https://www.who.int/teams/environment-climate-change-and-health/air-quality-and-health/health-impacts

WWF (2020) Living planet report 2020—bending the curve of biodiversity loss. In: Almond REA, Grooten M, Petersen, T (eds) WWF, Gland, Switzerland

WHO (World Health Organization) (2023) Air Quality and Health: Health Impacts. https://www.who.int/teams/environment-climate-change-and-health/air-quality-and-health/health-impacts

Yasseen BT, Abu-Al-Basal MA (2008) Ecophysiology of Limonium axillare and Avicennia marina from the coastline of Arabian Gulf-Qatar. J Coast Conserv 12, 35–42 https://doi.org/10.1007/s11852-008-0021-z

Zahedi SM, Karimi M, Venditti A (2021) Plants adapted to arid areas: specialized metabolites. Natural Product Research, 35(19):3314–3331. https://doi.org/10.1080/14786419.2019.1689500

Zhang Z, Doi K, Iwasaki A, Xu G (2020) Unsupervised domain adaptation of high-resolution aerial images via correlation alignment and self training. IEEE Geoscience and Remote Sensing Letters, 18(4): 746-750

Zomer, RJ, Xu J, Trabuco A (2022) Version 3 of the global aridity index and potential evapo-transpiration database. Scientific Data 9, 409. https://www.nature.com/articles/s41597-022-01493-1

Open Access This chapter is licensed under the terms of the Creative Commons Attribution 4.0 International License (http://creativecommons.org/licenses/by/4.0/), which permits use, sharing, adaptation, distribution and reproduction in any medium or format, as long as you give appropriate credit to the original author(s) and the source, provide a link to the Creative Commons license and indicate if changes were made.

The images or other third party material in this chapter are included in the chapter's Creative Commons license, unless indicated otherwise in a credit line to the material. If material is not included in the chapter's Creative Commons license and your intended use is not permitted by statutory regulation or exceeds the permitted use, you will need to obtain permission directly from the copyright holder.

Chapter 5
Urban Ecology, and Agriculture

S. Abdul Majid and T. Karanisa

Why a rich and diverse urban ecology matters to Qatar: connecting urban green spaces with urban agriculture to promote biodiversity in the Qatari Metropolis and beyond.

Abstract Urban landscapes are often densely built and ecologically simplified, typically lacking in native biodiversity. However, urban green spaces, including urban agriculture, can restore vegetation and reintroduce biodiversity, not just within their boundaries but also in surrounding areas through a "spill-over" effect. This dynamic is crucial for supporting wildlife in human-dominated environments. Green spaces such as parks, gardens, landscaped roadsides, and small-scale agriculture provide a wide range of benefits:

- They regulate microclimates, significantly reducing urban heat, especially when mature trees are introduced.
- They help capture and store carbon, contributing to climate change mitigation.
- Green areas enhance air quality, release oxygen, and reduce noise pollution, especially when vegetation is diverse and well-structured.
- Vegetation supports soil stability, preventing erosion and suppressing dust.
- Urban biodiversity is maintained through plant variety, which in turn supports insect and animal life.
- Access to green spaces improves mental and physical well-being, reduces stress, and supports social interaction.
- Proximity to green areas can uplift property values and foster a sense of place and belonging.
- They reconnect city residents with nature, nurturing environmental awareness and stewardship.

S. Abdul Majid (✉)
Eartha Centre for a Sustainable Future, Doha, Qatar
e-mail: sabdulmajid@qf.org.qa

T. Karanisa
Ministry of Rural Development and Food, Athens, Greece
e-mail: tkaranisa@minagric.gr

© The Author(s) 2026

A. D. Chatziefthimiou et al. (eds.), *Pathways to Nature Conservation and Resilience in Hot and Arid Lands*, Advances in Geographical and Environmental Sciences,
https://doi.org/10.1007/978-3-032-11046-6_5

- Urban agriculture and home gardening can boost food security and promote local food systems.
- Emerging studies suggest green spaces may also play a role in strengthening human immune health through contact with environmental microbes.

This chapter explores the current landscape of green spaces and urban food systems in Qatar, highlights key challenges, and offers recommendations to promote ecological and food diversity in Doha and beyond.

5.1　Urban Green Spaces in Qatar

The climate of the Arabian Peninsula, which includes Qatar, results in hyper-arid desert, and thus forms a unique desert landscape and ecology. Open vistas with sparse vegetation are the norm. Interestingly, the design of many parks in the region often aspires to the lushness of North American, European, and Mogul precedents, rather than embracing recreational spaces that reflect the desert ecology and landscapes native to their surroundings. According to Moosavi (2015), landscape architects and designers, involved in Middle East projects, frequently encounter resistance from clients to mimic desert ecology in urban situations as they favor the creation of more lush and green urban landscapes. A number of scholars have pointed out that green spaces are synonymous with paradise in the Islamic architectural conception, explaining the aspiration to verdant gardens (Al Mohanadi et al. 2021).

In Qatar, various organizations collaborate to create urban green spaces: the Public Parks Department (PPD) focusses on the design, vegetation selection, execution and monitoring of public parks. Ashghal (Public Works Authority) focusses on urban beautification of roads and public places, and the Urban Planning Department (Ministry of Municipality) on spatial planning of green spaces (Marthya 2022). The above organizations are working toward the Qatar National Vision 2030 and the Qatar National Masterplan.

The PPD has the predominant aim of greening the Qatari metropolitan and improving the quality of life; they have a motto in Arabic "Let's make Qatar a big garden" (Trees of Qatar 2016). Qatar municipalities/cities have recently been accredited with the World Health Organization "Healthy Cities" stamp, in part for their large green space to inhabitant ratio (QNA 2022).

The PPD plant palette is selected based on three factors, including survivability to the climate, water requirement and aesthetic beauty (Pers Comm. Al Khouri 2016). The PPD prefers flowering plants over fruiting plants as they are more appealing. They do use native plants especially on roadsides (such as *Ziziphus* and *Prosopis* spp), however, most of their selected species are exotic (e.g., *Adenium obesum, Azadirachta indica, Cupressus sempervirens, Delonix regia, Sphagneticola trilobata, Lantana camara, Ruellia brittoniana, and Muhlenbergia rigens*), and are imported from different geo-climatic zones.

The PPD issued a publication in Arabic titled "Trees of Qatar" that highlights the trees and shrubs that are suitable for planting in the Qatari environment. Each plant is accompanied by a plant description, seed treatment recommendations, home of origin, adaptability to the Qatari climate, and beautification value (Trees of Qatar 2016). The book is aimed at the Qatari population as a guide to identify trees and plants in the city and for gardening enthusiasts.

Public parks in the country have undergone massive developments in the past decade. In 2015, Hashem (2015) pointed out that there was both a shortage and an uneven distribution of public parks in the Greater Doha area. Doha's Park provision ratio was 4 m^2 per inhabitant, which falls well below the World Health Organization's (WHO) recommended 9 m^2 of green space per urban resident. As of 2022, there has been a 164% increase in public parks in the Qatar since 2010, as their number increased from 56 parks in 2010, to 143 parks in 2022. The per capita share of green space increased from 1 m^2 per inhabitant in 2010 to 16.2 m^2 in 2022, with a 16-fold increase (Qatar Tribune 2022). Tannous et al. 2021 have looked at the quality and quantity of green spaces available to Doha's inhabitants and have shown that there is spatial and social logic to the physical and spatial characteristics of open green spaces in Doha above 20 acres, while smaller green spaces tend to be more random, mostly due to problems with land availability in private developments (Tannous et al. 2021). Another study investigating the spatial equality and equity of public park distribution in Greater Doha and Al Daayen Municipalities in Qatar, in terms of walkability and ease of pedestrian access, found that there is an absence of distributional fairness in park provisions, unintentionally targeting the low wage migrant expatriates (Marthya 2022).

In recent years, the PPD has introduced sustainable design practices aiming to meet sustainability criteria without compromising on aesthetics. In their report on Sustainable Design Practices (PPD 2017), the PPD addresses 5 design initiatives including: compliance to GSAS/QSAS (i.e., solar powered systems, LED lighting, green roofs for buildings, and low flow, low volume water features to reduce water and energy consumption), the use of a drought-tolerant plant palette (incorporating native plants against PPD's plant palette), recycling resources such as using TSE for irrigation and sewage sludge pellets as fertilizer, execution of agroforestry parks, and specialized desert eco-parks (design concepts/locations are included in the report) and connecting green spaces to reduce the urban heat-island effect while providing urban ecosystem services and promoting biodiversity (Fig. 5.1). Recently, a sustainably designed park, Umm Al Seneem, received a Guinness World Record for the longest (1,143 m in length) air-conditioned pedestrian and jogging path utilizing the technology of air regeneration using solar energy, coupled with trees strategically placed to dramatically reduce urban heat island effect (Quirino, 2022; Tribune 2022).

Ambitious landscaping initiatives such as planting one million trees around Qatar, using primarily TSE were also introduced by the Ministry of Municipality in 2020, under the country's commitment to the Paris Agreement on Greenhouse Gas Emissions Reduction (Ataullah 2019). As of November 2022, the total number of trees that have been planted at the state level within the initiative, reached 1 million and 193 thousand and 665 trees (Qatar Tribune 2022). In addition, a GreenBelt Zone

Fig. 5.1 A proposed scheme for connected green paces in the Greater Doha put forward by a planning consultant for the PPD (Report on Sustainable Design Practices 2017)

encompassing the city of Doha and its suburbs is to be established as part of the Qatar Second National Development Strategy (2018–2022) (QNDS 2018). This is expected to reduce the effects of desertification by reducing dust-packed winds, sand encroachment and conserving the wild desert beyond the green belt, in addition to introducing an aesthetic character to these green spaces (Peninsula 2022).

5.2 Urban Food Production in Qatar

Qatar is highly urbanized with 99% of its population living in urban areas in 2021 (World Bank 1, 2021). It has the highest GDP per capita globally, reaching $61,276 in 2021 (World Bank 2, 2021). Over the past 20 years, Qatar's population has had an annual growth rate of approximately 1.7%. However, the United Nations projects that there will be a considerable reduction in the growth rate with an average of between zero and one percent being sustained from 2030 to the end of the century (United Nations, 2022).

Qatar is largely dependent on food imports to feed its current population (June 2022) of 2.66 million (Planning and Statistics Authority in Qatar 2022) and this reliance affects and compromises the long-term food security of the country.

Qatar's high import dependency as about 90% of consumed food is imported (Miniaoui, Irungu, & Kaitibie 2018) is primarily a consequence of its harsh climate, which in turn leads to unsuitable soils. However, in recent years, domestic food production in Qatar has expanded as a response to increased food demand and because of food security issues, these last were mainly caused by shocks to the food supply chain.

A severe shock occurred in June 2017 when a blockade was imposed upon Qatar by its neighboring countries and others within the MENA region causing an abrupt dislocation of Qatar's food supply, the nation's nutritional lifeline being 90% dependent on imports, which unsurprisingly caused a great deal of public alarm and considerable subsequent economic disturbance (Miniaoui et al. 2018).

A second severe shock was caused by the outbreak of the COVID-19 pandemic in early 2020, which including the consequent lockdowns, again disrupted Qatar's international food supply chain, and local distribution in a similar way as the previous embargo did. However, this time the pandemic was global in scale and potentially if the resultant mortality had been higher, could have led to severe food shortages in Qatar. Thus, for Qatar, the COVID-19 outbreak of 2020 raised national awareness of the vital importance of maintaining food security for urban dwellers.

Nonetheless, the Qatari government responded rapidly to these two emergencies, by taking bold measures, thus ensuring good levels of food security for its citizens. Qatar is ranked 30th out of 113 participating countries in the Global Food Security Index 2022 (Global Food Security Index, 2022) based on four indicators including food affordability, availability, sustainability and adaptation, as well as quality and safety (Table 5.1). Moreover, as shown in Table 5.1, Qatar ranked second among the Gulf Cooperation Council (GCC) countries in the same index in 2022. This position highlights that the various policies instigated and their strategic implementation by the government in response to the shock of the blockade and COVID-19 pandemic, have been very successful in making the country now among the most food-secure/resilient countries in the world.

Stimulated by the 2017 blockade, Qatar made significant advances in local food production, increasing the domestic production (self-sufficiency) of the main food produce groups, including vegetables, poultry, dairy products, and fish. Subsequently,

Table 5.1 Qatar's ranking in the global food security index (2022)

Country	Global rank	Rank in the GCC countries
Qatar	30	2
UAE	23	1
Oman	35	3
Bahrain	38	4
Saudi Arabia	41	5
Kuwait	50	6

self-sufficiency in these food products increased (Fig. 5.2). In an aspirational declaration, the Qatari government set the target of producing approximately 6% of its own food by 2023. This will include all dairy produce, fresh foods (vegetables, fruit, etc.), and meat as shown in Fig. 5.2 (Ministry of Municipality and Environment 2020; Planning and Statistics Authority in Qatar 2020a, b). It is noteworthy that Qatar achieved over 100% self-sufficiency in its fresh poultry and dairy product consumption in 2019, 124%, and 106% respectively, and this trend continued throughout 2020 and 2021 despite the COVID-19 pandemic.

The use of the urban environment for agriculture has several benefits such as improving food quality and accessibility and alleviating food deserts, thereby raising food security (Duchemin et al. 2008). Urban agriculture also provides multiple social, environmental, and economic development benefits for urban populations. Some of these social benefits include the creation of united, self-sufficient communities with access to fresher and healthier local foods, and enhancement of social interactions between community members, thus supporting mental health (Al-Mayahi et al. 2019). Furthermore, community members that grow their own vegetables and fruits ensure a high-quality diet which is one of the determinants of a healthy lifestyle.

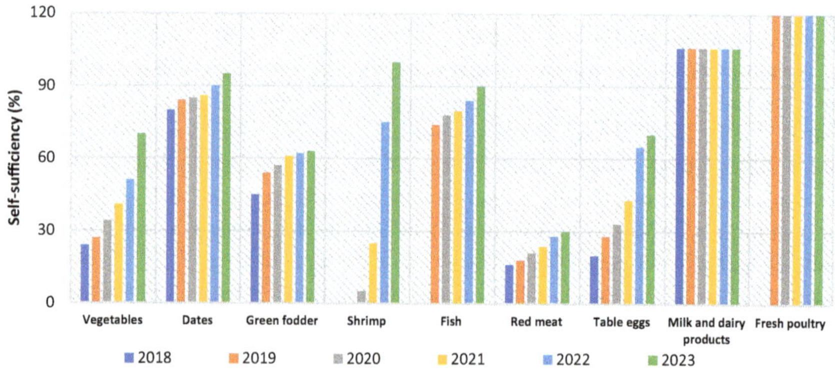

Fig. 5.2 Self-sufficiency (i.e., local production available for home consumption as a percentage of demand) of major food groups (vegetables, dates, fodder, shrimp, fish, red meat, eggs, milk, dairy, and fresh poultry) in Qatar (2018–2023) (Ministry of Municipality and Environment 2020; Planning and Statistics Authority in Qatar 2020a, b)

The potential benefits of urban agriculture to improve people's diet, and thus their health, and overall well-being are especially important in Qatar where childhood and adult obesity rates are of increasing concern. To elaborate, the childhood obesity of children under 5 years of age was 6% and for 13–18-year-olds was 22.3%. In 2016, adult obesity for males was 33.5% while for females it was 44.6% (Global Obesity Observatory, 2022). Furthermore, all the four main Non-Communicable Diseases (NCD)—cardiovascular diseases, neoplasms, diabetes, and respiratory diseases, are the main contributors to death (males 59.6% and females 60.7%) in the country (Qatar Ministry of Public Health, 2016) are all adversely affected by poor diet.

Urban agriculture also improves "urban living". Urban gardens provide amenity spaces that greatly increase urban outlook/neighborhood landscape aesthetic, leisure, and relaxation, and promote outdoor physical activities, allowing communities to connect with nature (Duchemin et al. 2008). They also provide urban residents with improved ecosystem services, such as water and air purification, temperature moderation and the creation of large green areas which contribute to the protection of urban biodiversity, and the reduction of a city's carbon footprint. This last is achieved by reducing food miles (which is significant in Qatar as a great deal of its food produce is imported via air transport) (van Veenhuizen and Danso 2007). In addition, climate change adaptation measures will become increasingly important in Qatar where annual average temperatures are predicted to rise above 50°C. Along with other "green-intensive" urban landscape measures, urban agriculture has the potential to improve the urban climate by decreasing the urban heat island effect and consequently reducing energy consumed by cooling (de Zeeuw 2011).

While urban and peri-urban agriculture provides so many benefits to the local population it is still seen as an emerging activity; often regarded as a hobby for small gardeners rather than an important aspect of a sustainable urban environment. Its potential contribution to Qatar's food security is, as yet, unrecognized.

However, the practice of urban agriculture in Qatar has recently gained popularity as social media groups, private clubs, and government reports demonstrate the potential and benefits of locally grown food. Self-grown and locally grown food seems to have become a growing trend, which fortuitously echoes the government's vision to become a self-sustaining nation in terms of food production. Moreover, in this effort to become more self-sufficient and resilient, there is the added benefit of considerably reducing the carbon footprint of Qatar's food consumption. Indeed, recent research has shown that food produced locally in Qatar has less CO_2 emissions than imported foods, although exact reductions are dependent on food type, country of origin, means of transport, and production system employed. However, as a guide, reducing imports in Qatar by 50% reduces CO_{2e} emissions from transportation by 450 kg CO_2 per capita^{-1} per year^{-1}. And this is usually combined with a reduction in packaging which in turn leads to a reduction in plastic waste entering the environment and further CO_{2e} reductions (Vincente and Piorr 2021).

The potential for Qatar to initiate large-scale urban farms based on modern horticultural systems such as hydroponics, aquaponics, and vertical farming in real-estate development projects is starting to be recognized, as the space availability is not a constraint. According to the Planning and Statistics Authority in Qatar and the Qatar

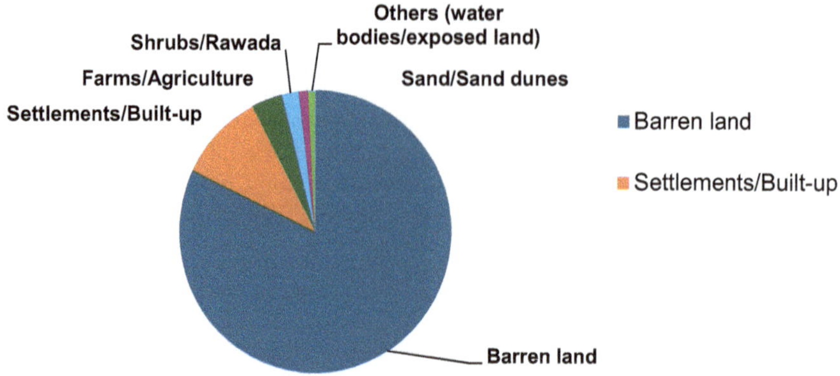

Fig. 5.3 Land cover for Qatar in 2020 (Planning and Statistics Authority, 2020)

Atlas for 2020, the state had in terms of land cover approximately 10,2% settlements and built-up (1.181,95 km^2), 3,7% farms and agriculture (427,79 km^2), 2% shrubs and rawada (233,26 km^2), 0,1% mangroves and forests (9,68 km^2), 1,2% sand and sand dunes (134,07 km^2), 82,1% barren land (9.528,52 km^2) and 0,8% other land covers such as water bodies/exposed land (94,10 km^2) of the total country area, as shown in Fig. 5.3.

The barren land includes other land use categories like residential, commercial, governmental sports, and religious complexes, etc.

Urban food production in Qatar could be the key catalyst that results in.

- increasing food availability, food access and utilization, as well as supply stability to urban dwellers;
- a bulwark that protects Qatar from food supply disruption, e.g., the COVID-19 pandemic and other shocks;
- provision for the increased food demand required as Doha and other urban areas have grown into large metropolitan areas;
- reducing the carbon footprint of Qatar's food supply chain; (Vincente and Piorr 2021); and
- progress toward the Qatari government's goal of increasing food security by 2023 and beyond and resilience (Ministry of Municipality and Environment 2020; Planning and Statistics Authority in Qatar 2020a, b).

5.3 Challenges and Recommendations

5.3.1 Biodiversity

It is widely known that urbanization is associated with biodiversity loss (Liu et al. 2025). Permanently replacing natural habitats leads to considerable changes in the composition and abundance of species assemblages (Geschke et al. 2018 and McKinney, 2002). Native species' richness typically declines, with the degree of urbanization (McKinney 2006). Urban areas usually support some invasive species, with the number of invasive species usually increasing with the magnitude of urbanization (McKinney 2006). Urban areas can also lead to phenotypic adaptations yielding rapid eco-evolutionary change (Alberti et al 2017).

While urbanization presents a significant threat to biodiversity, cities can support numerous species, which occasionally have larger populations, higher productivity, and faster growth rates, inside cities than outside of them (Spotswood et al. 2021). According to researchers (Spotswood et al. 2021) there are five pathways in which cities in deserts can promote regional biodiversity including.

- freeing species from threats in the wild,
- acting as migratory stopovers,
- increasing regional habitat heterogeneity and genetic diversity,
- pre-adapting organisms to climate change,
- and improving public environmental stewardship.

These pathways may be helpful if incorporated into urban greening strategies in Qatar as they are tailored for desert environments and can support local and regional biodiversity conservation.

At small spatial scales, plant diversity is a principal predictor of insect diversity (Southwood et al. 1979). Studies of backyard home gardens in Toronto, (Sperling and Lortue 2010) Leon (Nicaragua) (Gonzalez-Garcia and Sal 2008) and Santarem (Brazil) (WinklerPrins & Oliveira 2010) that possessed a reasonable variety of vegetation showed that insect diversity was greatly enhanced by plant diversity within the small spatial scales of the backyard home gardens studied (Southwood et al. 1979). Their study went on to show that insect abundance and diversity were correlated with plant diversity and that these backyard home gardens even had a greater abundance of winged flying invertebrates when compared to urban forests and grasslands (Sperling & Lortue 2010). Another study showed that community gardens in New York City, USA, providing food produce as well as ornamental plants, supported urban pollinators; among which there were 54 species of bees. Another study looked at home gardens with stratified vegetation, i.e., layered vegetation according to height of species, resembling agroforestry systems (systems combining trees, crops, or livestock to increase biodiversity, productivity, profitability, and environmental stewardship) (WinklerPrins and Oliveira 2010) and observed high levels of biodiversity (WinklerPrins 2002; Udawatta et al. 2019).

Bird and vertebrate abundance and diversity may be increased through the use of wildlife-friendly features (Goddard et al. 2013). Some of these features are as follows (Good 2000):

- planting fruit and/or seed-bearing trees;
- forming compost heaps;
- constructing bird tables and insect hotels; and
- restricting the use of pesticides and herbicides.

Further studies linking urban gardens, their plant diversity to greater biodiversity generally are.

- In Stockholm, Sweden studies showed that garden heterogeneity increased the biodiversity of insectivorous birds (Andersson et al. 2007), and
- In Leon, Nicaragua, garden area and tree height have been found to positively correlate to the presence of iguanas in urban green spaces (González-García et al. 2009).

These findings demonstrate that urban green spaces including urban agriculture can have very positive outcomes on insect biodiversity in urban settings. A large variety of vegetation can be planted in urban areas in Qatar, a combination of native and introduced plant species as well as food crops, allowing for considerable variation in plant diversity. Such variation in vegetation can benefit biodiversity by promoting insect and vertebrate diversity. In addition to habitat heterogeneity, habitat connectivity is a crucial feature for ground-dwelling animals to carry on in urban spaces (Braaker et al. 2014). Therefore, strategies such as movement corridors (i.e., connected to other green areas to permit movement), providing food and nesting resources to sustain vertebrate diversity in urban areas (Lin et al. 2015) may be helpful to biodiversity when planning urban green spaces.

Heterogeneity of plant and animal communities in urban green spaces is crucial as it influences urban ecosystem services such as pollination, climate regulation, and pest control. Urban agriculture is able to support a large assemblage of bees and butterflies when incorporated with a variety of flowering plants (Matteson & Langellotto 2010). This is invaluable for crop production and fruit setting. Studies (Cane 2001) suggest that a network of small scale, natural habitats interspersed in urban areas can support pollinator foraging and dispersal needs. Urban heat island mitigation has been discussed. Natural pest control, whereby pest populations are controlled through the use of other organisms, has been used for centuries in agricultural systems (Flint and Dreistadt 1998). This could potentially permit sustainable crop production without the use of toxic pesticides, especially important in urban areas whereby human exposure can be risky (Robbins 2001). However, such a management strategy requires extensive knowledge of natural enemies. These services protect production values in US crops by over $57 billion annually (Loreau 2001, Losey and Vaughan 2006, Daily 1997).

To the best knowledge of the authors, no studies were conducted on the link between biodiversity and urban ecology and agriculture in Qatar. While the work of others on urban green spaces and improved diversity demonstrates promising results, more research must be conducted in Qatar or similar desert environments.

5.3.2 *Connectivity*

Urban development increases fragmentation and isolation of habitats which may threaten biodiversity (Hobbs and Saunders 1990). Urban greening is key to maintaining biodiversity in urban areas. In order to maintain biodiversity, it is crucial to develop a comprehensive ecological network that can maintain landscape-scale connectivity at the urban planning level (WinklerPrins 2002; Udawatta et al. 2019). Developing such networks entails the protection of existing green spaces and the creation of links that restore and maintain connectivity among green spaces (Pauliet & Duhme 2000).

The literature shows that in urban environments, there's usually a large green space or mother node that tends to have high biodiversity and provide important habitat for species within, as well as peripheral species and transients due to its size (Rudd et al. 2002). The mother node is usually connected to smaller green spaces or satellite nodes ranging in size from 0.1 to over 100 ha (Rudd et al. 2002). These cannot support a large number of species on their own and are partially or completely dependent on individuals coming in from the mother node (Hansson 1991). As a result, they experience higher extinction rates than the mother nodes and require continuous repopulation, which depends on accessibility and proximity to the mother node (Van Apeldoorn et al. 1992). Urban areas are becoming increasingly more fragmented, satellite nodes are getting smaller and farther away from the mother node, making dispersal even more difficult. To maintain the biological integrity of a landscape, networks and habitat matrices should be established to facilitate species dispersal between green spaces.

There are several models that help generate and evaluate potential networks including Paul Revere model, Traveling Salesman model and the Least Cost to User model (Linehan et al. 1995). Researchers have recommended the Least Cost to User model as it is the most complex whereby all nodes are directly connected to each other (Rudd et al. 2002; Kong et al. 2010). Researchers also recommend using the gravity model to evaluate the level of interaction between the nodes in terms of species movement along connected corridors (Linehan et al. 1995) and the node-network analysis to evaluate species that move across corridors to patches (Forman 1999). Typically, nodes have a greater interaction when they are bigger and closer together (Linehan et al. 1995).

While models are helpful, connecting urban spaces is not a straightforward task for urban planners. Various decisions must be made such as the following:

- Must corridors connect to the wild scape? Is it a good idea to bring in wild animals to urban areas?
- Should corridors be natural or semi-natural?

- When locating corridors do, we assess the needs of 1 urban species or various for these corridors? Which species to focus on?
- Assess the minimum habitat size requirement for urban species.

Designing corridors utilizing the connectivity analysis is increasingly more effective than randomly choosing networks (Rudd et al. 2002). It is proven that a network that not only connects parks but also backyard habitats and food gardens, boulevards and utility rights-of-way would yield a great matrix of connections that increases biodiversity in cities and enhances the quality of life for residents (Rudd et al. 2002). It would also be beneficial to monitor the quality of the network over time as some may not provide suitable habitat for flora and fauna. More research on connected green spaces is recommended for Qatar and desert environments.

5.3.3 Natives Versus Exotics

While the Public Parks Department predominantly utilizes exotic species in green urban spaces, globally, there is a continuing debate between ecologists on the value of native versus exotic vegetation. Traditionally, natural areas with native communities were considered to have greater resilience in the face of global changes and that any restoration effort should resemble such reference systems. However, scientists started to debate whether biodiversity is the principal determinant for ecosystem integrity for ecological restoration projects. The focus has shifted from native to adaptable species (Simonson et al. 2021). Various scientists believe that functional diversity is the most significant factor for ecosystem integrity. This idea champions plant traits over taxonomic composition (Esperon-Rodriguez et al. 2020; Wood & Dupras 2021; Ossola & Lin 2021).

Arguments in Favor of Exotic Species:

- Exotic species occasionally surpass the native vegetation as they have higher foliar biomass, offer a denser canopy, and contribute significantly to the species' diversity (Riley et al. 2018).
- In extremely degraded areas such as landfills, former mining plots or abandoned urban areas, exotic pioneer species can colonize under harsh conditions and enrich native plant communities by providing shelter for native species to thrive (Oduor et al. 2018; Martins et al. 2020).
- Exotic species can act as a replacement for lost native species in urban green spaces, helping to maintain vital levels of functional diversity (Gonçalves et al. 2021). The key is to maintain a specific assemblage of plants that exhibit the essential functional traits of lost native species (Aquilué et al. 2021).
- Exotic species are preferred in domestic gardens or public spaces and are associated with cultural landscapes by people (Hoyle et al. 2017). These may include edible fruit trees, herb or vegetable gardens, that not only provide a habitat for urban species but also nutritional value to the inhabitants of the city.

- In most cases, exotics excel over natives, in the efficient capture and use of resources, particularly in scarcity contexts, which is frequently associated with their invasive behavior (Sun et al. 2022; Díaz-Barradas et al. 2020; Zunzunegui et al. 2020).
- Exotic species have a better ecosystem adaptive capacity as they have better chances of coping with the adverse effects of climate change (Esperon-Rodriguez et al. 2020; Wood and Dupras 2021; Ossola & Lin 2021), showing a better capacity for colonization in places where natives cannot subsist and for adjustment to changing conditions, such as hydrologic alterations due to rising sea levels, extended drought periods, or higher temperatures, which are expected to worsen with climate change (Goa et al. 2018).

Arguments in Favor of Native Species:

- Native species largely exceed exotics in the support of native animal species, the association with soil biota and in the harbor of other plant species (Berthon et al. 2021; Barrico et al. 2018; Prendergast et al. 2022).
- Native species are better at interacting with abiotic factors, such as the soil nutrient and water cycles (Pérez-Corona et al. 2021; Vasquez-Valderrama et al. 2020) while exotic species have negative impacts on the nutrients cycle, resource consumption, and levels of soil pH (Zhang et al. 2021; Kotzen et al. 2020; Gasteur et al. 2021).
- Native vegetation provides better ecological performance especially in arid and semi-arid environments. Native species have high plasticity and adaptive traits that allow them to handle the environmental stress in their harsh environments (Esfahani et al. 2022; Guillen-Cruz et al. 2021).
- Natural areas with native communities have better resilience and capacity for the supply of ecosystem services, especially when faced with global biodiversity alterations and climate change. Therefore, restoration interventions should mimic such reference natural systems. Moreover, in urban green spaces that are exposed to additional stresses, natives can be crucial as they also provide various regulating ecosystem services (Esperon-Rodriguez et al. 2020).
- In spite of the adaptive potential shown by exotic species, research points to natives as a viable option and states that projects should favor the evolutionary development of native communities (Rice and Emery 2003). These are some strategies to enhance the resilience and adaptive ability of native species and communities:

  - Increasing genetic diversity between species (as it can enhance the adaptivity of a native community) and within an individual species, for example by utilizing seeds from various ends of the species distribution range (Rice and Emery 2003).
  - Using rare, specialist native plants in their ecological optimum can lower biodiversity loss and enhance ecosystem stability which in turn can increase ecosystem service supplies (Jensen et al. 2021).
  - The increase in the number of species and the enlarging of habitat areas both fortify ecosystem integrity and ecological connectivity, enhancing the

ecosystem's ability to harbor biological diversity in the face of climate change (Maxwell et al. 2019).

- The ecosystems adaptive capacity can be supported by intervention strategies such as genetic strategies (assisted migration), or silvicultural interventions (density management), to lower resource competition, fire risk, or the spread of disease (Safford et al. 2019).
- Diverse native communities are thoroughly prepared to survive adverse events due to climate change, storms, floods, drought, increasing temperatures, and fires, while exotic communities, particularly invasive ones, progressively deteriorates the system (Galleguillos et al. 2021; Hamburg et al. 2020; Vivian et al. 2020).

One study (Carvalho et al. 2022) looked at 164 papers on the potential of native and exotic species to enhance ecological integrity in urban green spaces. The literature overwhelmingly shows that native species provide significant ecological advantages across biodiversity conservation, ecosystem performance, resilience to climate change, ecosystem services, and cultural perception, without documented disadvantages. While exotic species occasionally outperform natives in resource-scarce or degraded environments, their use often results in ecological trade-offs, including biodiversity loss, nutrient cycle disruption, and increased ecosystem vulnerability. Although urban and disturbed landscapes sometimes benefit from exotics, the consensus favors native vegetation for sustainable and resilient ecosystem management.

When designing urban green spaces, a serious challenge to using numerous exotic species is the introduction of an invasive one. Invasives could potentially harm native ecosystems and be detrimental to the delivery of ecosystem services (Blitzer et al. 2012). For example, in Qatar, the introduction of the invasive *Prosopis juliflora* tree has caused detrimental competition to the native *Prosopis cineraria* population (Bibi and Abu-Dieyeh 2016). *Prosopis juliflora* is highly aggressive and grows so densely that it crowds out native vegetation (El Keblawy and Al Rawai 2007). It significantly reduced the evenness, richness and density of the associated plants beneath, compared with open places beyond their canopies (El-Keblawy et al. 2014). In the UAE's artificial forests, planting exotic trees like Eucalyptus and *Prosopis juliflora* has led to notable declines in the richness and density of understory plants when compared to native species such as *P. cineraria* and *Acacia arabica* (El-Keblawy and Ksiksi 2005).

Aggarwal et al. (1976) observed that *Prosopis juliflora* canopies supported significantly fewer understory species than four other forestry species studied, while the native *P. cineraria* was linked to the highest diversity. The invasive success of *P. juliflora* is mainly due to its prolific seed production and effective dispersal (Shiferaw et al. 2004). Its spread leads to reduced grass cover and lower livestock carrying capacity. The species negatively impacts plant biodiversity by forming dense thickets that block light, physically obstruct seedling growth, deplete groundwater, and release allelopathic chemicals that harm native vegetation (Shiferaw et al. 2004).

In addition to invasive species, there are other challenges to urban green spaces such as the introduction of weeds, pathogens and/or pests. When designing an ecological landscape, care must be taken to not introduce pests. For example, the Red Palm Weevil (RPW) is a devastating pest of palm species in the Middle East. The RPW grubs are the most destructive stage of the weevil's lifecycle as they chew on the palm's soft tender tissues disrupting its vascular system. Their destructive effect is considerable. The Arabian Peninsula accounts for 30% of global date production. Researchers (Elsabea et al. 2009) estimated the economic losses of approximately 5.18 to 25.92 million US dollars due to the infestation of RPW (Hussain et al. 2013).

5.3.4 Urban Crops

Urban agriculture concentrates largely on crops that do not require large-scale land-holding and require limited inputs. The products are often perishable for example vegetables, fruits, herbs, and spices as well as small livestock. In regions with harsh climatic conditions, urban farmers regularly create a favorable microclimate, ameliorating temperature, humidity, wind, and shading to provide conditions that are favorable for cultivation and animal husbandry. Creating a suitable microclimate is even more important in desert climates and can be readily achieved in urban agricultural settings, where optimal conditions for food production can be established sustainably (Tchakerian 2015).

According to feedback derived from groups of very experienced local gardeners in Qatar, there are more than 70 types of vegetables, herbs, and fruits that have been regularly grown in Qatar's urban areas, as well as rearing a few animal species for meat and egg production for many years (Krishiyidam 2022). Although numerous plant species can be cultivated in Qatar, special attention should be given to the resources available for their growth, as some may be too resource-demanding (e.g., rice and wheat, etc.), and therefore cannot be grown sustainably in the urban environment. Moreover, urban farming could have a positive effect on biodiversity in urban areas as it offers diverse vegetation and more foraging opportunities than bare land or any typical urban land cover. However, according to researchers, this enhancement is species and location-specific and needs more empirical research (Claucas et al. 2018). Appropriate plant selection is therefore the key; the plant species under selection for urban farming for food production could be, as possible.

- native species;
- species already adapted and grown in the Qatari environment;
- diverse/mix and match species that attract pollinators and other animal species;
- short growth cycle such as small fruit varieties, cherry tomatoes and leafy greens;
- drought tolerant and pest resistant;
- non-resource demanding for their growth; and
- non-traditional plant species like halophytes and saline crops.

Recent studies around the world have shown that the yields from urban agriculture can be as productive, if not more so, than established agricultural methods (Orisini et al. 2013, FAO 2010; Baudoin 2010). According to the FAO, garden plots can be up to 15 times more productive than rural holdings and an area of just one square meter can provide 20 kg of food a year (FAO 2010). Nonetheless, yields vary among growers as they are highly dependent on the cultivation system employed and the skills of the farmers, as well as the quality of the agricultural inputs, the season's weather which can be very variable, and the exact crop varieties where small differences can matter.

In Qatar, informal surveys carried out with some of the local urban growers revealed that they can grow enough food in their backyard to feed a family of four, during Qatar's traditional growing season, typically from October to May each year. However, no matter how skilled their husbandry is, they recognize that their products cannot cover all their dietary requirements, as not all food types can be grown under Qatari climatic conditions (Pers. Com. Adukkalathottam 2022).

Concerning how much food can be produced in Qatar, it seems that the country could potentially be 100% self-sufficient for certain fresh items that can be grown under local conditions. High yields could be achieved with the utilization of modern production techniques such as hydroponics, aquaponics, aquaculture, and vertical farming, as was already discussed. Some of these fresh products are vegetables such as tomato, cucumber, eggplant, pepper, zucchini, melon, and beans. Many other plant species can be grown under the local pedoclimatic conditions, but several factors should be considered apart from the feasibility of production itself. Important factors are the production cost, storage/logistics cost as well as environmental aspects such as the sustainable resource use such as water and energy that could prove the production of certain food items nor profitable or sustainable.

5.3.5 *Water*

Water use, especially in arid lands, is another serious challenge that has to be confronted when it comes to irrigating urban green spaces. Qatar is deficient in freshwater resources due to extremely low rainfall and lack of freshwater bodies. There are 3 water resources currently used in the country including, desalinated water, recycled water and groundwater. The Qatar Second National Development Strategy (QNSDS, 2018–2022) acknowledges that the 3 resources available are inefficient but necessary for water security (QSNDS 2018). Desalinated water is costly to produce and comes with a multitude of environmental marine impacts. In 2016, the groundwater consumption rate was 5 times higher than the recharge rate (QSNDS 2018). The agricultural sector in Qatar utilizes 91% of all renewable freshwater resources (aquifers) in the state (Planning & Statistics Authority, 2020), and groundwater is becoming increasingly saline. This excessive extraction is linked to the lack of water tariffs and poor management of groundwater extraction on farms (QSNDS, 2018). Therefore, the available water for irrigation is not only limited but also in many cases the abstracted groundwater must be desalinated before use as it is highly saline, which could negatively impact plant growth.

In 2015, generated recycled water amounted to 194 million m^3 (at a fourth of the treatment cost of that of desalinated water). Around 34% was used for agriculture, 16% for landscape irrigation, 29% injected into deep non-freshwater aquifers, and 20% discharged into manmade lakes (QSNDS 2018). The use of Treated Sewage Effluent (TSE) for irrigation in Qatar is cheaper and more sustainable than the use of desalinated water (Jasim et al. 2016; Côté et al. 2005; and Missimer et al. 2014), however TSE supply and delivery systems need expansion in the country (QNSDS 2018). In the summer months, TSE supply for irrigation of landscaped areas in Doha is not sufficient to meet demand (Per Comm. Lawler 2022).

Consequently, sustainable water resource management and groundwater conservation policies are imperative. These policies could include the.

- diversification of water sources for irrigation like the use of Treated Sewage Effluent (TSE), recycled water, or even desalinated seawater than abstracted freshwater;
- adaptation of water-efficient irrigation practices, e.g., drip irrigation as well as optimal use of irrigation water;
- promotion of water-efficient vegetations and crops;
- promotion of water-saving food growing systems like hydroponics and aquaponics;
- awareness enhancement of water resources and management to the gardeners and farmers; and
- installation of meters on groundwater wells in farms to inform water management and rationalization programs.

In particular, hydroponics promotes extreme efficiency of water use, in addition to the possibility of being applied to non-arable or marginal lands, such as dry or urban areas (Jones 2014). Hydroponic plant production uses approximately 70% less water (Okemwa 2015) than traditional agriculture; the exact reduction in irrigation is dependent on the crop variety. There is also the possibility of recirculation of both water and nutrient solution in "closed systems" where there is minimal evaporation as the same nutrients or nutrient solutions are repeatedly recycled and re-applied (Abusin and Mandikiana 2020). Similarly, as aquaponics combines hydroponics and aquaculture (fish farming) into one production system the use of inputs such as water is optimized as the plants filter and clean the water so that it can be recycled back into the fish tank creating a symbiotic environment in which the water recirculates (Goddess et al. 2019).

5.3.6 Soil

Qatar is a small country covering approximately 11,600 km^2, of which 65,000 ha are suitable for arable cultivation and of this area a total of 13,430 hectares are already being cultivated (green fodder, vegetables, date palm, cereals and fruits), leaving 51,570 hectares to potential farmland in 2021 (Planning and Statistics Authority in Qatar 2021). Most Qatar soils are characterized by sandy coarse texture, shallow soil, low water retention, and slow soil formation and are therefore considered a challenge for landscaping and/or growing crops. While landscaping may require fewer soil amendments, depending on the choice of vegetation, in this section we will focus on arable soil. As already mentioned, arable soil is limited; however, soil cultivation can be feasible when appropriate farming methods such as enhancing soil fertility and improving its properties are applied. Some of these techniques to enhance soil health and increase productivity include the.

- use of substrates like coco peat;
- addition of compost, animal manure, biochar;
- imported soils;
- crop rotation;
- low tillage;
- long-term farming; and
- appropriate farming techniques.

According to researchers (Koberl et al. 2011), long-term farming can enhance desert soils and thus promote plant health and biodiversity. After 30 years of organic desert farming, studies on long-term agriculture of medical plants such as *Matricaria chamomilla* L. (german chamomile), *Calendula officinalis* L. (pot marigold), and *Solanum distichum* (Schumach. and Thonn.) on desert soil in Sekem (Egypt) found that bacterial communities in agricultural soil showed a higher diversity and a better ecosystem function for plant health. When coupled with the indigenous desert microorganisms, enhanced plant growth in desert agroecosystems was also seen (Koberl et al. 2011). In addition, indoor soil cultivation minimizes the risk of crop failure due to harsh climates and offers a better opportunity for plant pest control.

Additionally, agricultural soils in Qatar have recorded high levels of arsenic (As), chromium (Cr), and nickel (Ni), thus posing potential risks to human health (Alsafran et al., 2021a, b). According to research carried out in 2021, the consumption of some leafy vegetables like rocket, coriander, and parsley grown in open irrigated farms in Qatar posed a significant health risk (both noncarcinogenic and carcinogenic) to consumers (Alsafran 2021b). Limitations of soil cultivation in Qatar apart from the unsuitable soils and the scarcity of land is the harsh climate. When soil cultivation is practiced outside, it also faces the challenge of Qatar's climate which according to Köppen's classification (Kottek et al. 2006) is characterized as a hot desert (BWh) with hot humid summers and mild winters. Consequently, the traditional planting season is limited to the eight months between October and May. Thus, harsh summer climatic conditions limit the growing season to two-thirds of the year apart from a few species such as curry leaves, ladyfingers (okra), sweet potatoes, some varieties of cucumbers, and Indian spinach that can still be cultivated outdoors.

As soil cultivation for plant production can be practiced either in open fields or in greenhouses such as net houses, or plastic greenhouses, the limitations of soil cultivation can be overcome by.

- resorting to soilless cultivation techniques and
- using innovative agricultural technologies, often within greenhouses (where super-controlled environments can be established) that optimize the utilization of available resources (Karanisa et al. 2021).

These innovations include hydroponics, aquaponics, as well as vertical farming systems which can be deployed in small urban-scale greenhouses that can be easily integrated or adapted to existing buildings.

Besides, using the urban environment, urban-scale greenhouses and soilless cultivation methods for agriculture and food production there is no extra land taken and Qatar's undeveloped hinterland is preserved without depleting its unique habitat's biodiversity, endangering native flora and fauna.

Moreover, while extensive agriculture practiced in Qatar's hinterland would deplete its unique habitat's biodiversity and ecosystem services, regenerative agriculture and permaculture can restore and maintain natural systems. Regenerative agriculture is a set of approaches that restores and maintains soil health, protects natural resources and biodiversity (McLennon et al. 2021). Likewise, permaculture is an agricultural transdisciplinary system focusing on ecosystem health, individual responsibility as well as locally adapted and small-scale solutions (Didarali and Gambiza, 2019). These farming systems minimize the dependence on external inputs which restores and maintains natural systems such as improving soil health, ecosystem biodiversity, land and resource conservation and ecosystem services) enhancing agricultural and food systems sustainability. A successful example of a permaculture farm in Qatar is the Heenat Salma project, intergrading holistic agricultural methods, local architecture and community development. The project started in 2019, aiming to become a center for regenerative agriculture, hospitality and education through active community engagement (Heenat Salma, 2023).

Nonetheless, traditional agricultural practices have been always focused on enhancing productivity while maintaining soil health and preserving the ecosystem. According to researchers, local communities have already extensive knowledge and experience in sustainable farming practices that could be used by scientists, stakeholders and commercial companies. Indigenous people have always been part of the ecosystem; they have controlled and shaped the landscapes and have known about the plant species that thrive in their region for hundreds or thousands of years, being the biodiversity maintainers (Antonelli 2023). Thus, traditional knowledge incorporated into farming practices and food systems overall can enhance sustainability.

5.4 Best Ways to Move Forward

Although general literature demonstrates that urban green spaces promote biodiversity, literature on biodiversity and ecosystem services in urban green spaces in the Arabian Gulf and arid regions is scarce. Therefore, more research needs to be conducted in arid regions and in Qatar specifically. Our recommendations are as follows:

- Carry out an up-to-date country-wide baseline survey of flora and fauna—to account for rare/important/threatened species as well as identifying hot spots and biodiversity zones. Share this information with organizations and the public.
- Create a plant palette suitable for landscaping the country using native plants, regional plants, arid area plants and avoiding plants that require large quantities of water.
- Create urban green spaces using an ecosystem specific management scheme. It is not a one-size-fits-all approach. Thus, we propose that green spaces be developed not by design or aesthetics, chance or seasonal fashion but by science. A multidisciplinary science including ecologists, planners, social scientists and community members. Special attention should be given to the potential introduction of invasive species, weeds, pathogens, and/or pests.
- Design connection corridors utilizing the connectivity analysis. Networks can include parks, backyard habitats, urban and peri-urban farms, boulevards and utility rights-of-way which would yield a great matrix of connections that increases biodiversity in cities. Connecting large green spaces is a great way to increase biodiversity. It would also be beneficial to monitor the quality of the network over time as some may not provide suitable habitat for flora and fauna.
- Promote the use of native species and communities in the urban landscape as they not only present high adaptive capacity and resilience, but also provide additional benefits for biodiversity, ecosystem integrity, and people. Exotic species can be used for increasing biodiversity and painting an evergreen lush appearance.
- Adapt sustainable water resource management and groundwater conservation policies focusing on alternative water sources, agronomic aspects, and public awareness enhancement.
- Enhance the health and fertility of the arable soil to make plant cultivation feasible and productive. Long-term farming can enhance desert soils consequently promoting plant health and biodiversity. Soil improvement can be achieved through the use of soil amendments and appropriate cultivation techniques.
- Promote urban farming to enhance the biodiversity of plant and animal species in urban areas via the appropriate plant selection. This needs empirical research and a multidisciplinary approach to allow optimal selection of plant species for each location and desired result.
- Promote sustainable farming systems such as permaculture, regenerative agriculture and utilize indigenous knowledge that enhances soil health, ecosystem biodiversity, land and resource conservation and ecosystem services.

References

Abusin SAA, Mandikiana BW (2020) Towards sustainable food production systems in Qatar: Assessment of the viability of aquaponics. Glob Food Secur, 25. https://doi.org/10.1016/j.gfs.2020.100349

Adukkalathottam Doha community of urban growers in Qatar (2022) Personal correspondence

Aggarwal RK, Gupta JP, Saxena SK, Muthana KD (1976) Studies on soil physico-chemical and ecological changes under twelve years old desert tree species of Western Rajasthan. Indian Forester 102:863–872

Ahmed MT (2002) Millennium ecosystem assessment. Environ Sci Pollut Res 9(4):219–220. https://doi.org/10.1007/BF02987493

Al Khouri MA (2016) Public parks department interview

Alberti M et al (2017) Global urban signatures of phenotypic change in animal and plant populations. Proc Natl Acad Sci USA 114:8951–8956. https://doi.org/10.1073/pnas.1606034114

Al-Mayahi A et al (2019) Home gardening in Muscat, Oman: Gardeners' practices, perceptions and motivations. Urban Forestry & Urban Greening 38:286–294

AlMohannadi M, Zaina S, Zaina S, Furlan R (2015) Integrated approach for the improvement of human comfort in the public realm: The case of the Corniche, the linear urban link of Doha. Am J Sociol Res 5(4):89–100. https://doi.org/10.5923/j.sociology.20150504.01

AlMohannadi A, Sinclair B, Furlan R, Al-Matwi R (2021) From String of Pearls to String of Parks - Exploring Potency + Prospective of Landscape Design in the Islamic City: The Compelling Case of Doha, Qatar. ARCC 2021 International Conference Architectural Research Centers Consortium, Performative Environments, Virtual: Tucson, Arizona USA. https://www.academia.edu/46584522/From_string_of_pearls_to_string_of_parks_The_compelling_case_of_Doha_Qatar

Alsafran M, Usman K, Rizwan M, Ahmed T, Al Jabri H (2021b) The carcinogenic and non-carcinogenic health risks of metal(oid)s bioaccumulation in leafy vegetables: a consumption advisory. Front Environ Sci 9:742269. https://doi.org/10.3389/fenvs.2021.742269

Alsafran M, Usman K, Al Jabri H, Rizwan M (2021a) Ecological and health risks assessment of potentially toxic metals and metalloids contaminants: a case study of agricultural soils in qatar. Toxics, 9(35). https://doi.org/10.3390/toxics9020035

Andersson E, Barthel S, Ahrné K (2007) Measuring social–ecological dynamics behind the generation of ecosystem services. Ecol Appl 17(5):1267–1278. https://doi.org/10.1890/06-1116.1

Antonelli A (2023) Indigenous knowledge is key to sustainable food systems. Nature, 613(7943), 239–242. https://doi.org/10.1038/d41586-023-00021-4

Aquilué N, Messier C, Martins KT, Dumais-Lalonde V, Mina M (2021) A simple-to-use management approach to boost adaptive capacity of forests to global uncertainty. Forest Ecology and Management, 481. https://doi.org/10.1016/j.foreco.2020.118692

Ataullah S (2019). Project to plant 1 million trees in Qatar kicks off. Peninsula Newspaper. https://thepeninsulaqatar.com/article/20/03/2019/Project-to-plant-1-million-trees-in-Qatar-kicks-off

Attwell K (2000) Urban land resources and urban planting—case studies from denmark. Landsc Urban Plan 52:145–163. https://doi.org/10.1016/S0169-2046(00)00129-8

Barrico L, Castro H, Coutinho A, Gonçalves MT, Freitas H, Castro P (2018) Plant and microbial biodiversity in urban forests and public gardens: Insights for cities' sustainable development. Urban Forestry and Urban Greening 29:19–27. https://doi.org/10.1016/j.ufug.2017.10.012

Baudoin W (2010) Programme for Urban and Peri-urban Horticulture UPH. Plant Production and Protection Division. Factsheet 6. FAO

Berthon K, Thomas F, Bekessy S (2021) The role of 'nativeness' in urban greening to support animal biodiversity. Landsc Urban Plan 205:103959. https://doi.org/10.1016/j.landurbplan.2020.103959

Bibi S Abu-Dieyeh MH (2016) Allelopathic effects of the invasive prosopis juliflora (sw.) DC. on seed germination of selected Qatari native plant species. Hamad bin Khalifa University Press (HBKU Press), 20. https://doi.org/10.5339/qproc.2016.qulss.20

Blitzer EJ et al (2012) Spillover of functionally important organisms between managed and natural habitats. Agr Ecosyst Environ 146(1):34–43. https://doi.org/10.1016/j.agee.2011.09.005

Bolitzer B, Netusil NR (2000) The Impact of Open Spatials on Property Values in Portland, Oregon. J Environ Manage 59:185–193. https://doi.org/10.1006/jema.2000.0351

Bolleter J (2009) Para-scape: landscape architecture in Dubai. J Landsc Arch 4(1):28–41. https://doi.org/10.1080/18626033.2009.9723411

Braaker S, Ghazoul J, Obrist MK, Moretti M (2014) Habitat connectivity shapes urban arthropod communities: The key role of green roofs. Ecology, 95. 110–21. https://doi.org/10.1890/13-0705.1

Cane JH (2001) Habitat fragmentation and native bees: A premature verdict? Conserv Ecol, 5(1), 3. [online] URL: https://www.consecol.org/vol5/iss1/art3/

Carvalho CA et al (2022) Native or exotic: A bibliographical review of the debate on ecological science methodologies: Valuable lessons for urban green space design. Land 11(8):1201. https://doi.org/10.3390/land11081201

Clucas B, Parker ID, Feldpausch-Parker A (2018) A systematic review of the relationship between urban agriculture and biodiversity. Urban Ecosystems 21(4):635–643. https://doi.org/10.1007/s11252-018-0748-8

Côté P, Siverns S, Monti S (2005) Comparison of Membrane-based Solutions for Water Reclamation and Desalination. Desalination 182:251–257. https://doi.org/10.1016/j.desal.2005.04.015

Daily G (1997) Nature's services: societal dependence on natural ecosystems. Island Press, Washington, DC

de Zeeuw H (2011) Cities, Climate Change and Urban Agriculture. Urban Agriculture Magazine 25(8):39–42

Díaz-Barradas MC, Gallego-Fernández JB, Zunzunegui M (2020) Plant response to water stress of native and non-native *Oenothera drummondii* populations. Plant Physiol Biochem 154:219–228. https://doi.org/10.1016/j.plaphy.2020.06.001

Didarali Z, Gambiza J (2019) Permaculture: Challenges and benefits in improving rural livelihoods in South Africa and Zimbabwe. Sustainability 11(8):2219. https://doi.org/10.3390/su11082219

Duchemin E, Wegmuller F, Legault AM (2008) Urban agriculture: multi-dimensional tools for social development in poor neighbourhoods. Field Actions Sci Rep, l(1)

El-Keblawy A, Ksiksi T (2005) Artificial forests as conservation sites for the native flora of the UAE. For Ecol Manage 213(1–3):288–296. https://doi.org/10.1016/j.foreco.2005.03.058

El-Keblawy A, Abdelfattah MA, Khedr AHA (2015) Relationships between landforms, soil characteristics and dominant xerophytes in the hyper-arid northern United Arab Emirates. J Arid Environ 117:28–36. https://doi.org/10.1016/j.jaridenv.2015.02.008

El-Keblawy A, Al-Rawai A (2007) Impacts of the invasive exotic Prosopis Juliflora (Sw.) D.C. on the Native Flora and Soils of the UAE. *Plant Ecology*, 190(1), 23–35. https://doi.org/10.1007/s11258-006-9188-2

El-Keblawy A, Abdelfatah MA (2014) Impacts of native and invasive exotic Prosopis congeners on soil properties and associated flora in the arid United Arab Emirates. J Arid Environ 100–101:1–8. ISSN 0140-1963. https://doi.org/10.1016/j.jaridenv.2013.10.001

El-Sabea AMR, Faleiro JR, Abo El Saad MM (2009) The threat of red palm weevil *Rhynchophorus ferrugineus* to date plantations of the Gulf Region of the Middle East: An economic perspective. Outlooks on Pest Management 20:131–134. https://doi.org/10.1564/20jun11

Elsabea AI, Faleiro JR, Abo-El-Saad M (2009) The threat of red palm weevil Rhynchophorus ferrugineus to date plantations of the Gulf region in the Middle East: An economic perspective. Outlooks Pest Manag. 20, 131-134. https://doi.org/10.1564/20jun11

Esfahani RE, Paço TA, Martins D, Arsénio P (2022) Increasing the resistance of Mediterranean extensive green roofs by using native plants from old roofs and walls. Ecol Eng 178:106576. https://doi.org/10.1016/j.ecoleng.2022.106576

Esperon-Rodriguez M, Rymer PD, Power SA, Challis A, Prokopavicius RM, Tjoelker MG (2020) Functional adaptations and trait plasticity of urban trees along a climatic gradient. Urban Forestry & Urban Greening 54:126771. https://doi.org/10.1016/j.ufug.2020.126771

FAO Micro Gardens (2010) With micro-gardens, urban poor "grow their own". http://www.fao.org/ag/agp/greenercities/en/microgardens/index.html. (Accessed 15 December 2021).

Ferwati S, Skelhorn C, Shandas V, Makido Y (2019) A Comparison of Neighborhood-Scale Interventions to Alleviate Urban Heat in Doha. Qatar. Sustainability 11(3):730. https://doi.org/10.3390/su11030730

Flies E et al (2017) Biodiverse green spaces: A prescription for Global Urban Health. Front Ecol Environ 15(9):510–516. https://doi.org/10.1002/fee.1630

Flint ML, Dreistadt SH (1998) Natural enemies handbook: the illustrated guide to biological pest control. University of California Press, Oakland

Forman RTT (1999) Spatial models as an emerging foundation of road system ecology and a handle for transportation planning and policy. In, G.L. Evink, P. Garrett & D. Zeigler (Eds.). In: Proceedings of the third international conference on wildlife ecology and transportation (pp. 119–124). Florida Dept of Transportation, Tallahassee, Florida

Galleguillos M, Gimeno F, Puelma C, Zambrano-Bigiarini M, Lara A, Rojas M (2021) Disentangling the effect of future land use strategies and climate change on streamflow in a Mediterranean catchment dominated by tree plantations. J Hydrol 595:126047. https://doi.org/10.1016/j.jhydrol.2021.126047

Gao X, Zhao X, Li H, Guo L, Lv T, Wu P (2018) Exotic shrub species (*Caragana korshinskii*) is more resistant to extreme natural drought than native species (*Artemisia gmelinii*) in a semiarid revegetated ecosystem. Agric for Meteorol 263:207–216. https://doi.org/10.1016/j.agrformet.2018.08.029

Gastauer M, Ramos SJ, Caldeira CF, Siqueira JO (2021) Reintroduction of native plants indicates the return of ecosystem services after iron mining at the Urucum Massif. Ecosphere 12(11):e03762. https://doi.org/10.1002/ecs2.3762

Geschke A, James S, Bennett AF, Nimmo DG (2018) Compact cities or sprawling suburbs? Optimising the distribution of people in cities to maximise species diversity. J Appl Ecol 55:2320–2331. https://doi.org/10.1111/1365-2664.13183

Goddard MA, Dougill AJ, Benton TG (2013) Why garden for wildlife? Social and ecological drivers, motivations and barriers for biodiversity management in residential landscapes. Ecol Econ 86:258–273. https://doi.org/10.1016/j.ecolecon.2012.07.016

Goddek S, Joyce A, Kotzen B, Dos-Santos M (2019) Aquaponics and Global Food Challenges. In Goddek S, Joyce S, Kotzen B, Burnell GM (Eds). Aquaponics Food Prod Syst. Springer, Cham. https://doi.org/10.1007/978-3-030-15943-6_1

Gonçalves P, Vierikko K, Elands B, Haase D, Catarina Luz A, Santos-Reis M (2021) Biocultural diversity in an urban context: An indicator-based decision support tool to guide the planning and management of green infrastructure. Environmental and Sustainability Indicators 11:100131. https://doi.org/10.1016/j.indic.2021.100131

González-García A, Sal A (2008) Private urban greenspaces or patios as a key element in the urban ecology of tropical central America. Hum Ecol 36(2):291–300. https://doi.org/10.1007/s10745-007-9155-0

González-García A, Belliure J, Gómez-Sal A, Dávila P (2009) The role of urban greenspaces in fauna conservation: The case of the iguana *Ctenosaura similisin* the 'patios' of León city. Nicaragua. Biodiversity and Conservation 18(7):1909–1920. https://doi.org/10.1007/s10531-008-9564-4

Good R (2000) The value of gardening for wildlife-what contribution does it make to conservation? British Wildlife 12(2):77–84

Guillen-Cruz G, Rodríguez-Sánchez AL, Fernández-Luqueño F, Flores-Rentería D (2021) Influence of vegetation type on the ecosystem services provided by urban green areas in an arid zone of northern Mexico. Urban Forestry & Urban Greening 62:127135. https://doi.org/10.1016/j.ufug.2021.127135

Habib S, Al-Ghamdi SG (2021) Estimation of Above-Ground Carbon- Stocks for Urban Greeneries in Arid Areas: Case Study for Doha and FIFA World Cup Qatar 2022. Front Environ Sci 9:635365. https://doi.org/10.3389/fenvs.2021.635365

Hamberg LJ, Fraser RA, Robinson DT, Trant AJ, Murphy SD (2020) Surface temperature as an indicator of plant species diversity and restoration in oak woodland. Ecol Ind 113:106249. https://doi.org/10.1016/j.ecolind.2020.106249

Hansson L (1991) Dispersal and connectivity in metapopulations. Biol J Lin Soc 42:89–103

Hashem N (2015) Assessing spatial equality of urban green spaces provision: a case study of Greater Doha in Qatar. Local Environ 20. https://doi.org/10.1080/13549839.2013.855182

Hobbs RJ, Saunders DA, Hussey BMT (1990) Nature conservation: The role of corridors. Ambio 19(2):94–95

Hoyle H, Hitchmough J, Jorgensen A (2017) Attractive, climate-adapted and sustainable? Public perception of non-native planting in the designed urban landscape. Landsc Urban Plan 164:49–63. https://doi.org/10.1016/j.landurbplan.2017.03.009

Hussain A et al (2013) Managing invasive populations of Red Palm Weevil: A Worldwide Perspective. J Food Agric Environ 11(2):456–463

Global Food Security Index 2022: Country Profiles. Economist Impact 2022. https://impact.econom ist.com/sustainability/project/food-security-index/explore-countries (accessed on 8 December 2022).

Jasim SY, Saththasivam J, Loganathan K, Ogunbiyi OO, Sarp S (2016) Reuse of Treated Sewage Effluent (TSE) in Qatar. J Water Process Eng 11:174–182. https://doi.org/10.1016/j.jwpe.2016. 05.003

Jensen DA, Rao M, Zhang J, Grøn M, Tian S, Ma K, Svenning JC (2021) The potential for using rare, native species in reforestation—A case study of yews (Taxaceae) in China. For Ecol Manage 482:118816. https://doi.org/10.1016/j.foreco.2020.118816

Jones BJ (2014) Complete Guide for Growing Plants Hydroponically. CRC Press Taylor & Francis Group, Boca Raton, Florida, USA

Karanisa T, Amato A, Richer R, Abdul Majid S, Skelhorn C, Sayadi S (2021) Agricultural production in qatar's hot arid climate. Sustain 13:4059. https://doi.org/10.3390/su13074059

Kaur R, Gonzales WL, Llambi LD, Soriano PJ, Callaway RM et al. (2012) Community Impacts of Prosopis juliflora Invasion: Biogeographic and Congeneric Comparisons. PLoS ONE 7(9), e44966. https://doi.org/10.1371/journal.pone.0044966

Kleyn L, Mumaw L, Corney H (2020) From green spaces to vital places: connection and expression in urban greening. Aust Geogr 51(2):205–219. https://doi.org/10.1080/00049182.2019.1686195

Koberl M, Müller H, Ramadan EM, Berg G (2011) Desert farming benefits from microbial potential in arid soils and promotes diversity and plant health. PLoS ONE, 6(9), e24452. https://doi.org/ 10.1371/journal.pone.0024452.

Kong F, Yin H, Nakagoshi N, Zong H (2010) Urban green space network development for biodiversity conservation: Identification based on graph theory and gravity modeling. Landsc Urban Plan 95(1–2):16–27. https://doi.org/10.1016/j.landurbplan.2009.11.001

Kottek M, Grieser J, Beck C, Rudolf B (2006) World map of the Köppen-Geiger climate classification updated. Meterorologische Z 15(3):259–263

Kotzen B, Branquinho C, Prasse R (2020) Does the exotic equal pollution? Landscape methods for solving the dilemma of using native versus non-native plant species in drylands. Land Degrad Dev 31:2925–2935. https://doi.org/10.1002/ldr.3650

Lal R (2020) Home gardening and urban agriculture for advancing food and nutritional security in response to the COVID-19 pandemic. Food SecUrity 12:871–876. https://doi.org/10.1007/s12 571-020-01058-3

Laurance FW, Engert J (2022) Sprawling cities are rapidly encroaching on Earth's biodiversity. In: Proceedings of the national academy of sciences, 119(16). https://doi.org/10.1073/pnas.220224 4119

Lin BB, Fuller RA (2013) Sharing or sparing? How should we grow the world's cities? J Appl Ecol 50(5):1161–1168

Lin BB et al (2015) The future of urban agriculture and biodiversity-ecosystem services: Challenges and next steps. Basic Appl Ecol 16(3):189–201. https://doi.org/10.1016/j.baae.2015.01.005

Linehan J, Gross M, Finn J (1995) Greenway planning: developing a landscape ecological network approach. Land—Scape and Urban Planning 33:179–193. https://doi.org/10.1016/0169-2046(94)02017-A

Liu N, Liu Z, Wu Y (2025) Direct and indirect impacts of urbanization on biodiversity across the World's cities. Remote Sens 17(6):956. https://doi.org/10.3390/rs17060956

Loreau M et al (2001) Biodiversity and ecosystem functioning: Current knowledge and future challenges. Sci 294(5543):804–808

Losey JE, Vaughan M (2006) The economic value of ecological services provided by insects. Bioscience 56(4):311–323. https://doi.org/10.1641/0006-3568(2006)56[311:TEVOES]2.0.CO;2

Marthya KLAR (2022) Spatial logic of park access in Qatar. Qatar University Digital Hub. Available at: https://qspace.qu.edu.qa/handle/10576/26363

Martins PLSS, Furtado SG, Menini Neto L (2020) Could epiphytes be xenophobic? Evaluating the use of native versus exotic phorophytes by the vascular epiphytic community in an urban environment. Community Ecol 21:91–101. https://doi.org/10.1007/s42974-020-00001-y

Matteson K, Langellotto G (2010) Determinates of inner city butterfly and bee species richness. Urban Ecosyst 13:333–347. https://doi.org/10.1007/s11252-010-0122-y

Matthys B et al (2010) Spatial dispersion and characterisation of mosquito breeding habitats in urban vegetable-production areas of Abidjan, Côte d'Ivoire. Ann Trop Med Parasitol 104(8):649–666. https://doi.org/10.1179/136485910X12851868780108

Maxwell SL, Reside A, Trezise J, McAlpine CA, Watson JE (2019) Retention and restoration priorities for climate adaptation in a multi-use landscape. Glob Ecol Conserv, 18. https://doi.org/10.1016/j.gecco.2019.e00649

McKinney ML (2002) Urbanization, biodiversity, and conservation: the impacts of urbanization on native species are poorly studied, but educating a highly urbanized human population about these impacts can greatly improve species conservation in all ecosystems. BioScience 52(10):883–890. https://doi.org/10.1641/0006-3568(2002)052[0883:UBAC]2.0.CO;2

McKinney ML (2006) Urbanization as a major cause of biotic homogenization. Biol Cons 127:247–260. https://doi.org/10.1016/j.biocon.2005.09.005

McLennon E, Dari B, Jha D, Sihi D, Kankarla V (2021) Regenerative agriculture and integrative permaculture for sustainable and technology driven global food production and security. Agron J 113:4541–4559. https://doi.org/10.1002/agj2.20814

Miniaoui H, Irungu P, Kaitibie S (2018) Contemporary issues in qatar's food security. Middle east insights, 185. Middle East Institute: Singapore

Ministry of Municipality and Environment. Qatar National Food Security Strategy (2018–2023); Ministry of Municipality and Environment: Doha, Qatar, 2020

Ministry of Public Health in Qatar (2016) Qatar Health Report 2014–2016. Ministry of Public Health in Qatar, Doha, Qatar

Missimer TM, Maliva RG, Ghaffour N, Leiknes T, Amy GL (2014) Managed aquifer recharge (MAR) economics for wastewater reuse in low population wadi communities. Kingdom of Saudi Arabia. Water 6(8):2322–2338. https://doi.org/10.3390/w6082322

Moosavi S, Makhzoumi J, Grose M (2015) Landscape practice in the Middle East between local and global aspirations. Landsc Res 41(3):265–278. https://doi.org/10.1080/01426397.2015.1078888

Mougeot LJA (2005) Agropolis: the social, political and environmental dimensions of urban agriculture (1st ed.). Routledge. https://doi.org/10.4324/9781849775892

Nadeem H (2015) Assessing spatial equality of urban green spaces provision: a case study of Greater Doha in Qatar. Local Environ 20(3):386–399. https://doi.org/10.1080/13549839.2013.855182

Nastiti F, Giyarsih S (2019) Green open space In urban areas: a case In the government office of boyolali, Indonesia. RegNal Sci Inq, Hell Assoc RegNal Sci, 0(2), pages 19–28

Oduor AMO, Long H, Fandohan AB, Liu J, Yu X (2018) An invasive plant provides refuge to native plant species in an intensely grazed ecosystem. Biol Invasions 20:2745–2751. https://doi.org/10.1007/s10530-018-1757-5

Okemwa E (2015) Effectiveness of aquaponics and hydroponic gardening to traditional gardening. Int J Sci Res Innov Technol 2:2313–3759

Oliveira J, Biondi D, Nunho A (2022) The role of urban green areas in noise pollution attenuation. Dyna (Medellin, Colombia), 89(220), 210–215. https://doi.org/10.15446/dyna.v89n220.95822

Orsini F, Kahane R, Nono-Womdim R, Gianquinto G (2013) Urban agriculture in the developing world: a review. Agron Sustain Dev 33:695–720. https://doi.org/10.1007/s13593-013-0143-z

Ossola A, Lin BB (2021) Making nature-based solutions *climate-ready* for the 50 °C world. Environ Sci Policy 123:151–159. https://doi.org/10.1016/j.envsci.2021.05.026

Pauleit S, Duhme F (2000) Assessing the environmental performance of land cover types for urban planning. Landsc Urban Plan, 52, 1–20. https://doi.org/10.1016/S0169-2046(00)00109-2.

Peninsula, Qatar (2022) Qatar making great efforts to double green spaces. The Peninsula Qatar thepeninsulaqatar.com/article/17/06/2022/qatar-making-great-efforts-to-double-green-spaces

Pérez-Corona ME, Pérez-Hernández MdC, Medina-Villar S, Andivia E, Bermúdez de Castro F (2021) Canopy species composition drives seasonal soil characteristics in a Mediterranean riparian forest. Eur J For Res, 140, 1081–1093. https://doi.org/10.1007/s10342-021-01387-8

Planning and Statistics Authority in Qatar. Economic and Agricultural Statistics, 2020; Planning and Statistics Authority: Doha, Qatar, 2020

Planning and Statistics Authority in Qatar. Qatar Atlas, 2020; Planning and Statistics Authority: Doha, Qatar, 2020. https://gis.psa.gov.qa/QatarAtlas/StatisticalSector.aspx?Left_Menu_DIV= Landuse&Left_Menu_DIV_SubItem=MenuItem_LU_3

Planning and Statistics Authority in Qatar. Economic and Agricultural Statistics, 2021; Planning and Statistics Authority: Doha, Qatar, 2021

Planning and Statistics Authority in Qatar. Population Statistics Quarterly Bulletin, 2nd Quarter 2022; Planning and Statistics Authority in Qatar: Doha, Qatar, 2022

PPD (2017) Sustainable design practices qatar public parks. rep Doha, Qatar: Public Parks Department

Prendergast KS, Tomlinson S, Dixon KW, Bateman PW, Menz MHM (2022) Urban native vegetation remnants support more diverse native bee communities than residential gardens in Australia's southwest biodiversity hotspot. Biol Cons 265:109408. https://doi.org/10.1016/j.biocon.2021. 109408

Krishiyidam Qatar-Malayalam (2022) Facebook private group. Available at: https://www.facebook. com/groups/955834331176365/

QNA, Doha (2022) A Tale of Qatar's Healthy Cities. Gulf Times, www.gulf-times.com/story/725 600/a-tale-of-qatars-healthy-cities

QNDS (2018) Qatar second national development strategy 2018–2022. rep. Doha, Qatar: Minist Dev Plan Stat

Quirino T (2022) Qatar's newly opened umm Al seneem park receives guinness world record. www. iloveqatar.net/news/general/qatar-umm-al-seneem-park-receives-guinness-world-record.

Reeve N (2014) Rocky desert digital image. A Brief Trip to Qatar. Retrieved May 10, 2023, from https://hedgehogsite.wordpress.com/2014/04/09/a-brief-trip-to-qatar/

Rice KJ, Emery NC (2003) Managing microevolution: Restoration in the face of global change. Front Ecol Environ 1(9):469–478. https://doi.org/10.1890/1540-9295(2003)001[0469:MMRITF]2.0. CO;2

Riley CB, Herms DA, Gardiner MM (2018) Exotic trees contribute to urban forest diversity and ecosystem services in inner-city Cleveland, OH. Urban For & Urban Green, 29, 367–376. https:// doi.org/10.1016/j.ufug.2017.01.004

Robbins P, Polderman A, Birkenholtz T (2001) Lawns and toxins—An ecology of the city. Cities 18(6):369–380. https://doi.org/10.1016/S0264-2751(01)00029-4

Rudd H, Vala J, Scharfer V (2002) Importance of backyard habitat in a comprehensive biodiversity conservation strategy: a connectivity analysis of urban green spaces. Restor Ecol 10(2):368–375. https://doi.org/10.1046/j.1526-100X.2002.02041.x

Safford HD, Vallejo VR.(2019) Chapter12—Ecosystem management and ecological restoration in the Anthropocene: Integrating global change, soils, and disturbance in boreal and Mediterranean forests. In Developments in Soil Science, pp. 259–308. Busse M, Giardina CP, Morris DM, Page-Dumroese DS (Eds). Global Change and Forest Soils, Elsevier.

Heenat Salma (n.d.) Heenat Salma Farm. Retrieved March 31, 2023 from https://heenatsalma.earth/

Shiferaw H, Teketay D, Nemomissa S, Assefa F (2004) Some biological characteristics that foster the invasion of *Prosopis juliflora* (Sw.) DC. at Middle Awash Rift Valley Area, north-eastern Ethiopia. J Arid Environ, 58(2), 135–154. https://doi.org/10.1016/j.jaridenv.2003.08.011

Simonson WD, Miller E, Jones A, García-Rangel S, Thornton H, McOwen C (2021) Enhancing climate change resilience of ecological restoration—A framework for action. Perspectives in Ecology and Conservation 19(3):300–310. https://doi.org/10.1016/j.pecon.2021.05.002

Southwood TRE, Brown VK, Reader PM (1979) The relationships of plant and insect diversities in succession. Biol J Lin Soc 12(4):327–348. https://doi.org/10.1111/j.1095-8312.1979.tb00063.x

Sperling C, Lortie C (2010) The importance of urban back gardens on plant and invertebrate recruitment: A field microcosm experiment. Urban Ecosystems 13(2):223–235. https://doi.org/10.1007/s11252-009-0114-y

Spotswood EN et al (2021) The biological deserts fallacy: Cities in their landscapes contribute more than we think to regional biodiversity. Bioscience 7(2):148–160. https://doi.org/10.1093/biosci/biaa155

Sun F, Zeng L, Cai M, Chauvat M, Forey E, Tariq A, Graciano C, Zhang Z, Gu Y, Zeng F et al (2022) An invasive and native plant differ in their effects on the soil food-web and plant-soil phosphorus cycle. Geoderma 410:115672. https://doi.org/10.1016/j.geoderma.2021.115672

Tannous H, Major M, Furlan R (2021) Accessibility of green spaces in a metropolitan network using space syntax to objectively evaluate the spatial locations of parks and promenades in Doha, State of Qatar. Urban Forestry & Urban Greening 58:1618–8667. https://doi.org/10.1016/j.ufug.2020.126892

Tchakerian VP (2015) Deserts and desertification. In G.R. North, J Pyle F Zhang (Eds). Encycl Atmos Sci (2nd ed., vol. 3, pp. 185–192). John Wiley & Sons, New York

The World Bank Data. GDP per capita (current US$)—Qatar. Available at: https://data.worldbank.org/indicator/NY.GDP.PCAP.CD?locations=QA (accessed on 8 December 2022).

The World Bank Data. Urban population (% of total population)—Qatar. Available at: https://data.worldbank.org/indicator/SP.URB.TOTL.IN.ZS?locations=QA (accessed on 8 December 2022)

Trees of Qatar (2016) Doha: Public parks department

Qatar Tribune (2022) Qatar's green space increased 10 times since 2010: Minister. Qatar Tribune. Qatar Tribune. Available at: https://www.qatar-tribune.com/article/28820/latest-news/qatars-green-space-increased-10-times-since-2010-minister

Udawatta PR, Rankoth L, Jose S (2019) Agroforestry and Biodiversity. Sustain 11:2879. https://doi.org/10.3390/su11102879

United Nations. Department of economic and social affairs population dynamics. Qatar Popul Prospect. Available at: https://population.un.org/wpp/ (accessed on 28 November 2022)

Van Apeldoorn RC, Oostenbrink WT, van Winden A, van der Zee FF (1992) Effects of habitat fragmentation on the bank vole, *Clethrionomys glareolus*, in an agricultural landscape. Oikos 65:265–274. https://doi.org/10.2307/3545018

Vasquez-Valderrama M, González-M R, López-Camacho R, Baptiste MP, Salgado-Negret B (2020) Impact of invasive species on soil hydraulic properties: Importance of functional traits. Biol Invasions 22:1849–1863. https://doi.org/10.1007/s10530-020-02222-8

van Veenhuizen R, Danso G (2007) Profitability and sustainability of urban and peri-urban agriculture. Agricultural management, marketing and finance occasional paper. FAO, Rome, Italy, 19

Vicente-Vicente JL, Piorr A (2021) Can a shift to regional and organic diets reduce greenhouse gas emissions from the food system? A case study from Qatar. Carbon Balanc Manag, 16(2). https://doi.org/10.1186/s13021-020-00167-y

Vieira J, Matos P, Mexia T, Silva P, Lopes N, Freitas C, Correia O, Santos-Reis M, Branquinho C, Pinho P (2018) Green spaces are not all the same for the provision of air purification and climate regulation services: The case of urban parks. Environ Res 160:306–313. https://doi.org/10.1016/j.envres.2017.10.006

Vivian LM, Greet J, Jones CS (2020) Responses of grasses to experimental submergence in summer: Implications for the management of unseasonal flows in regulated rivers. Aquat Ecol 54:985–999. https://doi.org/10.1007/s10452-020-09788-4

WinklerPrins AGA (2002) House-lot gardens in Santarém, Pará, Brazil: Linking rural with urban. Urban Ecosystems 6(1–2):43–65. https://doi.org/10.1023/A:1025914629492

WinklerPrins A, Oliveira PS, de S (2010) Urban agriculture in santarém, Pará, Brazil: diversity and circulation of cultivated plants in urban homegardens. Bol Do Mus Para Emílio Goeldi Ciências Humas, 5(3). https://doi.org/10.1590/S1981-81222010000300002

Wolch J, Jerrett M, Reynolds K, McConnell R, Chang R, Dahmann N, Brady K, Gilliland F, Su JG, Berhane K (2011) Childhood obesity and proximity to urban parks and recreational resources: a longitudinal cohort study. Health Place, 17(1), 207–14. http://org.https://doi.org/10.1016/j.healthplace.2010.10.001

Wood, S. L. R.; & Dupras, J. (2021). Increasing functional diversity of the urban canopy for climate resilience: Potential tradeoffs with ecosystem services? *Urban Forestry & Urban Greening*, 58. https://doi.org/10.1016/j.ufug.2020.126972

World obesity. Global Obesity Observatory. Ranking (% obesity by country). Available at: https://data.worldobesity.org/rankings/?age=a&sex=f (accessed on 9 December 2021).

Yousef AH, Abdalla OA, Akbar MA (2020) Progress report for the national report of the state of qatar on the UNCCD implementation. Minist Munic Aff Agric, Dep Agric Water Res—Soil Res Sect. Qatar: State of Qatar

Zhang Y, Yu C, Xie J, Du S, Feng J, Guan D (2021) Comparison of fine root biomass and soil organic carbon stock between exotic and native mangrove. Catena, 204. https://doi.org/10.1016/j.catena.2021.105423

Zunzunegui M, Ruiz-Valdepeñas E, Sert MA, Díaz-Barradas MC, Gallego-Fernández JB (2020) Field comparison of ecophysiological traits between an invader and a native species in a Mediterranean coastal dune. Plant Physiol Biochem, 146, 278–286. https://doi.org/10.1016/j.plaphy.2019.11.032

Open Access This chapter is licensed under the terms of the Creative Commons Attribution 4.0 International License (http://creativecommons.org/licenses/by/4.0/), which permits use, sharing, adaptation, distribution and reproduction in any medium or format, as long as you give appropriate credit to the original author(s) and the source, provide a link to the Creative Commons license and indicate if changes were made.

The images or other third party material in this chapter are included in the chapter's Creative Commons license, unless indicated otherwise in a credit line to the material. If material is not included in the chapter's Creative Commons license and your intended use is not permitted by statutory regulation or exceeds the permitted use, you will need to obtain permission directly from the copyright holder.

Chapter 6
Coastal Marine Ecology

C. Reeves, R. Riera⬡, P. Range, R. Ben-Hamadou, A. Leitão, J. MK. Wong, B. W. Giraldes, B. Böer, D. Abdelwahed, T. R. R. Bontognali, K. T. Bron, A. Amato, and A. D. Chatziefthimiou

Abstract Qatar's coastal and marine ecosystems comprise a mosaic of coral reefs, seagrass meadows, mangroves, oyster beds, and sabkhas supporting high biodiversity and providing critical ecosystem services. These include fisheries production, carbon sequestration, shoreline stabilization, and habitat provision for endangered species such as dugongs, whale sharks, and hawksbill turtles. However, these ecosystems are increasingly threatened by climate change, habitat degradation, coastal development, and pollution. Coral reefs have experienced substantial declines due to marine heatwaves and anthropogenic stressors. Seagrass meadows and mangroves, which underpin carbon storage and nursery functions, are under pressure from fragmentation and disturbance. Oyster beds, once central to regional economies and ecological function, have largely collapsed. Sabkhat represent unique hyper-saline ecosystems of geobiological, scientific, educational, and global heritage importance.

C. Reeves (✉)
Sustainable Oceans Management Limited, Ripon, North Yorkshire, UK
e-mail: katiereeves@someconsulting.co.uk

R. Riera (✉)
BIOCON (Biodiversity and Conservation), ECOAQUA, University of Las Palmas de Gran Canaria, Canary Islands, Las Palmas de Gran Canaria, Spain
e-mail: rodrigo.riera@ulpgc.es

P. Range · A. Leitão · B. W. Giraldes
Environmental Science Center, Qatar University, Doha, Qatar
e-mail: prange@qu.edu.qa

A. Leitão
e-mail: Alexandra.Hamadou@qu.edu.qa

B. W. Giraldes
e-mail: bweltergiraldes@qu.edu.qa

R. Ben-Hamadou
Earthna, Center for a Sustainable Future, Qatar Foundation, Doha, Qatar
e-mail: rbenhamadou@qf.org.qa

J. MK. Wong
Ministry of Environment and Climate Change, Doha, Qatar
e-mail: jmwong@mecc.gov.qa

© The Author(s) 2026
A. D. Chatziefthimiou et al. (eds.), *Pathways to Nature Conservation and Resilience in Hot and Arid Lands*, Advances in Geographical and Environmental Sciences,
https://doi.org/10.1007/978-3-032-11046-6_6

Recent mapping and biodiversity assessments highlight the urgency of conservation measures. To preserve the ecological integrity and services of these ecosystems, immediate and coordinated action is required. Action includes evidence-based habitat restoration, strengthened marine governance including a dedicated Marine Spatial Plan, and the expansion and enforcement of Marine Protected Areas (MPAs) to meet the ambitious global target of the CBD's Post-2020 Framework UN of at least 30% MPA's by 2030 in Qatar's Exclusive Economic Zone. In addition to MPAs other conservation management tools will be required such as Other area based Effective Conservation Means (OECM) to contribute to the health of Qatar's marine ecosystems.

Keywords Marine ecosystems · Coral reefs · Mangroves · Seagrass · Biodiversity · Coastal conservation · Marine protected areas · Blue economy

6.1 Obligations for Conservation of Qatar's Biodiversity

Qatar has a range of legal and ethical obligations to conserve its marine biodiversity, stemming from international agreements like the CBD (Convention on Biological Diversity) and CITES (Convention on International Trade in Endangered Species of Wild Fauna and Flora), as well as its commitment to sustainable development under the UN's 2030 Agenda. These obligations are further reinforced by Qatar's own national initiatives, including the Qatar National Vision 2030 and the National

B. Böer
UNESCO South-Asia Office in New Delhi, New Delhi, India
e-mail: b.boer@unesco.org

D. Abdelwahed
UNESCO Office for Gulf States & Yemen, UNESCO, Doha, Qatar
e-mail: d.abdelwahed@unesco.org

T. R. R. Bontognali
Space Exploration Institute, Neuchâtel, Switzerland

Department of Environmental Sciences, University of Basel, Basel, Switzerland

T. R. R. Bontognali
e-mail: tomaso.bontognali@space-x.ch

K. T. Bron
GeoTree Consulting, Melbourne, Australia
e-mail: treenabron@gmail.com

A. Amato · A. D. Chatziefthimiou
Earthna, Center for a Sustainable Future, Qatar Foundation, Doha, Qatar
e-mail: aamato1uk@mac.com

A. D. Chatziefthimiou
e-mail: a.d.chatziefthimiou@gmail.com

Biodiversity Strategy and Action Plan 2015 to 2025, which is currently being updated with the aim to align the new NBSAP to the 2050 goals and 2030 targets of the Global Biodiversity Framework (GBF), as well as to the goals and targets of the Ministry of Environment and Climate Change's strategic goals and those of the Qatar National Development Strategy version 3. Qatar's efforts in this regard are vital for the protection of its marine ecosystems and the long-term well-being of its environment and people.

Qatar's journey towards biodiversity conservation and the establishment of Marine Protected Areas is a testament to the nation's commitment to preserving its natural heritage. The legal framework and guidelines provide a solid foundation for future conservation efforts, highlighting the nation's commitment to the United Nations Convention on Biological Diversity (UNCBD) and its objectives: the conservation of biological diversity, sustainable use of biological components, and the fair sharing of benefits from genetic resources. In the realm of biodiversity conservation and the establishment of Marine Protected Areas (MPAs), Qatar has embarked on a journey marked by significant milestones and a strong commitment to environmental preservation, and its ability to adapt to the evolving environmental challenges of the modern world.

In 1978, Qatar took its initial stride towards environmental protection by issuing Decree 55. This decree was aimed at safeguarding the marine environment from pollution, underscoring the nation's early recognition of the need for conservation efforts. The year 2002 marked a turning point in Qatar's conservation initiatives. Law 4 was introduced, ushering in regulations that effectively banned hunting activities in protected areas. This legal measure was a clear demonstration of the government's commitment to environmental conservation. Simultaneously, Law 30 for Environmental Protection was enacted in 2002, further solidifying the nation's dedication to environmental safeguards. In 2005, Qatar unveiled its first National Biodiversity Strategy & Action Plan, a comprehensive document outlining the country's strategy for biodiversity conservation this was renewed in 2015 for another decade of action. This pivotal year also witnessed the enactment of legal instruments, Law 7 and Law 8, designating Al Reem and Al Wusail as protected areas for terrestrial and marine animals, respectively. Another critical milestone in 2006 was the passage of Law 6, which was designed to protect the mangroves and the habitat of 134 bird species within the Al Thakira Reserve. Qatar's dedication to conservation continued in 2007 with the establishment of Khor Al Adaid as a protected area under Law 1. This area encompassed lagoons, sand dunes, or wadis, among others, and was home to various wildlife, including salkha, gazelle, hare, hedgehog, and fox. The year 2008 marked a visionary leap for Qatar with the publication of the Qatar National Vision for 2030. This vision set the stage for a Protection Area Action Plan spanning from 2008 to 2013. A central aspect of this plan was the emphasis on the designation of protected areas. The Ministry of Municipality and Environment (MME) was empowered as a pivotal agency to prohibit and control activities that posed threats to endangered species in their natural habitats. In 2011, Qatar's commitment to biodiversity conservation was further cemented when biodiversity protection became an integral part of the National Development Strategy for 2011–2016. By 2013, MME had declared

12 Protected Areas, covering 24% of the land area and 1.7% of the marine area. This step underscored Qatar's dedication to conserving its natural heritage. Looking towards the future, in 2021, an Integrated Coastal Zone Management Plan (ICZMP) was drafted, with a focus on environmental planning and sustainable development. In 2022, the Ministry of Environment and Climate Change (MoECC) adopted the Qatar National Marine Resource Conservation & Management Action Plan, signifying a renewed commitment to the protection of marine resources. In the same year, MoECC accepted a proposal for the designation of a Marine Protected Area, marking a significant milestone in the protection of Qatar's marine ecosystems.

Marine Protected Areas (MPAs) are widely recognized tools for conserving marine biodiversity in Qatar (Hilborn 2018; Pendleton et al. 2018). Their effective management is crucial for preserving coastal and marine ecosystems (Burt et al. 2017). According to the International Union for Conservation of Nature (IUCN), MPAs aim to protect biological diversity, natural resources, and cultural values through legal or effective means (Phillips 2002).

In Qatar, specific criteria guide MPA identification, including the diversity of natural themes, uniqueness of marine resources, degree of natural state, and educational value. The IUCN provides guidelines categorizing MPAs based on objectives, such as economic resource use, biodiversity conservation, and species protection (Wells et al. 2016). These categories range from strict protection to sustainable use of natural ecosystems. Nested MPAs with different categories allow distinct objectives within each area. Coastal conservation on land and marine protection at sea are vital for safeguarding marine wildlife, promoting fish stock recovery, and ensuring biodiversity.

Qatar's MPA network serves multiple purposes, such as protecting sensitive habitats, restoring damaged areas, rebuilding populations, and sustaining fishery production. The network also contributes to ecosystem services, research, education, and tourism. However, many MPAs in the Gulf Cooperation Council (GCC) region face challenges due to limited technical capacity and management plans (Van Lavieren and Klaus 2013). To effectively achieve MPA objectives, significant protection from extractive activities, especially fishing, is essential (Groud-Colvert et al. 2021). Strict protection, as per IUCN Category 1, is crucial for success, preventing MPAs from being symbolic without meaningful conservation impact (Rice et al. 2012; Belharet et al. 2020. The protected areas established in Qatar should have marine activities evaluated against IUCN Protected Area Management Categories (Ia to VI). The stricter categories indicating limited human use, while more flexible categories allow sustainable use or conditional access. Activities assessed include fishing, aquaculture, shipping, tourism, research, military use, and more. These are described more fully in Day et al. (2012).

Qatar's MPA network should prioritize high protection areas (IUCN Category 1a and 1b) to minimize human disturbance and allow natural recovery. Adequate funding, management, and permanence are vital for their success. Globally recognized best practices outlined in the IUCN's MPA Guidelines should inform permitted activities within MPAs. The establishment and effective management of MPAs are vital steps for preserving Qatar's marine biodiversity and ensuring the health of coastal and marine ecosystems.

6.2 Coastal Marine Habitats

Qatar, nestled on the northeastern coast of the Arabian Peninsula, is a nation intricately intertwined with the sea. With a coastline stretching for over 500 km, Qatar boasts a vast marine realm encompassing approximately 35,000 square km–a marine environment three times the expanse of its land. Along its western coastline, the coastal waters reveal a striking characteristic: extreme shallowness. Beneath these waters lie a dynamic tapestry of seabed sediments, forming a rich mosaic of habitats. These range from the pristine sands to mixtures of sand and mud that nurture seagrass meadows, to the rugged rock outcrops and hard substrates that provide a foundation for soft and hard corals (Butler et al. 2021). Yet, the marine life dwelling within these waters has to adapt to the distinctive and, at times, unforgiving conditions of the Arabian Gulf. With consistently high average temperatures and elevated salinity levels, this aquatic realm has become a unique crucible, shaping life in extraordinary ways.

The coastal marine environment of Qatar is more than just a scenic backdrop; it is the lifeblood of the nation's economy. These waters teem with a wealth of ecosystem services that underpin Qatar's prosperity. Yet, the very beauty and vitality of the marine life here face mounting threats from human activities and the spectre of climatic changes. Intriguingly, while approximately a quarter of Qatar's terrestrial expanse benefits from protection, a stark contrast emerges in the realm of the sea. Here, only about 2% of the marine environment enjoys the safeguard of protected status. This incongruity highlights the pressing need for concerted efforts to ensure the preservation of Qatar's unique coastal marine habitats, which play a pivotal role in shaping both the nation's natural heritage and its future.

6.2.1 Coral Reefs

The Gulf, a semi-enclosed marginal sea, stands separated from the Indian Ocean by the formidable Strait of Hormuz. Its unique characteristics, including shallow depth and limited water exchange, create extreme conditions, with hypersaline waters surpassing 40 PSU and seawater temperature fluctuations ranging from a scorching 36 °C in summer to a comparatively chilly 16 °C in winter (John et al. 1990; Sheppard et al. 2010). Remarkably, these conditions, which far exceed the physiological tolerance limits for corals in other regions, were historically conducive to extensive, thriving, and lush coral ecosystems along Qatar's coastline (Shinn 1976; Burt et al. 2021).

Despite the resilience of Gulf corals, they are not immune to a barrage of threats and disturbances. The primary culprits include climate change (Wabnitz et al. 2018), driven by the increasing frequency and duration of marine heat waves, attributed to rising global greenhouse gas emissions (Burt et al. 2019; Paparella et al. 2019). Localized degradation and loss have also occurred due to coastal development (Burt and

Bartholomew 2019), overfishing (El Sayed 1996), and marine pollution (Naser 2013), leading to algal overgrowth on dead coral skeletons (Bouwmeester et al. 2022). This algal encroachment further intensifies following the reduction of live coral cover, as algae compete with corals for light and space (McCook et al. 2001). The situation is exacerbated by the decline of herbivorous fish and invertebrates, crucial for facilitating coral recovery in degraded reefs (Bouwmeester et al. 2022). Recent surveys conducted in Qatar reveal a disheartening reality–most of the coastal reefs have succumbed to death or severe degradation, sustaining only a remnant community of stress-tolerant coral species amid escalating algal cover (Bouwmeester et al. 2022; Fanning et al. 2021). A comprehensive benthic habitat map, spanning 4,500 km^2 of Qatar's coastline from the intertidal zone to approximately 10 km offshore, unveiled the fact that only 7 km^2 of live coral habitat, a mere 0.21% of the total area, remain (Butler et al. 2020). Most of these surviving reefs are now concentrated in the north-eastern tip of the Qatari peninsula and around offshore seamounts and islands (Burt et al. 2017). While many of these reefs remain understudied, historical records report 55 scleractinian coral taxa in Qatar (Coles 2003). Recent surveys, however, confirm the presence of only 38 coral species, representing 58% of the known coral diversity in the Gulf (66 species; DiBattista et al. 2016). Remarkably, Qatar stands as the second most diverse nation in terms of coral species richness among the eight countries bordering the Arabian Gulf, surpassed only by Iran (Bouwmeester et al. 2020). Dominant families in Qatari reefs include Merulinidae, representing 64% of all scleractinian coral cover, and Poritidae, comprising 18% of total coral cover. Species such as *Platygyra daedalea*, *Dipsastraea pallida* (Merulinidae), and *Porites Lutea* (Poritidae) are the most widely distributed (Bouwmeester et al. 2022; Fanning et al. 2021).

In the face of this declining trend, it remains undeniable that coral reefs continue to be the most biodiverse, productive, and economically significant marine ecosystems in the Arabian Gulf (Feary et al. 2013; Burt et al. 2016). While coral diversity is naturally limited due to the extreme environmental conditions, these reefs support the most diverse communities of associated fish and invertebrate species. The significance of coral conservation is underscored by the impact of coral loss on reef-associated fish, which account for a substantial 95% of total fisheries landings in Qatar. Recent national fisheries statistics indicate a decline in the percentage of fish stocks within safe biological limits, from 72% in 2010 to 68% in 2015 (Ministry of Development Planning and Statistics 2017). The extensive overlap between coral reef habitats and fishing activities within Qatar's Exclusive Economic Zone, as well as the use of inadequate fishing gear (Fanning et al. 2021), are likely linked to the observed increase in algal cover within reef habitats during the same period. The ongoing degradation and loss of coral ecosystems pose a significant threat not only to marine biodiversity but also to fisheries stocks, the livelihoods of fishermen, and food security in Qatar (Riegl 2002; Buchanan et al. 2016).

While addressing the global challenge of climate change requires coordinated international efforts to reduce greenhouse gas emissions, there are local actions that can contribute to improving coral reef conditions and enhancing their resilience

(Gil et al. 2019). Qatar's National Biodiversity Strategy and Action Plan 2015–2025 has identified the preservation of coastal ecosystems and the sustainability of marine resources as a key goal, aligning with the Aichi targets of the Convention on Biological Diversity (Carr et al. 2020). Priority actions within this framework are critical for the preservation and restoration of coral reefs in Qatar. These actions include identifying key areas for coral conservation and designating them as marine protected areas, including "no take" reserves for areas with *Acropora* table corals. Comprehensive benthic habitat mapping efforts, especially in offshore areas, are essential, given that these reefs house most of the coral cover and diversity, yet their spatial extent remains largely unknown (Butler et al. 2020).

6.2.2 Seagrass Meadows

Seagrass habitats along the coast of Qatar represent a vital marine resource in the region (Abdelbary and Al Ashwal 2021). These habitats play a multifaceted role, supporting high primary production (Pergent et al. 1994), fostering a rich biodiversity of associated plant and animal species (Duffy 2006), and serving as crucial nursery grounds for various commercially valuable fish, penaeid shrimps, pearl oysters, and a plethora of other marine organisms, including sharks, rays, marine turtles, seahorses, and pipefish (Heck et al. 1997; Jackson et al. 2001; Unsworth et al. 2019). Notably, the extensive seagrass meadows surrounding Qatar, along with its neighboring UAE and Bahrain, house the world's second-largest population of approximately 5,800 dugongs (Preen 2004). These remarkable marine mammals rely almost exclusively on seagrasses as their primary food source (Erftemeijer and Shuail 2012). The seagrass meadows are exceptionally adapted to withstand the harsh environmental conditions of the Arabian Gulf, characterized by significant seasonal fluctuations in water temperature and salinity (Price and Coles 1992; Phillips 2003). These conditions are suitable for only four opportunistic seagrass species: *Halodule uninervis, Halophila stipulacea, H. ovalis* and Thalassia testudinum..

Recent high-resolution benthic habitat maps for Qatar's entire coastal zone unveiled a complex interplay of seagrass, macroalgal, and reef habitats (Butler et al. 2020). This comprehensive mapping is instrumental in recognizing sensitive areas that warrant consideration for marine protection. The mapping revealed significant seagrass meadows located off the northwest and central-east coasts of Qatar, with sparse seagrass accounting for 14% of the Qatari coastal shelf habitats and dense seagrass contributing 13%.

Notably, the seagrass meadows off the northwest coast, stretching from Bahrain to Qatar and extending south into the Gulf of Salwa, are among the densest and most expansive in the region. The embayments of Al Khor and Al Dhakira host dense seagrass meadows that extend over 15 km offshore, linking to reef complexes in Fasht al Hurabi. Similarly, a submarine bank southeast of Masaieed, spanning 10 km, showcases mosaic habitats of sparse and dense seagrass, macroalgae, and further dense seagrass meadows located at the entrance to the narrow channel of Khor al Adaid (Abdelbary and Al Ashwal 2021).

Despite their ecological importance, seagrass habitats are among the least protected coastal ecosystems (Henderson et al. 2019). They often face cumulative pressures from coastal development, nutrient run-off, and the effects of climate change (Adams et al. 2020). In Qatar, rapid industrial and coastal development for power generation, port facilities, desalination plants, and land-reclamation projects pose an unprecedented threat to seagrass habitats (Burt et al. 2017). These areas are further imperilled by unregulated fishing activities, the anchoring of commercial and recreational vessels, trampling, and marine litter deposition, leading to small areas of erosion that fragment the seagrass habitat. Seagrasses also serve as carbon sinks, effectively removing substantial quantities of carbon from the atmosphere and storing it in their biomass and associated sediments (Duarte et al. 2010; Fourqurean et al. 2012). In healthy habitats, the carbon sequestered in the sediment can be preserved for hundreds to thousands of years, contributing significantly to climate change mitigation (Duarte and Krause-Jensen 2017). The Qatari coastal zone holds substantial blue carbon potential. The mapping efforts not only offer insights into the spatial extent of blue carbon sinks but also identify areas with high blue carbon potential for field verification trials (Butler et al. 2020). Given the considerable carbon storage and sequestration capacity of seagrass ecosystems, they can be integrated into a nation's nationally determined contributions (NDCs), assisting in meeting targets under the Paris Agreement and the United Nations Framework Convention on Climate Change (UNFCCC). Restoring seagrass habitats presents Qatar with an opportunity to fulfil commitments made during the UN Decade on Ecosystem Restoration (UNEP 2020). The management, restoration, and protection of seagrass align with the new UN CBD targets (Kunming-Montreal Post-2020 GBF).

It is imperative that all dense seagrass areas and potentially restorable sparse areas be included in a suite of coastal marine protected areas (MPAs) (Fanning et al. 2021). These areas provide a wide range of ecosystem services, including their blue carbon potential (Nordlund et al. 2018). A national biodiversity action plan focused on seagrass can be developed, with targets for protecting all seagrass areas and restoring sparse areas, coupled with regular monitoring efforts. Monitoring can inform potential loss for habitat areas where seagrass migration up shorelines cannot occur due to coastal development. Management measures should encompass policies like no-anchoring in seagrass areas and the availability of seagrass-friendly mooring buoys without chains and raising awareness of the habitat with marine recreational and tourism sectors.

6.2.3 Oyster Beds

The pearl oyster holds a distinguished position in the Arabian Gulf, where it serves as an iconic ecosystem builder, forming extensive oyster beds that define one of the prominent seascapes in this semi-enclosed sea (Hightower 2013; Bento et al.

2022). Historically, the pearl oyster bed extended from Kuwait to the UAE on the western side of the Arabian Gulf, contributing significantly to the economies of the Gulf countries for centuries (Bowen 1951; Carter 2005). These oyster beds were a cornerstone of food security and economic income in the region (Carter 2005; Maslamani et al. 2018).

Despite its economic and social importance, the pearl oyster has not been immune to the environmental pressures resulting from the rapid industrial and urban growth of the Gulf countries in recent decades (Mateos-Molina et al. 2021). The entire marine biodiversity and ecosystems in the Gulf face multiple anthropogenic stressors, including domestic sewage inputs, eutrophication, overfishing, and coastal construction (Naser 2014), which are likely to be exacerbated by the effects of climate change. Balancing the demands of expanding economic growth with the need to maintain, restore, and protect marine biodiversity and ecosystem health is a complex challenge for the Gulf countries (Sheppard et al. 2010), and the decline of the iconic pearl oyster beds is one of the major consequences. A survey conducted on five historically renowned pearl oyster beds in Qatar by Smyth et al. (2016) revealed that only one out of the five sites assessed could still be characterized as oyster dominant, even though all were once highly productive oyster fisheries. The results underscore the profound impact of a combination of natural and anthropogenic hazards on these traditional pearl oyster beds, which were once an invaluable economic resource in the Arabian Gulf (Bishara et al. 2016). The pearl oyster beds continue to play a crucial role in maintaining the integrity of the marine environment in the region (Giraldes et al. 2023). These beds are not isolated but are part of the broader marine ecosystem (Grabowski et al. 2012). The decline of the pearl oyster ecosystem will contribute to the long-term loss of biological and economic functionality within key shallow-water marine environments. Oysters possess significant filtration potential (Jahromi et al. 2021), and oyster beds are recognized as natural filtering systems in marine environments (Grabowski and Peterson 2007). The substantial reduction in filtration services is a significant environmental concern in this semi-enclosed marine system, already facing extreme hydroclimatic conditions. The southwestern ecoregion of the Gulf features extensive and ideal marine sites for oyster beds within its shallow bathymetric zones. Recent studies with predictive modelling suggest that if these shallow areas were restored to their maximum densities, the entire volume of the Gulf could be filtered on a trimestral basis. Restoring the filtration functionality of these oyster seascapes and the associated recovery of the trophic chain could also support the revival of accompanying seascapes within the Gulf, such as coral reefs, seagrasses, and their associated fish stocks (Giraldes et al. 2023).

The continued loss of the ecosystem services provided by the oyster beds' habitat will result in the ongoing deterioration of the "health" of the Arabian Gulf's marine ecosystems (Samara et al. 2023). Considering the top-down and bottom-up trophic effects of oyster beds, the restoration of this ecosystem would have a positive impact on the recovery of associated seascapes, including coral reefs, seagrasses, and their associated fish stocks (Giraldes et al. 2023). Restoring this iconic natural resource in southwestern Gulf countries, such as Qatar, the UAE, and Bahrain, would also support the natural recovery of the biological and economic functionality of this marine resource. There is growing interest globally in oyster reef restoration, with good examples of oyster restoration projects across the globe with research and findings being published in site selection, substrate, structure and habitat value (Hughes et al. 2023. McLeod et al. 2019. Cole et al. 2024). Additionally, there is a need to place a particular emphasis on enhancing aquaculture knowledge and technologies for oyster production and reintroduction in strategically chosen sites.

6.2.4 Mangroves and Coastal Wetlands

The global remaining mangrove coverage is estimated at 152,000 km^2, with several countries, including Qatar, experiencing significant losses of more than 40% of their mangrove areas over a 25-year period (Romañach et al. 2018). In the Red Sea and the Arabian Gulf, only two mangrove species, *Avicennia marina* and *Rhizophora mucronata*, are found (Saifullah 1996; Khalil 2015). The region's harsh environmental conditions, characterized by high salinity levels, very low precipitation rates, low winter temperatures, and extremely high summer temperatures, make it a biogeographically interesting area (Vaughan et al. 2019). On land, vegetation cover is typically less than 1%, occasionally reaching up to 5%, and exceptionally up to 10% (Almahasheer 2018). However, in the intertidal zone, the vegetation cover of mangroves and salt marshes can reach up to 100%. This intertidal vegetation is dominated by salt tolerant *Avicennia marina*, often coexisting with salt marsh plants, e.g., *Arthrocnemum macrostachyum, Halcnemum strobilaceum, Halopeplis perfoliata, Salicornia* species, and *Limonium axillare*, among others (Al-Khayat and Jones 1999).

Qatar's mangroves, covering approximately 12.3 km^2 in the past, are primarily located in sheltered lagoons along the eastern shores, notably in Al Khor and Al Dhakeera (Fig. 6.1), with fewer stands to the south and north (Al-Khayat and Balakrishnan 2014). Hypersaline conditions along the western shore of Qatar, in the Gulf of Salwa, limit the growth of the seawater-tolerant mangrove species *Avicennia marina* in that area. The total mangrove coverage has significantly decreased over the years (Pernot et al. 2017) (Fig. 6.2).

Photograph showing the coastal landscape of Al Dhakeera, featuring dense green mangroves, adjacent salt marshes, and a winding tidal creek. The vegetation contrasts with the surrounding arid environment, highlighting the ecological richness of the area. Tidal water channels are visible flowing through the mangroves.

Qatar holds a significant position in the Arab Region in terms of mangroves (Almahasheer 2018). These ecosystems provide not only essential ecosystem services like food and support for tourism (Mitra and Mitra 2020) but are also recognized as a component of Blue Carbon (BC) (Alongi 2020). They serve as a nature-based solution for reducing greenhouse gas emissions by sequestering and storing substantial amounts of carbon. Given these ecological benefits, it is crucial to preserve Qatar's remaining Blue Carbon ecosystems and explore strategies for restoring or offsetting the lost mangroves, salt marshes, seagrass beds, and other wetlands. Earthna

Fig. 6.1 Al Dhakeera mangroves, salt marshes and creek. The image illustrates the interconnected coastal habitats that support biodiversity and provide ecosystem services such as shoreline protection and nursery grounds for marine species (*Photo* Entalek 2013)

begun in 2025 a Mangrove Restoration Project in the north of Qatar planting over 90,000 mangroves in a 50 hectares area, this is being funded by the MSC Foundation (Ben-Hamadou pers. comm.3).

Qatar's mangroves also offer numerous ecosystem services to both people and wildlife. They are primary producers at the beginning of the marine food chain, providing biomass for various marine species. These mangroves serve as habitats for a wide range of fauna, including birds, crustaceans, fish, and even camels, which browse on *Avicennia marina*, known locally as 'Qurm.' Moreover, they play a crucial role in marine productivity and carbon sequestration, making them ecologically and economically significant. People also use mangrove forests for recreational purposes and ecotourism, such as picnics and canoe trips.

Figure 6.2 contains a series of six photographs showing the transformation of the Al Maroona mangrove inlet due to coastal development. Images a, c, and e (© M. Aigner, 2013) depict the site before infill, featuring natural mangrove-lined shorelines, open water channels, and a healthy intertidal ecosystem. Images b, d, and f (© V. Johnson, 2023) show the same locations after infill and development, with visible land reclamation, loss of mangrove cover, altered coastline, and disrupted aquatic habitats—highlighting the degradation of what was once an important fish nursery area.

Fig. 6.2 Photographs showing the ecological loss due to coastal development at the Al Maroona mangrove inlet, previously a vital nursery habitat for fish. Images **a**, **c**, and **e** © M. Aigner (2013) illustrate the site before infill; images **b**, **d**, and **f** © V. Johnson (2023) show the same areas after infill and development

Efforts have been made in Qatar to conserve, study, and restore mangrove ecosystems (Al-Khayat and Balakrishnan 2014; Burt et al. 2017). In the 1980s, the government collaborated with experts on an afforestation campaign, planting *Avicenia marina* to rejuvenate existing mangrove forests and establish new ones (Al-Khayat and Balakrishnan 2014). Unfortunately, some areas in Qatar have lost their mangroves due to coastal development and urban encroachment, such as in Al Wakra, Al Maroona, Al Khor, and Al Dhakeera (Fig. 6.2). These losses are attributed to changes in coastal hydrology due to development impacts that can lead to oxygen deficiency or excessive salinity in the mangrove habitats. Superimposing satellite images on historical aerial photographs revealed a 46% reduction in mangrove cover

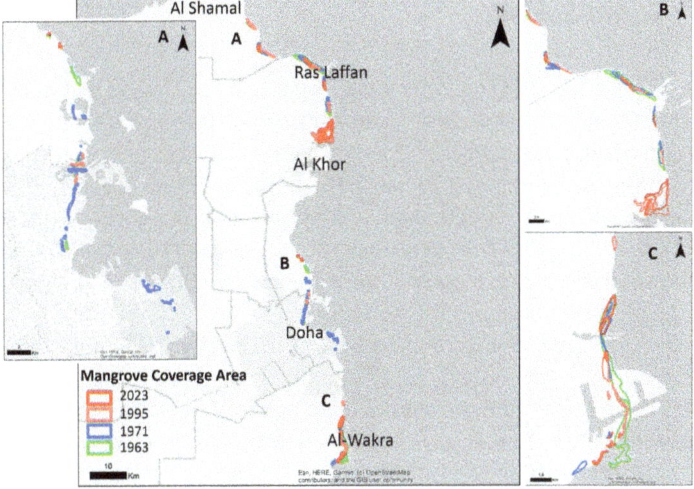

Fig. 6.3 Mangrove habitat loss along the East coast of Qatar, illustrating the extent of environmental change due to urban expansion, land reclamation, and coastal development pressures prepared by Dr Ruqaiya Yousif

on the east coast of Qatar from 1963 to 1995, excluding the recent loss in Al Maroona (Fig. 6.3). Therefore, the actual reduction in mangrove strands may be higher.

Qatar is home to a UNESCO Biosphere Reserve, 'Al Reem,' established in 2007, which includes coastal ecosystems. Another area, Khor Al Udayd, has received significant attention and support from UNESCO, possibly aiming for UNESCO designation. A recommendation is to designate the mangrove areas of Al Khor and Al Dhakeera as UNESCO Biosphere Reserves that would enable authorities to limit access to a core conservation area while allowing access to less vulnerable buffer zones (dedicated to research and monitoring) and transition areas (for awareness raising and ecotourism). Globally and within the MENA region, community-based initiatives have proven to be effective in restoring and protecting mangroves. Numerous advanced restoration methods and project examples (Lovelock et al. 2022), including those supported by UNESCO, can provide valuable insights and assistance to Qatar. Mangrove forests have a remarkable capacity to regenerate under the right conditions, and one approach could be non-intervention to allow natural restoration to occur (Sherman et al. 2000; Otero et al. 2020).

Figure 6.3 shows visual evidence of mangrove habitat loss along the East coast of Qatar. The image highlights areas where mangrove cover has been reduced or removed, often replaced by coastal infrastructure, dredged areas, or reclaimed lands. The contrast between remaining natural vegetation and developed or barren zones illustrates the scale and impact of habitat degradation.

6.2.5   Coastal Sabkhas

The sabkhas (salt flats) of Qatar offer a rare opportunity to observe and study geological and microbiological processes that hold relevance for various fields of scientific research (Elisabeth Krumbein et al. 2004). These arid sabkha environments, characterized by strong evaporation, give rise to the formation of carbonates, sulfates, and other evaporite minerals commonly found in ancient sedimentary sequences (Al-Amoudi et al. 1995). These minerals have significance as economically important gas and oil reservoir rocks, records of past climatic conditions, and repositories of some of the oldest evidence for life on Earth.

Since the 1960s, the sabkhas of Qatar have been investigated as modern analogues for ancient sedimentary sequences (Wells 1962; Alsharhan and Kendall 2003) (Fig. 6.4). The findings of these studies have provided valuable insights for developing sedimentological models of hydrocarbon reservoirs (Warren 2006). These reservoirs are primarily located deep below the Earth's surface, making it challenging to assess and predict their mineralogy, chemistry, and physical properties, which are crucial for optimizing the exploitation of oil and gas reservoirs, thereby holding importance for the energy industry. Recent research has unveiled the crucial role of microbes in the sabkhas in the formation of various carbonate and evaporite minerals, which were traditionally attributed solely to physicochemical processes

Fig. 6.4 A photograph of Khor Al Adaid sabkha (salt flat) in Qatar, showing an expansive, flat, and arid landscape covered by microbial mats. These dark, patchy microbial communities dominate the surface and play a key role in shaping the sedimentary environment by promoting the formation of evaporitic minerals and unique sedimentary structures such as polygonal cracking and layering characteristic of this extreme environment

(Bontognali et al. 2010; Brauchli et al. 2016; Perri et al. 2018). Despite the seemingly sterile and inhospitable appearance of Qatari sabkhas, they host an astonishing diversity of microorganisms. In the absence of competition from more advanced eukaryotic life forms, which cannot endure the extreme salinity and harsh temperatures of the sabkhas, these primitive microbial communities thrive and produce thick microbial mats. The study of these microbial mats has led to the discovery of unique microbial species capable of mediating mineral formation through processes relevant both in the field of material science and carbon sequestration, which can help mitigate climate change (DiLoreto et al. 2019, 2021) and be considered as blue carbon reserves (Hameed et al. 2024). Microbes isolated from Qatari sabkhas have been successfully cultured in the laboratory and are currently employed for mineral precipitation experiments (Al Disi et al. 2017, 2019).

For all former reasons, the sabkhas of Qatar represent unique ecosystems where multidisciplinary research holds great potential (Brauchli et al. 2016). These ecosystems, particularly the microbial mats with their rare and distinctive community composition, are highly sensitive to changes in water chemistry and physical disturbances. Therefore, they warrant preservation and active protection. Although Qatar's sabkhas may appear desolate at first glance, they should be regarded as a precious and natural heritage of the country. Sabkhat, when exposed to saline moisture, or freshwater moisture, do actually gradually convert into ecosystems with a vegetation cover, depending on salinity levels (Böer and Al Hajiri 2002. Norton et al. 2009).

6.3 Marine Species

Qatar's coastal marine habitats are not only breathtaking seascapes but also vital ecosystems that support a remarkable array of marine species (Carpenter et al. 1997). The marine waters of Qatar are teeming with life, offering refuge to a diverse community of creatures. These marine habitats, with their unique blend of environmental conditions, provide a home to both common and extraordinary species, many of which are found in few other places on our planet. The significance of Qatar's coastal marine environments extends far beyond their intrinsic beauty. They are sanctuaries for some of the most vulnerable and endangered species on Earth. Among these remarkable inhabitants are the annual aggregations of majestic whale shark, a species referred to as the gentle giants of the ocean. Qatar's waters also harbour populations of the scalloped hammerhead (*Sphyrna lewini*), the dugong (*Dugong dugon*), as well as sea turtles, dolphins, and the Socotra cormorants (Preen 2004; Pilcher et al. 2015).

In the early assessment of Qatar's marine biodiversity, the original Qatar National Biodiversity Action Plan (NBSAP) identified 955 known species in its waters, reflecting the rich tapestry of life within the coastal marine environments (SCENR 2004). These include over 500 fish species, 67 types of macroalgae, 39 species of coral, 12 commercially valuable crustaceans, and more than 70 mollusks. Recent studies have increased the number of species of several taxonomic groups such as fish

(Torquato et al. 2017). However, as we delve deeper into understanding the intricacies of these ecosystems, large gaps in our knowledge persist, particularly concerning the taxonomy, biology, and ecological needs of these species within Qatar's waters. With the passage of time, some of these marine species have faced growing threats, challenges and even potential extirpation. The 2004 NBSAP highlighted that out of the 22 species at risk within the region, a substantial portion of them were marine. Unfortunately, the current understanding of species at risk has grown limited, and a comprehensive register to determine conservation measures aimed at protecting them from the brink of extinction remains absent.

In this section, our focus will be on those marine species that not only hold the status of keystone species within these marine habitats but also captivate our hearts with their charismatic nature. By highlighting the significance of these species and their critical role in these marine ecosystems, we aim to engage the public and raise awareness about the challenges these creatures face, as well as the broader environmental issues that surround their habitat. The coastal marine species of Qatar, like the intricate mosaic of ecosystems they inhabit, deserve our attention and protection, not only for their intrinsic value but for the overall health and balance of our shared planet.

6.3.1 Marine Mammals

Marine mammals play a pivotal role in marine ecosystems worldwide. Their conservation is of utmost importance; however, there remains a substantial gap in our understanding of the diversity, distribution, density, and abundance of these animals within the Arabian Gulf (Rabaoui et al. 2021). The marine mammals recorded in Qatar are listed in Table 6.1. Remarkably, while the Dugong (*Dugong dugon*) has received some attention and research in Qatar, the same cannot be said for cetaceans, their habitats, and different life stages.

Efforts have been made by offshore operators to contribute to the knowledge of marine mammals in Qatar through a citizen science sightings scheme and data from such schemes should be put into a wider database for marine species in Qatar. Most dedicated surveys have primarily focused on the Dugong, with extensive surveys conducted in 2014–2015. These surveys consistently spotted a large herd during the winter months, with similar observations in 1986 and 1999 (Preen 2004) and this continues to 2018 (Marshall et al. 2018), where cetacean species were also documented concurrently.

Given the urgency of the matter, it is imperative to ensure the continued involvement of the industry in this work. Engaging in discussions on conservation efforts and contributing their valuable resources to the routine monitoring of marine mammal populations around Qatar is vital for the protection of these magnificent creatures and their habitats.

Table 6.1 Marine mammal species observed in Qatar and their associated IUCN Red List Status, indicating their global conservation concern. Statuses follow the IUCN classification system

Common name	Scientific name	Status in Qatar	IUCN Status
Dugong	*Dugong dugon*	Extant resident population, annual aggregation in NW Qatar, and found a lesser extent in south-eastern area.	VU (Vulnerable)
Indo-Pacific Bottlenose Dolphin	*Tursiops aduncus*	Sighting nearshore and offshore in Qatar. Photographic evidence of calves in NE Qatar (Catherine Reeves, pers. comm).	NT (Near Threatened)
Humpback Dolphin	*Sousa plumbea*	Common in Doha Bay until 2013 (John Reeves pers. comm.). Sighted further north from Fuwariat, Ras Laffan and south Sealine Resort.	EN (Endangered)
Spinner Dolphin	*Stenella longirostris*	Confirmed in north east offshore waters	LC (least concern)
Long beaked common dolphin	*Delphinus capensis*	Sightings SE and offshore in NE Qatar, with the latter having photographic evidence.	LC (least concern)
Pantropical Spotted dolphin	*Stenella attenuata*	Two records from south of the border in UAE. Also, likely to be found in Qatar.	LC (least concern)
Indo-Pacific Finless Porpoise	*Neophocaena phocaenoides*	Sightings E of Doha, however it is small, elusive marine mammal	VU (Vulnerable)
Brydes whale	*Balaenoptera edeni*	Confirmed offshore, beached specimen in September 2022.	LC (least concern)
Killer whale	*Orcinus orca*	Confirmed in video in NE Qatar, 2020 and 2021.	Data Deficient (DD)
False killer whale	*Pseudorca crassidens*	Two records in UAE. It is likely to be in Qatar but considered rare.	NT (Near Threatened)
Humpback whale	*Megaptera novaeangliae*	One sighting in 1959, unlikely to be present or resident	LC (least concern)

6.3.2 Sea Birds

The entire Gulf region is internationally recognized as a crucial area for both migratory and breeding seabirds (see Table 6.2). These seabird populations rely on various ecosystems, including freshwater and salt marshes, mangroves, islands, and intertidal sand and mudflats. Offshore islands serve as essential nesting grounds for numerous seabird colonies, including terns and the Socotra Cormorant *Phalacrocorax nigrogularis*, which is classified as vulnerable on the IUCN Red List. These seabird species predominantly nest in coastal sites that are particularly vulnerable to the effects of climate change. Additionally, recreational activities can lead to damage to nesting sites and disturbances in their feeding and breeding grounds. Key habitats for these seabirds encompass saltmarshes and mudflats, along with the adjacent coastal waters that provide them with food sources. Any alterations or disruptions to these habitats can have a significant impact on the availability of food supplies and suitable nesting areas for seabirds.

Table 6.2 Common seabird species recorded in Qatar. Continued conservation of coastal and marine habitats is vital to support the populations and ecological roles of these seabirds

Common name	Scientific name	IUCN Status
Greater Flamingo	*Phoenicopterus roseus*	LC (least concern)
Eurasian Oystercatcher	*Haematopus ostralegus*	NT (Near Threatened)
Black-winged Stilt	*Himantopus himantopus*	LC (least concern)
Pied Avocet	*Recurvirostra avosetta*	LC (least concern)
Common Ringed Plove	*Charadrius hiaticula*	LC (least concern)
Kentish Plover	*Charadrius alexandrinus*	LC (least concern)
Lesser Sand Plover	*Charadrius atrifrons*	LC (least concern)
Greater Sand Plover	*Charadrius leschenaultii*	LC (least concern)
Eurasian Curlew	*Numenius arquata*	NT (Near Threatened)
Bar-tailed Godwit	*Limosa lapponica*	NT (Near Threatened)
Ruddy Turnstone	*Arenaria interpres*	NT (Near Threatened)
Sanderling	*Calidris alba*	LC (least concern)
Little Stint	*Calidris minuta*	LC (least concern)
Common Sandpiper	*Actitis hypoleucos*	LC (least concern)
Common Redshank	*Tringa totanus*	LC (least concern)
Marsh Sandpiper	*Tringa stagnatilis*	LC (least concern)
Slender-billed Gull	*Larus genei*	LC (least concern)
Caspian Tern	*Hydroprogne caspia*	LC (least concern)
Great Cormorant	*Phalacrocorax carbo carbo*	LC (least concern)
Socotra Cormorant	*Phalacrocorax nigrogularis*	VU (Vulnerable)
Grey Heron	*Ardea cinerea*	LC (least concern)
Purple Heron	*Ardea purpurea*	LC (least concern)
Western Osprey	*Pandion haliaetus*	LC (least concern)
Little Egret	*Egretta garzetta*	LC (least concern)
Indian Reef Heron (Western reef-egret)	*Egretta gularis*	LC (least concern)

6.3.3 Marine Reptiles

There have been documented sightings of five species of marine turtles in the waters of Qatar and the broader Arabian Gulf (see Table 6.3, 6.4). Within the Gulf, only two sea turtle species are known to nest, and Qatar hosts one of these species: the hawksbill turtle (*Eretmochelys imbricata*). In Qatar, approximately 200 hawksbill turtles nest each year at locations including Fuwariat, Ras Laffan, and Halul Island, and additional nesting sites have been identified along the coastline (Pilcher et al. 2015). Although loggerhead turtles (*Caretta caretta*) have been recorded infrequently, there have been occasional sightings of leatherback turtles (*Dermochelys coriacea*) and even a single olive ridley turtle (*Lepidochelys olivacea*). The presence of leatherback and olive ridley turtles is considered incidental, with these individuals likely being stray turtles. However, there is mounting evidence of loggerhead turtles inhabiting the waters offshore in Northeast Qatar. Photographic evidence suggests that there may be a foraging population residing off the northern shores of Qatar and Bahrain, potentially originating from Oman (Pilcher et al. 2015). There was a Ministerial Resolution No. 37 of 2010 on the Conservation of Turtles and Seabirds from Extinction to protect the nests of both seabirds and the turtle population at Fuwairiat Beach from disturbance and poaching.

The conservation and protection of marine turtle species, especially the nesting sites and foraging areas, are of paramount importance to ensure their continued survival and well-being. It is essential to further study and monitor these species to better understand their behavior and needs in the region. The main threats to sea turtles in Qatar are loss of eggs after nesting due to predation, loss of the nesting beaches due to coastal development, reclamation, and sand removal resulting from coastal erosion due to climate change. Turtles have been entangled in fishing gear

Table 6.3 Sea turtle species present in Qatar, highlighting their conservation status within the region

Common name	Scientific name	Status in Qatar	IUCN Status
Green Turtle	*Chelonia mydas*	Most likely species encountered at sea and more numerous than Hawksbill but do not nest	LC (least concern)
Hawksbill Turtle	*Eretmochelys imbricata*	Nesting on beaches on the Eastern side of Qatar, foraging grounds	CE Critically Endangered
Loggerhead Turtle	*Caretta caretta*	Photographic evidence of numerous sightings in offshore NE area	VU Vulnerable
Oliver Ridley Turtle	*Lepidochelys olivacea*	Infrequent	VU Vulnerable
Leatherback Turtle	*Dermochelys coriacea*	Occasional infrequent	VU Vulnerable

Table 6.4 Sea snake species present in Qatar, providing information on species diversity within the region's marine habitats

Common name	Scientific name	IUCN Status
Stoke's sea snake	*Hydrophis stokesii*	LC (least concern)
Beaked sea snake	*Enhydrina (Hydrophis) schistosus*	LC (least concern)
Annulated sea snake	*Hydrophis cyanocinctus*	LC (least concern)
Arabian Gulf Sea snake	*Hydrophis lapemoides*	LC (least concern)
Ornate sea snake	*Hydrophis ornatus*	LC (least concern)
Yellow sea snake	*Hydrophis spiralis (spiralis)*	LC (least concern)
Spine-bellied (Short) (Shaw's) sea snake	*Lapemis (Hydrophis) curtus*	LC (least concern)
Graceful small-headed (Slender) sea snake	*Microcephalophis gracilis*	LC (least concern)
Yellow-bellied (Pelagic) sea snake	*Pelamis (Hydrophis) platurus*	LC (least concern)
Viperine sea snake	*Praescutata (Hydrophis) viperinus*	LC (least concern)

and drowned, and can be impacted by oil spills and plastic ingestion of which cause inability to absorb nutrients from stomach. There have been anecdotal evidence of marine traffic causing fatal collisions and propeller strikes.

The Arabian Gulf is home to ten species of sea snakes, which are detailed in Table 6.5. The snakes more frequently reported by the people in 2013 were *Hydrophis curtus* (18% of the observations), *H. cyanocinctus* (13%) and *H. spiralis* (12%). The snakes more frequently reported by the people in 2016 were *Hydrophis lapemoides* (20% of the observations), *Hydrophis platurus* (16%) and *H. spiralis* (12%) (Castilla et al. 2017). These highly adapted reptiles have evolved to thrive in the unique conditions of the Gulf, where they play vital roles in the marine ecosystem. Their presence is a testament to the region's ecological significance and highlights the need for continued research and conservation efforts to protect their habitat and ensure the sustainability of these intriguing sea snakes.

6.3.4 Elasmobranchs: Sharks and Rays

Elasmobranchs are important inhabitants of Qatar's marine ecosystems. These enigmatic creatures are finely adapted to the extreme environmental conditions found in the Arabian Gulf. Several elasmobranch species are known to frequent Qatar's waters, contributing to the region's unique marine biodiversity (see Table 6.5). The presence of elasmobranchs serves as an indicator of the health and balance of the marine environment, and should make them a focal point of scientific research and conservation efforts. According to the IUCN Red List of Threatened Species, it is estimated that over one-third of sharks and related species are threatened with extinction. Ensuring the well-being of these magnificent animals is not only important for

Table 6.5 Elasmobranch species recorded in Qatar by the Shark Conservation Society during their 2009 expedition and by subsequent researchers, highlighting species diversity and presence in the region

Common name	Scientific name	IUCN Status
Scalloped Hammerhead	*Sphyrna lewini*	CE (Critically Endangered)
Arabian Smoothhound	*Mustelus mosis*	NT (Near Threatened)
Hooktooth shark	*Chaenogaleus macrostoma*	VU (Vulnerable)
Slender Weasel shark	*Paragaleus randalli*	VU (Vulnerable)
Snaggletooth shark	*Hemipristis elongata*	VU (Vulnerable)
Pigeye shark	*Carcharhinus amboinensis*	VU (Vulnerable)
Whitecheek shark	*Carcharhinus dussumieri*	EN (Endangered)
Bull shark	*Carcharhinus leucas*	VU (Vulnerable)
Blacktip shark	*Carcharhinus limbatus*	VU (Vulnerable)
Milk Shark	*Rhizoprionodon acutus*	VU (Vulnerable)
Great Hammerhead shark	*Sphyrna mokarran*	CE (Critically Endangered)
Whale shark	*Rhincodon typus*	EN (Endangered)
Spottail shark	*Carcharhinus sorrah*	NT (Near Threatened)
Zebra shark	*Stegostoma tigrinum*	EN (Endangered)
Tawny Nurse shark	*Nebrius ferrugineus*	VU (Vulnerable)
Sandbar shark	*Carcharhinus plumbeus*	EN (Endangered)
Sliteye shark	*Loxodon macrorhinus*	NT (Near Threatened)
Guitarfish(es)	*Rhinobatiformes sp.indet*	Most species are critically endangered (CE)
Halavi Guitarfish	*Glaucostegus halavi*	CE (Critically Endangered)
Green Sawfish	*Pristis zijsron*	CE (Critically Endangered)
Wedgefish	*Rhinidae sp. indet*	Most species are critically endangered (CE)

preserving marine biodiversity but also for maintaining the overall health of Qatar's coastal ecosystems. Studying and protecting elasmobranch populations are essential steps towards maintaining a sustainable and thriving marine environment in the region.

Due to the decline in shark populations, illegal fishing, and the difficulties to implement sustainable shark fishing some countries have opted to ban commercial shark fishing and the export of shark products altogether. Countries have designated their EEZ's as shark sanctuaries, where shark fishing and related activities are often heavily regulated, or even prohibited, aim to help populations to recover. These sanctuaries raise awareness in the local population and can become ecotourism hotspots. There is still the need for further management measures to reduce shark by-catch, ensure shark sanctuary have conservation goals, and guidelines to evaluate effectiveness (Ward-Paige 2017).

The correct identification of species is essential to record what is in the national catch, and thereby develop a shark management plan. Sharks of the Arabian Seas (Jabado et al. 2015) identification guide is unique as it addresses only species confirmed from the Arabian Seas region to rapidly and accurately identify sharks encountered whilst at sea, at landing sites or domestic fish markets in the region. The taxonomic status of many elasmobranch species in the Qatar has been unclear and there are species that are difficult to identify due to similar body shape, colour and overlapping distributions. Scientific information on rays, guitarfishes and wedge fishes is inadequate even though sightings are received regularly both onshore and offshore. Monitoring with environmental DNA (eDNA) techniques has shown in Qatar, and elsewhere that it detects 44% more shark species than traditional underwater visual censuses and baited videos. Furthermore, eDNA analysis reveals the presence of previously unobserved shark species in human-impacted areas (Boussarie et al. 2018).

Environmental DNA could be used to further confirm the continued presence of the critically endangered Green Sawfish in Qatari waters. In 2010 the Qatar government issued a decree protecting the Green Sawfish in Qatari waters, the Shark Conservation Society reported the landing of a juvenile in 2011 accidentally caught in a net, it has also been reported in both Bahrain and the UAE near to the south-east border with Qatar (Jabado and Spaet 2017). They are highly susceptible to capture in gillnets and any ghost fishing gear. It would be beneficial is there was a method for fishers to report by-catch of the Green Sawfish and other by-caught marine species as well as being able to check video data from the offshore sectors in Qatar.

6.4 Ecosystem Services

The coastal marine species and ecosystems of Qatar play a vital role in providing a wide array of ecosystem services that benefit both the environment and human society. In the context of Qatar's coastal marine environments, these services are intricately linked to the diverse array of species and habitats found in the region.

Under the Millennium Ecosystem Assessment's, and for operational purposes, these services are divided into four main types:

(i) *Provisioning services* encompass the tangible resources provided by marine ecosystems, such as fish, mollusks, and crustaceans, which are essential for food security and economic livelihoods in Qatar. The bountiful fisheries and other marine resources are key components of these provisioning services, supporting local communities and contributing to the nation's economy;

(ii) *Regulating services* involve the natural processes that help maintain a stable and healthy environment. Coastal ecosystems like mangroves, salt marshes, and seagrass beds in Qatar act as buffers against coastal erosion, protect against storm surges, and play a crucial role in purifying water and sequestering carbon.

These regulating services are vital for mitigating the impacts of climate change and protecting coastal areas from natural disasters;

(iii) *Supporting services* encompass the underlying processes and functions that sustain life within marine ecosystems (Elliff and Kikuchi 2017; Pascal et al. 2016). These include nutrient cycling, primary production, and the critical role of various species in maintaining ecological balance. Qatar's marine species, from microscopic plankton to large predators, are integral to these supporting services, ensuring the continued health and productivity of the coastal marine environments;

(iv) *Cultural services* refer to the intrinsic value that people derive from the marine environment, including recreational activities, tourism, and the aesthetic and cultural significance of these coastal landscapes. Qatar's marine ecosystems provide opportunities for leisure and connection to nature, fostering a sense of place and pride among its inhabitants.

Linkages between ecosystem health, biodiversity and ecosystem services are an important scientific information that should be used to improve management decision making. Considering that such linkages were poorly understood for the Gulf as a whole (Sheppard et al. 2010) and in Qatar in particular, we report here the most critical ecosystem services provided by marine ecosystems in Qatar.

6.4.1 Provisioning Services

Fisheries

One of the most significant services provided by marine ecosystems in Qatar is their crucial contribution to national food security through fisheries. For example, fish landings reached 17,154 tonnes in 2019 (FAO 2022), and seafood consumption has increased to 24.5 kg per capita annually (Abusin et al. 2022). The top three fished and consumed species in Qatar are the emperor fish (*Lethrinus* spp., locally known as Sha'ari), narrow banded Spanish mackerel (*Scomberomorus commerson*, locally known as Kanad), and groupers (*Epinephelus* spp., locally called Hamour). Fishing activities primarily employ fish traps (gargour) and angling, and using an automatic satellite boat tracking system, offshore coral reefs have been identified as the major fishing grounds for commercial fisheries in Qatar. The overlap of these major fishing hotspots with offshore coral sites underscores the vital importance of these habitats to Qatar's fisheries sector. Surveys of fish diversity and abundance conducted in both coral reefs and seagrass meadows have revealed that several of the most abundant and widely distributed species, such as *Lethrinus nebulosus* (Sha'ari) and *Siganus luridus* (Safi), are particularly associated with healthier ecosystems (Bouwmeester et al. 2022). Notably, while juveniles were predominantly observed in seagrass meadows, emphasizing their role as nursery areas, adults were found in coral reefs, indicating the significance of ecosystem connectivity in the provision of fisheries products.

Genetic Resources

Marine genetic resources encompass the genetic information found in marine organisms, enabling them to produce a wide array of biochemical compounds (Jaspars et al. 2016). These resources hold significant potential for the benefit of humankind through various applications in biodiscovery, including the development of pharmaceutical compounds, cosmetics, food supplements, research tools, and industrial processes (Blasiak et al. 2020; Harden-Davies 2020). Certain coral species, for instance, are utilized in the pharmaceutical industry due to their anti-inflammatory, antimicrobial, anticoagulant, and anticancer properties. Additionally, corals are sought after in the ornamental trade (Moberg and Folke 1999). Moreover, there is growing evidence supporting the use of coral skeletons in bone graft operations (Pountos and Giannoudis 2016).

6.4.2 Regulating Services

Heat Regulation

Coastal ecosystems play a pivotal role in climate regulation on Earth, contributing significantly to the mitigation of both seawater and atmospheric warming. Coral reefs play a crucial role in this process by releasing biogenic aerosols (Jackson et al. 2020). These aerosols, generated through the breaking waves on coral reefs or the production of mucus, have a cooling effect on local temperatures. Sea spray aerosols reflect sunlight and enhance the Earth's surface reflectivity, reducing the absorption of solar energy and thus cooling the air. Moreover, they can act as nuclei for cloud condensation, further enhancing the cooling effect (Xu et al. 2022).

Similar to coral reefs, mangrove forests and seagrass meadows exert both direct and indirect effects on the local climate (Massetti et al. 2013; Duarte et al. 2005). Indirectly, these ecosystems influence the atmosphere through various mechanisms. Mangrove leaves, for instance, release volatile organic compounds (VOCs) that can react with other atmospheric compounds to form secondary organic aerosols (SOA). These SOAs can function as cloud condensation nuclei (CCN), affecting cloud formation and properties, akin to the aerosols produced by coral reefs. The presence of mangroves can also alter local wind patterns and moisture transport, influencing precipitation patterns and humidity levels (Gilma Beserra de Lima and Galvani 2013). Mangrove forests and seagrass meadows further regulate ocean temperatures, with mangroves absorbing and storing heat during the day and releasing it at night, creating thermal refuges for wildlife. They also contribute to higher humidity levels through evaporation from the soil and transpiration from the trees. Additionally, these ecosystems help maintain ocean temperatures by providing shade and reducing the penetration of sunlight into the water, preventing excessive warming (Massetti et al. 2013).

Pollution Buffering

The extensive and pervasive industrialization and coastal development in the countries surrounding the Arabian Gulf over the past few decades have resulted in a mounting influx of both organic and inorganic pollutants into the coastal environment, and Qatar is no exception to this trend. This surge in pollution poses a serious threat to the delicate marine ecosystems that line Qatar's shores. Marine ecosystems play a pivotal role in capturing and mitigating pollutants, a vital service they provide to safeguard the coastal environment. These pollutants encompass a range of harmful substances, including plastics, nitrogen, phosphorus, and various organic compounds. One noteworthy capability of marine ecosystems is their proficiency in capturing marine plastics, which have become a widespread concern. Seagrass meadows and mangroves are particularly effective in this regard. Seagrasses, for instance, can trap a remarkable 1,470 plastic items per kilogram of plant material, as demonstrated by Sánchez-Vidal et al. (2021). In addition, plastics are pervasive in mangrove forests worldwide, with densities reaching up to 533 items per square meter, as evidenced in a Jakarta mangrove in Indonesia (Luo et al. 2021). The role of these ecosystems in plastic capture is invaluable for protecting marine life and mitigating the ecological damage caused by plastics.

Furthermore, marine ecosystems exhibit an exceptional capacity to bioremediate and retain inorganic pollutants like nitrogen and phosphorus. These nutrients are often present in excessive amounts due to industrial runoff and agricultural activities, leading to water pollution and eutrophication. Fortunately, common coastal ecosystems in Qatar, including mangroves, salt marshes, and seagrass meadows, have been shown to act as efficient filters, cycling, and storing nutrients. They achieve this by absorbing and retaining nutrients through their leaves and intricate root systems (de los Santos et al. 2020). This crucial service helps maintain water quality, reduce the harmful effects of eutrophication, and ensure the health of Qatar's marine environment.

Carbon Sequestration

As the state with the highest per capita carbon emissions, Qatar faces a significant challenge in mitigating these emissions and reducing its overall carbon footprint. However, Qatar can leverage its marine ecosystems to play a vital role in achieving these goals by sequestering carbon and acting as effective carbon sinks. The carbon sequestered within marine sediments is commonly referred to as "blue carbon," and it holds enormous potential in the context of carbon reduction (Lincoln et al. 2021). Notably, Qatar's coastlines are rich in blue carbon ecosystems, including mangrove forests, microbial mats in saline mudflats (sabkhas), saltmarshes, and seagrass meadows. The significance of blue carbon ecosystems is garnering global recognition due to their remarkable capacity to store and sequester carbon. Recent research indicates that coastal ecosystems worldwide, encompassing mangroves, saltmarshes, and seagrass meadows, are responsible for storing approximately 81 million metric tons of carbon annually. This is roughly equivalent to the combined greenhouse gas emissions of Qatar and the United Arab Emirates (Bertram et al.

2021). The global ecosystem service provided by blue carbon is conservatively estimated to be valued at USD 190 billion per year (World Bank, 2023).

In Qatar, these coastal blue carbon ecosystems, comprising mangroves, tidal marshes, and seagrass meadows, efficiently sequester carbon dioxide. They offer a range of benefits, primarily in the form of regulating services. These ecosystems help safeguard shorelines from the impacts of storms and sea level rise, playing a crucial role in climate change mitigation and enhancing resilience against its effects. On the other hand, coral reefs, another essential marine ecosystem in Qatar, also contribute to carbon sequestration over geological timescales. The calcification process, where corals bind with calcium to form calcium carbonate, plays a fundamental role in the global calcium balance (Moberg and Folke 1999). These processes further underscore the invaluable ecosystem services provided by Qatar's marine environments in carbon sequestration and climate regulation.

6.4.3 Cultural Services (Recreation and Heritage)

Cultural services in Qatar are intricately linked to the rich history and heritage associated with the sea, especially the thriving pearling industry that once flourished in the region, including Qatar, during the 19th and early twentieth centuries. This historical connection underscores the cultural significance of Qatar's marine ecosystems. Coral reefs serve as unique places that not only offer natural beauty but also generate and support a wide range of human experiences. These ecosystems hold profound cultural importance and are deeply connected to the nation's historical identity. The rich marine biodiversity and vibrant coral reefs not only provide aesthetic pleasure but also support a variety of traditional livelihoods and associated characteristics. Moreover, Qatar's coral reefs offer abundant opportunities for education and research, further enriching the cultural fabric of the region. Reef-based tourism plays a vital role in this regard, encompassing both on-reef activities such as diving and snorkelling. These outdoor activities are immensely popular among Qatari citizens and residents, allowing them to connect with and appreciate the underwater world. The reefs also indirectly contribute to tourism through activities like beachcombing and enjoying scenic coastal views (Woodhead et al. 2019).

The cultural services provided by Qatar's marine ecosystems are deeply rooted in the nation's history, with the pearling industry serving as a testament to the enduring relationship between its people and the sea. Today, these services continue to offer opportunities for cultural enrichment, recreation, and education, further strengthening the bond between the people of Qatar and their marine environment. Recognizing and preserving these cultural services are essential for ensuring that future generations can continue to benefit from and celebrate this unique aspect of Qatar's heritage.

6.4.4 Supporting Services

Supporting services in the context of marine ecosystems are vital for the proper functioning of the marine environment and the continued provision of other ecosystem services. Two primary supporting services in Qatar's marine ecosystems are primary production and the nursery function.

Primary Production

Marine primary production is the essential process through which primary producers in the ocean, including phytoplankton and seagrass, harness energy from the sun through photosynthesis to generate organic matter. This organic matter forms the cornerstone of the marine food web, supplying energy and vital nutrients to sustain all other marine organisms, from fish and mammals to birds. Therefore, marine primary production serves as a pivotal supporting service in the marine environment. Its significance lies in its critical role in maintaining the functionality of marine ecosystems and preserving biodiversity. Without the continuous process of marine primary production, the entire marine food web would crumble, resulting in the loss of crucial ecosystem services such as fisheries, carbon sequestration, and oxygen production. Changes in the rate, timing, and spatial distribution of marine primary production can have substantial consequences for marine ecosystems in Qatar. These changes can lead to shifts in the distribution and abundance of marine species, modifications in nutrient cycling, and alterations in the ocean's ability to absorb atmospheric carbon dioxide. Along the coastline of Qatar, coastal mudflats, saltmarshes, seagrass meadows, and mangrove forests are the primary contributors to local coastal primary productivity, with rates varying from 1 to 10 kg C/m^2/yr, influenced by factors such as nutrient availability, light penetration, and temperature. These ecosystems are vital for sustaining the marine environment and the ecosystem services it provides (Rodrigues-Filho 2023; Ullah et al. 2018).

Nursery

The nursery function plays a pivotal role in certain marine and coastal ecosystems by providing a safe and conducive environment for the early life stages of various species. These ecosystems act as critical breeding and nursery grounds for a diverse array of fish, invertebrates, and other marine organisms. These young organisms rely on specific environmental conditions to ensure their survival and growth (Beck et al. 2001). Notably, in Qatar, examples of such ecosystems include seagrass beds, coral reefs, mangrove forests, and salt marshes, among others. These habitats provide essential features such as refuge from predators, a consistent food source, and ideal water conditions, including temperature and salinity, which are imperative for the successful growth and development of juvenile organisms. The nursery function is categorized as a supporting service because it lays the foundation for a spectrum of other ecosystem services that directly benefit human populations. These benefits encompass diverse sectors such as, fisheries, tourism, and shoreline protection (Pittman et al. 2017). In the absence of these ecosystems effectively functioning as nurseries, numerous marine species would struggle to reproduce and maintain their populations, setting off a chain reaction that would reverberate across the entire coastal environment (Ferreira et al. 2020). The preservation and conservation of these critical nursery habitats are thus essential for maintaining healthy marine ecosystems and the various services they provide.

6.5 Potential Marine Protected Areas in Qatar Exclusive Economic Zone (EEZ)

In the Qatar Exclusive Economic Zone (EEZ), several potential Marine Protected Areas (MPAs) have been identified by the MoECC, each serving a unique and crucial purpose in conserving marine biodiversity and promoting sustainable practices. These potential MPAs are as follows:

(i) *Western MPA for Seagrass, Dugong, Dolphin, and Sea Turtles*: This area is dedicated to the protection of seagrass meadows, essential habitats for dugongs, dolphins, and sea turtles. It is designed to safeguard the critical ecosystems and species that rely on seagrass beds along the western coast.

(ii) *Northeast Offshore MPA for Whale Sharks, Tuna Mackerel, Cetaceans, and Offshore Seabed Ecosystems*: This offshore MPA aims to preserve the habitats of whale sharks, tuna mackerel, cetaceans, and the delicate offshore seabed ecosystems, particularly areas with soft corals.

(iii) *Halul Island MPA for Hard and Soft Corals, Nesting Hawksbill Turtles, and Seabirds*: Halul Island serves as a vital location for the protection of hard and soft coral ecosystems, as well as providing nesting sites for hawksbill turtles and a habitat for seabirds.

(iv) *Al Thakira & Al Khor MPA for Mangroves, Seabirds, Fish Nursery Areas, and Marine Invertebrates*: This MPA focuses on conserving the rich mangrove ecosystems, important fish nursery areas, and various marine invertebrates that are integral to the coastal environment in Al Thakira and Al Khor.

(v) *Khor Al Adaid MPA for Natural Geological Features, Sand Dunes, Salt Marshes, Islands, Mangroves, Macroalgae, Coral Ecosystems, and Associated Species*: This area is dedicated to the protection of the diverse natural features found in Khor Al Adaid, including sand dunes, salt marshes, islands, mangroves, macroalgae, hard and soft coral ecosystems, as well as the associated species, including marine mammals and seabirds.

(vi) *Offshore Coral and Oyster Beds*: This MPA is established to safeguard offshore coral and oyster bed ecosystems, which play a crucial role in maintaining marine biodiversity.

(vii) *Umm Tays to Ras Laffan Sea Turtle Nesting Beaches and Coastal Mangrove, Seagrass, and Ecosystems*: This area focuses on the protection of sea turtle nesting beaches and the valuable coastal mangrove and seagrass ecosystems that provide essential habitats for various marine species.

(viii) *Al Wakra Coastal Mangrove, Sand Dunes, and Seagrass Meadows*: This MPA is designed to conserve the coastal mangrove, sand dune, and seagrass ecosystems found in Al Wakra.

UNESCO identified as Khor al Udyad, also known regionally as the 'Inland Sea', located in the south-east of the State of Qatar as a future UNESCO Global Goepark and World Natural Heritage Site (Licciardi et al. 2007). The area showcases a remarkable landscape shaped by a globally unique combination of geological and geomorphological features. These elements create a diverse and exceptionally beautiful natural scenery that remains largely undeveloped. Each landscape unit—such as the Arabian Gulf, expansive mobile calcium-carbonate sand dunes, tidal embayment systems, and both inland and coastal sabkhas—contributes to this distinctive environment.

The Ministry of Environment and Climate Change (MoECC) has already initiated the process of designating these new MPAs. The Qatar West Coastal MPA, aimed at safeguarding dugongs and seagrass meadows, the Qatar North MPA, and Qatar North-East offshore MPAs, dedicated to the protection of whale sharks, tuna mackerel, pelagic fish, and offshore island and coral ecosystems, are expected to be effective as the first batch of new marine protected areas. It is essential to acknowledge that the costs associated with establishing and maintaining a national MPA network are relatively small compared to the substantial economic benefits generated by fisheries. Even a modest 10% increase in fisheries yields could contribute significantly to covering the network's costs. Evidence from other regions demonstrate that the creation of MPAs can lead to substantial increases in catch rates. For example, in St Lucia, catch rates doubled within five years of MPA creation, while in the Philippines, they increased ten-fold after 20 years of protection (Gell and Roberts 2003).

As part of the CBD's Post-2020 Global Biodiversity Framework, a crucial goal is to protect at least 30% of the world's oceans by 2030 (MPA 2030). This program aims to provide the necessary information and knowledge to assist countries in achieving their 30% by 2030 targets. MPAs have a multifaceted role in protecting and restoring biodiversity, enhancing food security, supporting sustainable development, and promoting responsible fisheries management. As the global community works toward protecting 30% of the ocean, it is imperative that these MPAs not only conserve biodiversity but also contribute to food security and maintain these benefits in the face of global changes, including climate change. At the UN Oceans Conference in July 2022, the State of Qatar, represented by the Ministry of Environment and Climate Change, made a voluntary commitment to support and implement these MPA initiatives, underscoring the nation's dedication to marine conservation and sustainable ocean management. Qatar's ratification of the High Seas Treaty (BBNJ) in January 2026 helped bring it into force. This Treaty includes the first legally binding framework to conserve and sustainably use marine biodiversity in areas beyond national jurisdiction and enable the designation of MPAs in BBNJ.

6.6 Threats and Human Impacts on Marine Ecosystems in Qatar

There exists a complex interplay between human activities, marine ecosystems, and their profound impact on the condition of Qatar's oceans. Human actions are setting in motion extensive changes that reverberate through the physical, chemical, and biological systems of our seas. Importantly, these driving forces behind oceanic transformations originate largely from activities on land, mirroring the dynamics of many contributors to anthropogenic climate change. They encompass the escalating demand for sustenance required to feed burgeoning human populations, the global trade in agricultural products and commodities, and coastal degradation. Within the industrial sector, Qatar encounters multifaceted challenges arising from oil and mineral extraction, shipping, fisheries, tourism, coastal development, and the burgeoning aquaculture industry.

Human activities are instigating extensive alterations in the physical, chemical, and biological systems of the oceans. The primary forces driving these transformations in the marine environment originate primarily from terrestrial sources. Much like many of the leading contributors to anthropogenic climate change originate on land, the primary sources of amplified pressures on marine biodiversity and environmental quality can also be traced back to activities occurring on land. These encompass the demand for sustenance to feed growing human populations, the global trade in agricultural products and goods, and coastal degradation. Within the industrial sector, Qatar faces pressures arising from oil and mineral extraction, shipping, fisheries, tourism, coastal development, and aquaculture. The pursuit of profitability and cost-efficient production practices further contributes to pollution and contamination concerns (GRID-Arendal and UNEP 2016).

The intensified utilization of resources and oceanic areas is exerting detrimental effects on the state of the ocean. Across various indicators, the Arabian Gulf is undergoing significant transformations–the waters are undergoing warming and alarming acidification, commercial fish populations have been dwindling for extended periods, and coastal waters are facing an upsurge in contamination resulting from both onshore activities and marine industries such as aquaculture. Elevated levels of heavy metals and other noxious substances in certain marine mammals and fish render them unsuitable for human consumption, starkly underscoring the ongoing pollution of once-pristine seas. Many regions of the ocean are already severely degraded, and the imprint of human influence continues to expand. Failure to address these issues poses a substantial risk of triggering a pernicious cycle of deterioration, whereby the ocean can no longer offer many of the benefits currently enjoyed by humankind.

6.6.1 Pressures

Driven by the abovementioned global and regional forces, the pressures on the coastal and marine environment in the GCC region fall into one of the 3 main categories:

(i) *Pollution* including the introduction into the marine environment of chemical, physical and biological harmful components. The land- and sea-based pollutions are resulting mainly from an increasing regional desalination capacity; an ever-expanding urban and industrial marine and coastal activities; and oil and gas production and transportation. Marine traffic from and to the region's harbours is, for instance, a main gateway of invasive species through discharged ballast waters;

(ii) *Overexploitation* of living and non-living resources and habitat destruction is regarded as an overexploitation of space and the physical environment. This includes overfishing, harassing the natural carrying capacity and stock replenishment and the extraction of minerals (e.g., sand) at an unbalancing level of the coastal geochemistry;

(iii) Climate change-related stressors on the marine environment notably acting in the region are seawater warming, seawater acidification and sea level rise (Hosseini et al. 2021).

6.6.2 Fisheries and Aquaculture

Qatari fisheries represent a unique blend of artisanal methods operating on a commercially significant scale (Al-Abdulrazzak and Pauly 2013). The fishing fleet consists of two distinct vessel types: traditional Dhows (large boats) and Speedboats (small outboard-powered fiberglass vessels), both with an operational range of 60–100 km (Al-Ansi and Priede 1996). Speedboats, characterized by their small size and powerful engines, allow for rapid travel, usually within a day. In contrast, Dhows are larger boats, primarily constructed from wood or fiberglass, equipped with inboard powerful motors but are relatively slower, leading to longer fishing trips lasting on average 3 to 5 days (Stamatopoulos and Abdallah 2015). In 2016, the fishing fleet consisted of 624 boats, comprising 374 Dhows and 250 Speedboats, distributed across four ports: Al Khor, Doha, Al Wakra, and Al Shamal. Both vessel types are engaged in pelagic and demersal fishing activities, utilizing various fishing gear. The most common gear used is fish traps, locally known as "gargoor," followed by gillnets, handlines, and troll lines. Seasonal trolling and hand-lining for Spanish mackerel (*Scomberomorus commerson*) are also practiced (Al-Abdulrazzak and Pauly 2013). According to Stamatopoulos and Abdallah (2015), the top species contributing to Qatar's fish landings include the Spangled emperor (*Lethrinus nebulosus*) at 16.2% of total landings, the narrow-barred Spanish mackerel (*Scomberomorus commerson*) at 10.5%, the white-spot spinefoot (*Siganus canaliculatus*) at 8.3%, the pink ear emperor (*Lethrinus lentjan*) at 6.6%, and the orange-spotted grouper (*Epinephelus coioides*) at 6.0%.

Fishing activities in Qatari waters are concentrated primarily on the eastern side of the peninsula, focusing on offshore waters in the central Gulf, typically at depths of less than 50 m (Al-Ansi and Priede 1996). Demersal fishing efforts are often concentrated on traditional offshore fishing grounds, including the shallow 'hairãt' habitats, which are considered highly productive and support diverse benthic biodiversity (Egerton et al. 2018; Smyth et al. 2016). Many of these areas also harbor significant coral communities. As discussed in the coral reef section, there is extensive spatial overlap between reef habitats and fishing efforts. Recent analysis of AIS data (Automatic Identification System) has revealed that nine major hotspots for fishing activity in the EEZ of Qatar coincide with known offshore coral reefs (Fanning et al. 2021). This close association is reflected in national fisheries landings, where reef-associated fish species account for 90% of the total biomass (Grandcourt 2012). This underscores the critical importance of coral reefs to the fisheries sector in Qatar.

Regarding aquaculture in Qatar, the country has made significant strides in developing this sector. The Qatari government has been investing in aquaculture projects to enhance food security and reduce the reliance on wild-caught seafood. This initiative includes the farming of various marine species such as fish, shrimp, and mollusks in controlled environments. Aquaculture not only helps meet domestic seafood demand but also reduces pressure on natural fish stocks. Qatar's efforts in aquaculture align with global trends toward sustainable and responsible seafood production, ensuring a more resilient and diverse food supply. The aquaculture sector in Qatar is expected to grow, making it an essential component of the country's fisheries and food security strategy.

6.7 Protection of the Blue Economy

The protection of the blue economy is of paramount importance, not only for Qatar but for coastal and maritime nations worldwide. The blue economy encompasses a wide array of economic activities and sectors that are heavily reliant on the health and sustainability of marine ecosystems. To protect and promote the blue economy, several key considerations and actions are crucial:

i. *Sustainable Resource Management*: Effective management of living and non-living marine resources is fundamental. This involves implementing sustainable fishing practices, aquaculture management, and the responsible extraction of marine minerals and oil and gas resources. Sustainable management ensures that resources are not depleted to the point of no return, allowing future generations to benefit from them.

ii. *Coastal and Marine Habitat Protection*: Coastal and marine habitats, such as coral reefs, mangroves, seagrass meadows, and others, are the foundation of the blue economy. Protecting these habitats and restoring degraded ones is essential. Healthy habitats support fisheries, provide coastal protection, and serve as breeding and nursery areas for marine species.

iii. *Integrated and Strategic Planning*: Managing maritime developments, such as ports, shipping routes, and coastal infrastructure, must be done in an integrated and strategic manner. This approach considers the ecological, economic, and social aspects, ensuring that these developments are compatible with the long-term health of marine ecosystems.

iv. *Pollution Control*: Pollution, particularly plastic pollution, poses a significant threat to marine ecosystems and the blue economy. Implementing stringent pollution control measures, recycling programs, and reducing single-use plastics can help mitigate these threats.

v. *Climate Change Mitigation*: Climate change is a global challenge that directly impacts the oceans. Combating climate change through reducing greenhouse gas emissions and implementing strategies to adapt to its effects is essential for safeguarding the blue economy.

vi. *Education and Awareness*: Raising public and industry awareness about the importance of marine conservation and the blue economy is crucial. This can lead to behavioral changes that reduce the negative impacts of human activities on the marine environment.

vii. *Research and Innovation*: Investing in marine research, technology, and innovation can lead to sustainable solutions for blue economy sectors, from more efficient fishing practices to eco-friendly tourism and renewable energy sources.

viii. *International Collaboration*: Many marine issues transcend national borders. Collaborating with neighboring countries and international organizations can help address shared challenges and implement common solutions.

By focusing on these aspects, Qatar can protect and promote their blue economy while ensuring the long-term health and sustainability of marine ecosystems, which are the lifeblood of these economies. For example, the maritime economy of Qatar encompasses a broad spectrum of economic activities that rely on the marine environment, its assets, goods, and services. These activities span various sectors, including fisheries and aquaculture, marine biotechnology, maritime transport, the oil and gas industry, tourism, as well as ports and shipping. In this context, the concept of an ocean or sea economy can be categorized as either 'brown' or 'blue,' with the blue economy representing a sustainable approach where economic activities are in harmony with the long-term capacity of the ocean ecosystem to support these activities while maintaining resilience and health.

Factors promoting a blue economy emphasizing sustainable use of ocean resources for economic growth, improved livelihoods, and ocean ecosystem health are:

- Marine spatial planning;
- Integrated multi-level governance;
- Stakeholder Partnerships;
- Resources and capacity within the country and with support from international bodies;
- Science, research and technology development;

- Monitoring, control and surveillance;
- Financing & investment; and
- Adapting to the effects of climate change.

These can be seen in greater detail in UNESCO-IOC (2021b).

To realize the vision set forth by the Ministry of Environment and Climate Change (MoECC) for the State of Qatar, several key actions are essential. These actions involve safeguarding coastal and marine habitats and, where degradation has occurred, working to restore them. Additionally, it entails the sustainable utilization of living and non-living marine resources, the strategic and integrated management of maritime developments, and the elimination of pollution, including the pervasive problem of plastic pollution. This comprehensive approach, often referred to as ecosystem-based maritime and coastal management, is paramount in the context of Qatar's unique marine environment.

6.8 Ecosystem-Based Management for the Marine Environment

The Ecosystem-Based Approach (EBA) is a comprehensive strategy for the integrated management of land, water, and living resources. It is designed to promote the conservation and sustainable use of these resources in an equitable manner. The fundamental goal of the EBA is to ensure the sustainable utilization of ecosystem goods and services while preserving the overall health and integrity of the ecosystem. One of the distinctive features of the EBA is its ability to account for the cumulative impacts of various stressors on the environment. These stressors can include pollution, coastal development, excessive harvesting, and complex ecological interactions. By considering the combined effects of these influences, the EBA strives to strike a balance between competing uses and activities to prevent overexploitation and environmental degradation. In essence, the Ecosystem-Based Approach encourages a holistic perspective on natural resource management, concentrating on the long-term health of ecosystems. It recognizes that ecosystems are interconnected, and actions in one area can have far-reaching consequences. Moreover, the approach promotes resilience, adaptability, and sustainability. Integrated Coastal Zone Management (ICZM) shares similarities with the EBA, as both endorse holistic and adaptive approaches to natural resource management. However, ICZM places more emphasis on the institutional and governance aspects related to the coastal zone. It is inclined to focus on practical policies and actions that can be implemented to address immediate concerns.

Marine Spatial Planning (MSP) plays a crucial role in implementing the EBA. It ensures that planning considerations extend beyond administrative boundaries, allowing for the comprehensive assessment of cumulative impacts. A precautionary approach is also a key element of MSP, which means taking preventive measures to avoid potential harm to the marine environment. Additionally, MSP offers the flexibility to adapt to changing circumstances. Furthermore, a central aspect of MSP and the EBA is the incorporation of nature conservation policies in marine areas. This can include setting targets for marine conservation, like the objective to establish and effectively manage at least 30% of Marine Protected Areas (MPAs) by 2030. This approach not only safeguards marine biodiversity but also contributes to the overall health and sustainability of coastal ecosystems. MPAs are one component alongside other governance tools such as marine spatial planning, international maritime law, fisheries management, pollution mitigation, management and control, and climate change adaptation to improve ecosystem health.

EBM recognizes that the various components of the marine environment in Qatar are interconnected, and the well-being of one element is dependent on the condition of others. Here are some key ideas to further develop the concept of EBM in Qatar:

a. *Holistic Management*: EBM in Qatar involves taking a holistic view of the marine environment. It considers all aspects, from the physical environment (like coral reefs and mangroves) to the living resources (such as fish stocks and marine mammals). By looking at the entire ecosystem, management decisions can better account for potential impacts on various species and habitats.

b. *Sustainable Resource Use*: One of the primary goals of EBM is to ensure the sustainable use of marine resources. Qatar's rich marine biodiversity supports vital sectors like fisheries and tourism. EBM strives to manage these sectors in a way that ensures they can continue to thrive without compromising the long-term health of the marine environment.

c. *Ecosystem Health*: EBM focuses on preserving the overall health of ecosystems. This includes maintaining biodiversity, water quality, and habitat integrity. By keeping ecosystems healthy, they are better equipped to provide critical services like carbon sequestration, storm protection, and shoreline stability.

d. *Cumulative Impact Assessment*: EBM acknowledges that various human activities and stressors can impact marine ecosystems cumulatively. Whether its pollution, coastal development, or fishing practices, EBM aims to assess and manage these cumulative effects to prevent significant and lasting harm.

e. *Adaptive Management*: Marine ecosystems are dynamic and ever-changing. EBM incorporates adaptive management strategies that allow for flexibility in decision-making. This means that as new information becomes available or conditions change, management plans can be adjusted accordingly.

f. *Cross-Sector Collaboration*: Effective EBM often involves collaboration among various sectors, including government agencies, scientists, industries, and communities. By working together, stakeholders can develop comprehensive management plans that consider a broad range of interests and perspectives.

g. *Marine Protected Areas (MPAs)*: Establishing MPAs is a common strategy within EBM. These areas are designed to protect critical habitats, breeding grounds, and nursery areas for marine species. In Qatar, MPAs can play a significant role in conserving species like the dugong, sea turtles, and important fish stocks.

h. *Long-Term Vision*: EBM is not a short-term fix but rather a long-term vision for the marine environment. It seeks to ensure that future generations can continue to benefit from the services and resources provided by the sea.

i. *Monitoring and Research*: Ongoing monitoring and scientific research are essential components of EBM. By continually assessing the state of the marine environment and studying the impacts of human activities, decision-makers can make informed choices.

j. *Climate Resilience*: EBM also addresses climate resilience. Given the challenges of climate change, including rising sea levels and increased temperatures, EBM incorporates strategies to protect coastal areas from the impacts of a changing climate.

Ecosystem-Based Management in Qatar is a proactive and adaptable approach to safeguarding the marine environment while ensuring the well-being of coastal communities and supporting economic activities. It seeks to find a balance that allows for sustainable development and human well-being without jeopardizing the long-term health of the marine ecosystem.

6.9 Recommendations

Structural and Governance

To improve protective measures for marine resources Qatar policy decision makers, should consider the following;

a. *A Structural review*. Existing policy fishery and marine protection are under two separate Ministries. The Ministry of Municipality (MM) and the MoECC. In many countries both fishery and agriculture are under the same ministry with environment for effective management. Within the previous environmental ministry, the MME, there were too departments carrying out similar tasks. e.g., Protection and monitoring nature reserves department, wildlife development sections and biodiversity sections. Structural review within the ministries should be carried out.

b. *Development of a clear, coherent Qatar Marine Policy* to ensure a strategic direction for development, extraction, and conservation. This should include a legal duty of care to be placed on all State of Qatar Government and statutory bodies to protect the coastal and marine environment.

c. *Absence of regular multi-sectoral working groups or meetings of multi-disciplinary committees* provide consolidation and monitor follow up actions of projects and task. It is an effective means of management. Collective management is effective for development of any new projects including MPA programs or cumulative impact assessments (CIA) projects.

d. *A statutory body for nature conservation*—an 'ecological or environment agency—including the coastal and marine environment—is required with in-house monitoring, research and management team with marine science knowledge is required to ensure continuity of task and projects. For example, research has been carried out on many species for decades, yet limited government staff was trained to take up protection measures for these endanger species. Post training and capacity building for existing or newly recruits should be allowed and encouraged.

Marine Spatial Planning and Integrated Coastal Zone Management (ICZM) and Development

Planning in the Qatar EEZ has traditionally been done through sectoral policies that often have not considered the presence and impact of other sectors, resulting in a series of conflicts between users. Sectoral policies do not always consider spatial aspects, which can lead to conflicts over the use of the sea areas and resources. In addition, they have not considered the long-term cumulative impact of multiple developments in the marine zone of Qatar. There has been little 'space' for nature.

Marine Spatial Planning (MSP) is a comprehensive and strategic process to analyse and allocate the use of the sea areas to minimise conflicts between human activities and maximise benefits, while ensuring the resilience of marine ecosystems. It typically addresses many sectors, their interrelationships, and cumulative impacts, and provides for spatial and temporal measures to steer different uses of the sea areas or resources. Spatial measures can be, for instance, allocation of space for particular uses (and exclusion of uses) or place-specific or general conditions for the use of sea areas or resources (UNESCO-IOC/European Commission 2021a). From a development perspective, MSP is an inclusive process trying to address the needs of society within environmental limits. This is something that does not happen in Qatar at present. The MSP process takes sectoral management into account and may use it as a basis for planning provisions, but MSP does not replace single-sector management measures. For MSP to work there is the need to engage stakeholders actively to improve the legitimacy and quality of decision-making it builds trust and makes decisions more durable. In the context of MSP, stakeholders include representatives, workers or those affected by all maritime sectors, governmental and non-governmental marine related organisations, academia, and civil society, among others.

MSP serves as a decision-making process to determine the organisation of human activities and to facilitate the achievement of or shift towards sustainable paths in meeting the economic, environmental, and social needs of societies. MSP is considered the enabler of a Blue Economy because it:

a. Identifies sites for new and emerging uses, such as offshore renewable energy development, following an EBA approach,
b. Promotes multi-use spaces for coexistence and synergies.
c. Mitigates conflict.
d. Increases investor confidence by introducing transparency and predictability.

e. Facilitates knowledge gaps on the ocean, its health, and key sectors.
f. Can foster collaboration across borders.
g. Promotes capacity building, using new technologies.

There is now a global portal and guideline to MSP developed by UNESCO which has a step-by-step guide to developing a MSP system (UNESCO-IOC/European Commission 2021b). The authors point to this guide and international process to assist Qatar in developing an MSP for Qatar, with the following actions:

a. Develop a legal system to underpin Marine Spatial Planning in Qatar. Using area-based GIS approach. Ensure that all sector policies and expansion are mapped.
b. Ensure the ICZM plan is published, and a legal basis established.
c. Establish urgently a Qatar Coastal Forum
d. First role is to determine coastal indicators and ensure monitoring for the indicators occurs.
e. Ensure that all developments, including tourism, have to conduct an EIA. Update the EIA regulations to include fishery, aquaculture and tourism as requiring an impact assessment.
f. Develop a mechanism for cumulative impact assessments (CIA) in Qatar.

Biodiversity and MPAs

To develop an ecologically coherent network of Marine Protected Areas in Qatar in line with the GBF.

a. Establish new MPAs and provide a legal mechanism to identify, designate, manage and enforce protection of a network of nationally important MPAs, including highly protected marine reserves.
b. Qatar to designate 30% of waters as MPAs, Qatar to specify the target or percentage for highly protected areas.
c. To ensure management effectiveness in the MPAs ensure the statutory body for managing the network of MPAs has adequate human capacity as well as funding and clear boundaries.
d. An independent environmental fund specially allocated for management of MPAs is not present in Qatar. MoECC severely lacks resources required to administer proper monitoring of MPAs and rehabilitate endangered species that have required rescue operations. There is no environmental tax in Qatar. Many countries set up such tax to provide for the financial resource for environmental management.
e. Assessment of effectiveness of MPA management should be open both to in-house staff as well as outside institutions. Quantitative assessment of biological indicator species is the most direct mean of assessment. Strict documentation of procedures and recording data is also an important part of assessment.
f. Funding and a mechanism for State of Qatar marine and coastal National Biodiversity Action Plans; including identifying urgent areas for blue carbon and restoration projects such as oysters, seagrass and mangroves.

The effectiveness of MPAs and area-based measures may be limited for highly mobile species such as those seen in Qatari waters; the whale shark, marine turtles and cetaceans; little is known about their spatial and temporal movements. A recent study showed that the GBF 30% MPAs will be insufficient for marine megafauna conservation. Whilst MPAs are needed they need to be coupled with mitigation measures such as vessel separation and speed limit schemes, fishing regulation and limiting disturbance (Sequeira et al. 2025). For species protection measures the following are essential as a start for Qatar's ambition to protect marine species in Qatar:

g. Urgent research of existing data sets, surveys and monitoring work is required to provide data for marine species to identify key biologically important areas where aggregations of individuals of a species are known to display biologically important behaviour such as breeding, foraging, resting or migration. This should include e-DNA to monitor for species that might not be known to be present such as elasmobranchs.

h. Develop a list of marine species at risk, this should include both threatened and migratory species as specified under Annex I & II of the Bonn Convention (CMS) plus research into which marine invertebrates could be included.

i. Develop urgently a Marine Species Law to ensure that it is a legal requirement to not kill, maim, harass or disturb the list of marine species at risk. This should include all marine megafauna; cetaceans, dugongs, marine turtles, all shark species that are on the at-risk register.

j. An immediate step for Qatar and a key recommendation for species protection would be for Qatar to join the Bonn Convention on Migratory Species (CMS) as a signatory state, and the three key marine species Memorandum of Understanding (MoU).

    (a) The Memorandum of Understanding on the Conservation of Migratory Sharks (Sharks MOU).

    (b) The Memorandum of Understanding on the Conservation and Management of Marine Turtles and their habitats of the Indian Ocean and S-E Asia (IOSEA Marine Turtles MoU)

    (c) Memorandum of Understanding on Dugongs (Dugongs MoU).

k. Qatar could become a national shark sanctuary, with a total ban on the fishing of elasmobranchs (mandatory release for all species of shark, guitar and wedgefishes) due to nearly all species endangered or critically endangered status. This would be the first in the Red, Arabian and Gulf Seas.

Marine and Coastal Fisheries

Qatar currently manages the fishing industry through regulating the number of artisanal vessels, rather than limiting the catch size. No fisheries management plans exist and, as a result, policy directions for fisheries management are unclear and subject to frequent change. Qatar introduced a freeze on the issuing of new fishing licenses in 1998 and this has assisted in controlling fishing capacity. However, there were no restrictions introduced on fishing gear limitation or the size of vessels. As a result, replacement vessels of a larger size and carrying more fishing gear (particularly fish traps) and crew have entered the fishery in recent years. This has resulted in significant, but unmeasured, increases in fishing effort despite the restrictions on the issuing of new licenses. The subsidization of local fisheries remains an important part of the fisheries management policy. Fuel is tax-free and ice is delivered to the port at subsidised prices. The vessels are owned and financed by Qatari private entrepreneurs. The export of fish from Qatar is strictly controlled. Only Qatari nationals who are vessel owners benefit from these subsidies since foreign workers on the vessels are engaged on a contract basis (De Young 2006).

The national authority with responsibility for fisheries management is the Fisheries Department, Ministry of Municipality. The Fisheries department administers the basic national fisheries legislation in Qatar, which is contained in Law n°4 of 1983 for the use and conservation of marine resources. Emiri Decree n° 17 of 1993 subsequently amended this basic fisheries Law. Various Ministerial Decisions and directives implement specific actions under the basic law, including the Ministerial decision of 1992 to ban shrimp fishing in Qatari waters. Other competencies include: policy, licensing and collation and collection of statistics. As in other countries in the region, the fisheries department has no enforcement capabilities, which are concentrated in the Qatari coast guard, ministry of interior. Until mid-2012 statistical monitoring was limited to catch information collected at the Central Auction Market in Doha. Due to its limitations in data scope and coverage a new system, the National Fisheries Information System (NFIS), was implemented in 2012. The NFIS fishing statistics are collected for each boat-gear category and at all four fishing ports, including: catch or landing, fishing effort, catch per unit effort, fish size and its prices. The data is stored into an online database, later used for statistical analysis (Stamatopoulos and Abdallah 2015).

The United Nations Convention on the Law of the Sea (UNCLOS III) entered into force in 1994, and was ratified by the State of Qatar in December 2002. The Convention also includes sustainability indicators relating to fishing, including: (i) Yield-related indicators, e.g., catches, and (ii) Fishing capacity-related indicators such as, fishing effort and intensity. According to the classification in the latest fisheries statistics report, the proportion of fish stocks within biologically sustainable levels[1] has consistently recovered, from 59% in 2017 to 90% in 2020 (Planning and Statistics Authority 2021). Considering the previously mentioned dependency of fisheries landings in Qatar on reef-associated species and the strong decline in living coral cover in recent years, this is an unexpected result. Conversely, fishing activity is one of the most relevant pressures in offshore coral reefs in Qatar. In fact, despite the ban on bottom trawling, the prevalent use of the multi-hooked grapple known as "manshal", to retrieve the traditional fish traps "gargoors" from the seafloor, has a

---

[1] Indicator 14.4.1, Target 14.4 "Sustainable fishing", Goal 14 "Life below water" of the United Nations Sustainable Development Goals (SDGs); https://gis.psa.gov.qa/GISApps/SDG/Data/PDF/En/SDG14.pdf.

considerable detrimental impact on coral reef and other seabed habitats (Al Maslamani et al. 2018). Illegal fishing is another major problem for Qatar, given the small scale and economic importance of the industry. Illegal fishing (such as by driftnets) is common because enforcement agencies are unable to ensure compliance with regulations (De Young 2006). In addition, a growing and uncontrolled recreational fishing sector, estimated to deploy over 1,000 crafts, catches significant quantities of fish (De Young 2006). Given the country's rapidly growing population and the corresponding increase in recreational fishing, the catch of the recreational sector (which is unmonitored) could one day exceed that of the commercial sector.

Climate change is, however, the greatest challenge for fisheries in the region, as it is likely to push many commercial fish species to the limit of their physiological tolerance. Declines in commercial catches and changing species composition have been considered a severe risk for several Gulf countries, including Qatar (Bouwmeester et al. 2020; Lincoln et al. 2021).

Ultimately, effective fisheries management will require the adoption of an integrated approach that considers all major activities and their cumulative impacts on both the end users and the focal ecosystem services, as critical components of an informed decision-making process (Ecosystem-based management, Fanning et al. 2021).

The preservation of coastal ecosystems and sustainability of marine resources is one of the seven goals defined in the Qatar NBSAP. The national development strategy of the state of Qatar explicitly calls for the development of a national programme for the optimal and sustainable exploitation of fisheries stocks (Planning and Statistics Authority 2018). Some of the priority actions for achieving this goal include, but are not limited to, the following:

a. Eliminate fisheries subsidies and impose a moratorium on licensing of new fishing boats
b. Introduce new regulations, banning the use of destructive fishing practices, such as the "manshal" to protect benthic habitats.
c. For any new fisheries or aquaculture developments a full Environmental Impact Assessment (EIA) should be conducted.
d. Develop a training program for fishermen on marine environment protection, including reporting any by-catch of marine megafauna and develop reduction measures with fishers to reduce by-catch in the bottom-set gillnets.
e. Establish a fisheries observer program to monitor the enforcement of fishing regulations at sea
f. Implement regular stock assessments and impose catch quotas for the most important commercial species
g. Identify key areas that act as nurseries or spawning habitat for fish and introduce seasonal closures
h. Identify area-based fishery management measures such as areas for declaration as no take areas complementing the biodiversity MPAs, prohibition of specific gears in certain areas, regulation of fishing effort by area.

i. Develop a consumer awareness campaign on the status of local fish stocks, to encourage the purchase of sustainably fished species

Marine Pollution

Addressing marine pollution, particularly plastic pollution, requires a comprehensive and multi-faceted approach. By adopting the following policy recommendations, Qatar can make significant strides in mitigating marine pollution, particularly plastic pollution, and contribute to global efforts for a cleaner and healthier marine environment. Here are the key policy recommendations for Qatar:

a. Adopt and Implement UNEP National Guidance:

   (a) Embrace and implement the UNEP National Guidance for Plastic Pollution Hotspotting and Shaping Action as a foundational framework.
   (b) Use the common methodological framework provided by UNEP to assess and prioritize interventions to mitigate plastic pollution.

b. Reduction at Source:

   (a) Implement measures to reduce the use of single-use plastics and encourage the use of sustainable alternatives.
   (b) Target key sources of marine litter, such as fishing nets and microplastics from tires and cosmetics, through regulations and incentives for alternative materials.
   (c) Establish statutory routine monitoring of aquatic litter to assess the effectiveness of pollution reduction measures.

c. Educational Programs:

   (a) Develop and implement long-term educational programs to raise awareness about the impact of marine litter. Plus educational initiatives that focus on reducing litter both at the coast and at sea.
   (b) Target both coastal communities and the broader population to foster a sense of responsibility for resource management and littering.

d. Support for Volunteer Initiatives:

   (a) Provide support and resources for volunteer litter pickers, including the establishment of a structured program.
   (b) Explore opportunities to create a revenue stream for volunteer litter pickers, potentially through recycling incentives or government-funded initiatives.

e. Engage Fishers and Recreational Users:

   (a) Collaborate with fishers and recreational users to promote responsible waste disposal practices.
   (b) Introduce programs such as "fish for litter" with rewards to incentivize proper waste disposal and recycling.

(c) Conduct outreach campaigns to emphasize the importance of bringing litter back to shore. There are fishing for litter campaigns in different countries that could be used as examples to keep any litter brought up in fishing gear back to shore.

f. Port-Wide Waste Management:

(a) Implement comprehensive waste management systems at all ports, including fisher ports, to capture and manage marine litter.
(b) Ensure that waste management infrastructure is easily accessible and properly maintained.

g. Enforcement of Anti-Littering Laws:

(a) Strengthen and enforce laws against littering, with specific penalties for marine litter violations.
(b) Collaborate with law enforcement agencies to monitor and penalize illegal dumping and littering in coastal and marine areas.

h. International Collaboration:

(a) Collaborate with neighboring countries and international organizations to address transboundary marine pollution issues.
(b) Share best practices and lessons learned with the global community to contribute to a collective effort to combat marine pollution.

i. Research and Innovation:

(a) Invest in research and innovation to develop new technologies and solutions for monitoring, preventing, and cleaning up marine litter.
(b) Encourage the private sector to contribute to research and development efforts aimed at reducing plastic pollution.

References

Abdelbary EM, Al Ashwal AA (2021) Distribution and abundance of seagrasses in Qatar Marine Zone. The Arabian seas: biodiversity, environmental challenges and conservation measures, pp 327–362. https://doi.org/10.1007/978-3-030-51506-5_14
Abusin S, Mandikiana BW, Al Emadi N, Al-Boinin F (2022) Socioeconomic drivers of fish consumption in Qatar. Sustainability 14:10921. https://doi.org/10.3390/su141710921
Adams MP, Koh EJ, Vilas MP, Collier CJ, Lambert VM, Sisson SA, O'Brien KR (2020) Predicting seagrass decline due to cumulative stressors. Environ Model Softw 130:104717. https://doi.org/10.1016/j.envsoft.2020.104717
Al Disi ZA, Zouari N, Dittrich M, Jaoua S, Al-Kuwari HAS, Bontognali TRR (2019) Characterization of the extracellular polymeric substances (EPS) of Virgibacillus strains capable of mediating the formation of high Mg-calcite and protodolomite. Mar Chem 216:103693. https://doi.org/10.1016/j.marchem.2019.103693

Al-Abdulrazzak D, Pauly D (2013) From dhows to trawlers: a recent history of fisheries in the Gulf countries, 1950 to 2010. Fish Cent Res Rep 1(2):1–59

Al-Amoudi OSB, Abduljauwad SN (1995) Compressibility and collapse characteristics of arid saline sabkha soils. Eng Geol 39(3–4):185–202. https://doi.org/10.1016/0013-7952(95)00016-9

Al-Ansi M, Priede IG (1996) Expansion of fisheries in Qatar (1980–1992): growth of an artisanal fleet and closure of a trawling company. Fish Res 26(1–2):101–111. https://doi.org/10.1016/0165-7836(95)00396-7

Al-Disi ZA et al (2017) Evidence of a role for aerobic bacteria in high magnesium carbonate formation in the evaporitic environment of Dohat Faishakh Sabkha in Qatar. Front Environ Sci 5:1. https://doi.org/10.3389/fenvs.2017.00001

Al-Khayat J, Balakrishnan P (2014) *Avicennia* marina around Qatar: tree, seedling and pneumatophore densities in natural and planted mangroves using remote sensing. Int J Sci 5. https://ssrn.com/abstract=2573637

Al-Khayat JA, Jones DA (1999) A comparison of the macrofauna of natural and replanted mangroves in Qatar. Estuar Coast Shelf Sci 49:55–63. https://doi.org/10.1016/S0272-7714(99)80009-2

Almahasheer H (2018) Spatial coverage of mangrove communities in the Arabian Gulf. Environ Monit Assess 190(2):85. https://doi.org/10.1007/s10661-018-6472-2

Al-Maslamani I, David S, Giraldes B, Chatting M, Al-Mohannadi M, Le Vay L (2018) Decline in oyster populations in traditional fishing grounds; is habitat damage by static fishing gear a contributory factor in ecosystem degradation? J Sea Res 140:40–51

Alongi DM (2020) Global significance of mangrove blue carbon in climate change mitigation. Science 2(3):67. https://doi.org/10.3390/sci2030067

Alsharhan AS, Kendall CGSC (2003) Holocene coastal carbonates and evaporites of the southern Arabian Gulf and their ancient analogues. Earth Sci Rev 61(3–4):191–243. https://doi.org/10.1016/S0012-8252(02)00110-1

Beck MW, Heck KL Jr, Able KW, Childers DL, Eggleston DB, Gillanders BM, Orth RJ (2001) The identification, conservation, and management of estuarine and marine nurseries for fish and invertebrates: a better understanding of the habitats that serve as nurseries for marine species can help to conserve these areas for future generations. Bioscience 51(8):633–641. https://doi.org/10.1641/0006-3568(2001)051[0633:TICAMO]2.0.CO;2

Belharet M, Di Franco A, Calò A, Mari L, Claudet J, Casagrandi R, Melià P (2020) Extending full protection inside existing marine protected areas, or reducing fishing effort outside, can reconcile conservation and fisheries goals. J Appl Ecol 57(10):1948–1957. https://doi.org/10.1111/1365-2664.13688

Bento R, Jabado RW, Sawaf M, Bejarano I, Samara F, Yaghmour F, Mateos-Molina D (2022) Oyster beds in the United Arab Emirates: important fishing grounds in need of protection. Mar Pollut Bull 182:113992. https://doi.org/10.1016/j.marpolbul.2022.113992

Bertram C, Quaas M, Reusch TBH et al (2021) The blue carbon wealth of nations. Nat Clim Chang 11:704–709. https://doi.org/10.1038/s41558-021-01089-4

Beserra de Lima, N. G., & Galvani, E. (2013). Mangrove microclimate: A case study from southeastern Brazil. Earth Interactions, 17(2), 1-16.

Bishara FA, Haykel B, Hertog S, Holes C, Onley J (2016) The economic transformation of the Gulf. Emerge Gulf States Stud Mod Hist 187. http://hdl.handle.net/10871/20537

Blasiak R, Wynberg R, Grorud-Colvert K et al (2020) The ocean genome and future prospects for conservation and equity. Nat Sustain 3:588–596. https://doi.org/10.1038/s41893-020-0522-9

Böer B, Al Hajiri S (2002) The coastal and sabkha flora of Qatar: an introduction, pp 63–70. In: Barth H-J, Böer B (eds) 2002: Sabkah ecosystems vol I: The Arabian Peninsula and adjacent countries. Tasks for Vegetation Science 36. Kluwer Academic Publishers

Bontognali TRR, Vasconcelos C, Warthmann RJ, Bernasconi SM, Dupraz C, Strohmenger CJ, McKenzie JA (2010) Dolomite formation within microbial mats in the coastal sabkha of Abu Dhabi (United Arab Emirates): Sedimentology 57(3):824–844. https://doi.org/10.1111/j.1365-3091.2009.01121.x

Boussarie G, Bakker J, Wangensteen OS, Mariani S, Kiszka JJ, Manel MS, Robbins WD, Vigliola L, Mouillot D (2018) Environmental DNA illuminates the dark diversity of sharks. Sci Adv 4(5). https://doi.org/10.1126/sciadv.aap9661.

Bouwmeester J, Ben-Hamadou R, Range P, Al Jamali F, Burt J (2022) Spatial patterns of reef fishes and corals in the thermally extreme waters of Qatar. Front Mar Sci 9:989841. https://doi.org/10.3389/fmars.2022.989841

Bouwmeester J, Riera R, Range P, Ben-Hamadou R, Samimi-Namin K, Burt JA (2020) Coral and reef fish communities in the thermally extreme Persian/Arabian Gulf: insights into potential climate change effects. Perspectives on the marine animal forests of the world, pp 63–86. https://doi.org/10.1007/978-3-030-57054-5_3

Bowen RL (1951) The pearl fisheries of the Persian Gulf. Middle East J 5(2):161–180

Brauchli M, McKenzie JA, Strohmenger CJ, Sadooni F, Vasconcelos C, Bontognali TR (2016) The importance of microbial mats for dolomite formation in the Dohat Faishakh sabkha, Qatar. Carbonates Evaporites 31:339–345. https://doi.org/10.1007/s13146-015-0275-0

Buchanan JR, Krupp F, Burt JA, Feary DA, Ralph GM, Carpenter KE (2016) Living on the edge: vulnerability of coral-dependent fishes in the Gulf. Mar Pollut Bull 105(2):480–488. https://doi.org/10.1016/j.marpolbul.2015.11.033

Burt JA, Bartholomew A (2019) Towards more sustainable coastal development in the Arabian Gulf: opportunities for ecological engineering in an urbanized seascape. Mar Pollut Bull 142:93–102. https://doi.org/10.1016/j.marpolbul.2019.03.024

Burt JA, Smith EG, Warren C, Dupont J (2016) An assessment of Qatar's coral communities in a regional context. Mar Pollut Bull 105(2):473–479. https://doi.org/10.1016/j.marpolbul.2015.09.025

Burt JA, Ben-Hamadou R, Abdel-Moati MA, Fanning L, Kaitibie S, Al-Jamali F, Warren CS (2017) Improving management of future coastal development in Qatar through ecosystem-based management approaches. Ocean Coast Manag 148:171–181. https://doi.org/10.1016/j.ocecoaman.2017.08.006

Burt JA, Paparella F, Al-Mansoori N, Al-Mansoori A, Al-Jailani H (2019) Causes and consequences of the 2017 coral bleaching event in the southern Persian/Arabian Gulf. Coral Reefs 38:567–589. https://doi.org/10.1007/s00338-019-01767-y

Burt JA, Range P, Claerboudt M, Al Mealla R, Salimi MA, Salimi PA, Ben-Hamadou R, Bolouki M, Bouwmeester J, Taylor O, Wilson S (2021) Chapter 4. Status and trends of coral reefs of the ROPME Sea Area. In: Status of coral reefs of the world: 2020. International Coral Reef Initiative (ICRI), Global Coral Reef Monitoring Network (GCRMN), Australian Institute of Marine Science (AIMS), United Nations Environment Programme (UNEP), p 12

Butler JD, Purkis SJ, Yousif R, Al-Shaikh I, Warren C (2020) A high-resolution remotely sensed benthic habitat map of the Qatari coastal zone. Mar Pollut Bull 160:111634. https://doi.org/10.1016/j.marpolbul.2020.111634

Butler JD, Purkis LM, Purkis SJ, Yousif R, Al-Shaikh I (2021) A benthic habitat sensitivity analysis of Qatar's coastal zone. Mar Pollut Bull 167:112333. https://doi.org/10.1016/j.marpolbul.2021.112333

Carpenter KE, Krupp F, Jones DA, Zajonz U (1997) FAO species identification field guide for fishery purposes. The living marine resources of Kuwait, Eastern Saudi Arabia, Bahrain, Qatar, and the United Arab Emirates. FAO species identification field guide for fishery purposes. The living marine resources of Kuwait, Eastern Saudi Arabia, Bahrain, Qatar, and the United Arab Emirates

Carr H, Abas M, Boutahar L, Caretti ON, Chan WY, Chapman AS, Zivian A (2020) The aichi biodiversity targets: achievements for marine conservation and priorities beyond 2020. Peer J 8:e9743. https://doi.org/10.7717/peerj.9743

Carter R (2005) The history and prehistory of pearling in the Persian Gulf. J Econ Social History of the Orient 48(2):139–209. https://www.jstor.org/stable/25165089

Castilla AM, Riera R, Humaid MA, Garland T Jr, Alkuwari A, Muzaffar S, Valdeon A (2017) Contribution of citizen science to improve knowledge on marine biodiversity in the Gulf Region. J Assoc Arab Univ Basic Appl Sci 24:126–135. https://doi.org/10.1016/j.jaubas.2017.06.002

Cole VJ, Harasti D, Kirk Dahle S, Russell K (2024) Determining the best practice for Sydney rock oyster, Saccostrea glomerata, reef restoration and enhanced ecological benefits. BMC Ecol Evol 24:114

Coles SL (2003) Coral species diversity and environmental factors in the Arabian Gulf and the Gulf of Oman: a comparison to the Indo-Pacific region. Atoll Res Bull 507:1–19. https://doi.org/10.5479/si.00775630.507.1

Day J, Dudley N, Hockings M, Holmes G, Laffoley D, Stolton S, Wells S (2012) Guidelines for applying the IUCN protected area management categories to marine protected areas. IUCN, Gland, Switzerland, p 36. iucn_categoriesmpa_eng.pdf

de los Santos C, Scott A, Arias-Ortiz A, Jones B, Kennedy H, Mazarrasa I, McKenzie LJ, Mtwana Nordlund L, de la Torre-Castro M, Unsworth RKF, Ambo-Rappe R (2020) Seagrass ecosystem services: assessment and scale of benefits. In Potouroglou M, Grimsditch G, Weatherdon L, Lutz S (eds) Out of the blue: the value of seagrasses to the environment and to people. UN Environment Programme, Nairobi, Kenya, pp 21–35

De Young C (ed) (2006) Review of the state of world marine capture fisheries management: Indian Ocean (n 488). Food & Agriculture Organization

Di Battista JD, Roberts MB, Bouwmeester J et al (2016) A review of contemporary patterns of endemism for shallow water reef fauna in the Red Sea. J Biogeogr 43:423–439. https://doi.org/10.1111/jbi.12649

DiLoreto ZA, Bontognali TR, Al Disi ZA, Al-Kuwari HAS, Williford KH, Strohmenger CJ, Dittrich M (2019) Microbial community composition and dolomite formation in the hypersaline microbial mats of the Khor Al-Adaid sabkhas, Qatar. Extremophiles 23:201–218. https://doi.org/10.1007/s00792-018-01074-4

DiLoreto ZA, Garg S, Bontognali TRR, Dittrich M (2021) Modern dolomite formation caused by seasonal cycling of oxygenic phototrophs and anoxygenic phototrophs in a hypersaline sabkha. Sci Rep 11(1):1–13. https://doi.org/10.1038/s41598-021-83676-1

Duarte CM, Krause-Jensen D (2017) Export from seagrass meadows contributes to marine carbon sequestration. Front Mar Sci 4:13. https://doi.org/10.3389/fmars.2017.00013

Duarte CM, Middelburg JJ, Caraco N (2005) Major role of marine vegetation on the oceanic carbon cycle. Biogeosciences 2(1):1–8. https://doi.org/10.5194/bg-2-1-2005

Duarte CM, Marbà N, Gacia E, Fourqurean JW, Beggins J, Barrón C, Apostolaki ET (2010) Seagrass community metabolism: Assessing the carbon sink capacity of seagrass meadows. Glob Biogeochem Cycles 24(4). https://doi.org/10.1029/2010GB003793

Duffy JE (2006) Biodiversity and the functioning of seagrass ecosystems. Mar Ecol Prog Ser 311:233–250. https://doi.org/10.3354/meps311233

Egerton JP, Al-Ansi M, Abdallah M, Walton M, Hayes J, Turner J, Le Vay L (2018) Hydroacoustics to examine fish association with shallow offshore habitats in the Arabian Gulf. Fish Res 199:127–136. https://doi.org/10.1016/j.fishres.2017.12.002

El Sayed AFM (1996) Effects of overfishing and abandoning bottom trawling on Qatar's fisheries. Qatar Univ Sci J 16(1):173–178

Elisabeth Krumbein W, Gorbushina AA, Holtkamp-Tacken E (2004) Hypersaline microbial systems of sabkhas: examples of life's survival in" extreme" conditions. Astrobiology 4(4):450–459. https://doi.org/10.1089/ast.2004.4.450

Elliff CI, Kikuchi RKP (2017) Ecosystem services provided by coral reefs in a Southwestern Atlantic Archipelago. Ocean Coast Manag 136:49–55. https://doi.org/10.1016/j.ocecoaman.2016.11.021

Erftemeijer P, Shuail D (2012) Seagrass habitats in the Arabian Gulf: distribution, tolerance thresholds and threats. Aquat Ecosyst Health Manage 15:73–83. https://doi.org/10.1080/14634988.2012.668479

Fanning LM, Al-Naimi MN, Range P, Ali ASM, Bouwmeester J, Al-Jamali F, Ben-Hamadou R (2021) Applying the ecosystem services-EBM framework to sustainably manage Qatar's coral reefs and seagrass beds. Ocean Coast Manag 205:105566. https://doi.org/10.1016/j.ocecoaman.2021.105566

FAO (2022) GLOBEFISH highlights–international markets on fisheries and aquaculture products. Quarterly update. 1st issue 2022, with Jan.–Sep. 2021 Statistics–Globefish Highlights No. 1–2022. Rome. https://doi.org/10.4060/cc0222en

Feary DA, Burt JA, Bauman AG, Al Hazeem S, Abdel-Moati MA, Al-Khalifa KA, Wiedenmann J (2013) Critical research needs for identifying future changes in Gulf coral reef ecosystems. Mar Pollut Bull 72(2):406–416. https://doi.org/10.1016/j.marpolbul.2013.02.038

Ferreira GVB, Lima ARA, Barletta M (2020) Environmental factors affecting the nursery function for fish in the main estuaries of the Gulf of Cadiz (south-west Iberian Peninsula). Mar Environ Res 162:105115

Fourqurean JW, Duarte CM, Kennedy H, Marbà N, Holmer M, Mateo MA, Serrano O (2012) Seagrass ecosystems as a globally significant carbon stock. Nat Geosci 5(7):505–509. https://doi.org/10.1038/ngeo1477

Gell FR, Roberts CM (2003) Benefits beyond boundaries: the fishery effects of marine reserves. Trends Ecol Evol 18:148–155. https://doi.org/10.1016/S0169-5347(03)00189-7

Gil JD, Daioglou V, van Ittersum M, Reidsma P, Doelman JC, van Middelaar CE, van Vuuren DP (2019) Reconciling global sustainability targets and local action for food production and climate change mitigation. Glob Environ Chang 59:101983. https://doi.org/10.1016/j.gloenvcha.2019.101983

Gilma Beserra de Lima N, Galvan E (2013) Mangrove microclimate: a case study from Southeastern Brazil. Earth Interact 17(2):1. https://doi.org/10.1175/2012EI000464.1

Giraldes BW, Smyth D, Chatting M, Al-Ashwel AA, Al-Omary NH, Mello L, Leitão A (2023) Increasing knowledge to restore oyster beds and related services in the Arabian-Persian Gulf. Reg Stud Mar Sci 66:103172. https://doi.org/10.1016/j.rsma.2023.103172

Grabowski JH, Peterson CH (2007) Restoring oyster reefs to recover ecosystem services. Ecosyst Eng Plants Protists 4:281–298

Grabowski JH, Brumbaugh RD, Conrad RF, Keeler AG, Opaluch JJ, Peterson CH, Smyth AR (2012) Economic valuation of ecosystem services provided by oyster reefs. Bioscience 62(10):900–909. https://doi.org/10.1525/bio.2012.62.10.10

Grandcourt E (2012) Reef fish and fisheries in the Gulf. In: Coral reefs of the Gulf: adaptation to climatic extremes. Springer Netherlands, Dordrecht, pp 127–161

Groud-Colvert K, Sullivan-Stack J, Roberts C, Constant V, Horta e Costa B, Pike EP, Lubchenco J (2021) The MPA guide: a framework to achieve global goals for the ocean. Science 373(6560):eabf0861. https://doi.org/10.1126/science.abf0861

Hameed A, Böer B, Clüsener-Godt M, Gul B (2024) Sabkha ecosystems as blue carbon reserves. In: Clüsener-Godt M, Matsuda H, Böer B, Loughland R (eds) Blue carbon mangrove ecosystems. Springer Nature, pp 59–71

Harden-Davies H (2020) The exploitation of deep-sea biodiversity: components, capacity and conservation. In: Baker M, Ramírez-Llodra E, Tyler P (eds) Natural capital and exploitation of the deep ocean. Oxford University Press, pp 127–139

Heck Jr KL, Nadeau DA, Thomas R (1997) The nursery role of seagrass beds. Gulf Mexico Sci 15(1):8. https://doi.org/10.18785/goms.1501.08

Henderson CJ, Stevens T, Lee SY, Gilby BL, Schlacher TA, Connolly RM, Olds AD (2019) Optimising seagrass conservation for ecological functions. Ecosystems 22:1368–1380. https://doi.org/10.1007/s10021-019-00343-3

Hightower VP (2013) Pearls and the Southern Persian/Arabian Gulf: a lesson in sustainability. Environ Hist 18:44–59. https://www.jstor.org/stable/41721477

Hilborn R (2018) Are MPAs effective? ICES J Mar Sci 75(3):1160–1162. https://doi.org/10.1093/icesjms/fsx068

Hosseini, H., Saadaoui, I., Moheimani, N., Al Saidi, M., Al Jamali, F., Al Jabri, H., & Hamadou, R. B. (2021). Marine health of the Arabian Gulf: Drivers of pollution and assessment approaches focusing on desalination activities. Marine Pollution Bulletin, 164, 112085.

Hughes A, Bonačić K, Cameron T, Collins K, da Costa F, Debney A, van Duren L, Elzinga J, Fariñas-Franco JM, Gamble C, Helmer L, Holbrook Z, Holden E, Knight K, Murphy AJ, Pogoda B, Pouvreau S, Preston J, Reid A, Reuchlin-Hugenholtz E, Sanderson WG, Smyth D, Stechele B, Strand A, Theodorou JA, Uttley M, Wray B, zu Ermgassen PSE (2023) Site selection for European native oyster (Ostrea edulis) habitat restoration projects: an expert-derived consensus. Aquat Conserv Mar Freshw Ecosyst 33:721–736. https://doi.org/10.1002/aqc.3917

Illing LV, Wells AJ, Taylor JCM (1965) Penecontemporary dolomite in the Persian Gulf. In: Pray LC, Murray LC (eds) Dolomitazation and limestone diagenesis. Volume SEPM special publication 13, pp 89–111

Jabado RW, Al Ghais SM, Hamza W, Shivji MS, Henderson AC (2015) Shark diversity in the Arabian/Persian Gulf higher than previously thought: insights based on species composition of shark landings in the United Arab Emirates. Mar Biodivers 45:719–731. https://doi.org/10.1007/s12526-014-0275-7

Jabado RW, Spaet JL (2017) Elasmobranch fisheries in the Arabian seas region: characteristics, trade and management. Fish Fisheries 18(6):1096–1118

Jackson EL, Rowden AA, Attrill MJ, Bossey SJ, Jones MB (2001) The importance of seagrass beds as a habitat for fishery species. Oceanogr Mar Biol 39:269–304

Jackson R, Gabric AJ, Cropp R (2020) Coral reefs as a source of climate-active aerosols. PeerJ 8:e10023. https://doi.org/10.7717/peerj.10023

Jahromi FA, Keshavarzi B, Moore F, Abbasi S, Busquets R, Hooda PS, Jaafarzadeh N (2021) Source and risk assessment of heavy metals and microplastics in bivalves and coastal sediments of the Northern Persian Gulf, Hormogzan Province. Environ Res 196:110963. https://doi.org/10.1016/j.envres.2021.110963

Jaspars M, de Pascale D, Andersen JH, Reyes F, Crawford AD, Ianora A (2016) The marine biodiscovery pipeline and ocean medicines of tomorrow. J Mar Biol Assoc UK 96:151–158. https://doi.org/10.1017/S0025315415002106

Jabado, R. W., & Spaet, J. L. (2017). Elasmobranch fisheries in the Arabian Seas Region: Characteristics, trade and management. Fish and Fisheries, 18(6), 1096-1118.

John VC, Coles SL, Abozed AI (1990) Seasonal cycles of temperature, salinity and water masses of the western Arabian Gulf. Oceanol Acta 13(3):273–281

Khalil AS (2015) Mangroves of the red sea.The Red Sea: the formation, morphology, oceanography and environment of a young ocean basin, pp 585–597. https://doi.org/10.1007/978-3-662-45201-1_33

Licciardi G, Böer B, Schwarze H, Aspinall S (2007) Tentative list of world heritage sites in the State of Qatar. UNESCO, Doha, p 23p

Lincoln S, Buckley P, Howes EL, Maltby KM, Pinnegar JK, Ali TS, Alosairi Y, Al-Ragum A, Baglee A, Balmes CO, Hamadou RB, Burt JA, Claereboudt M, Glavan J, Mamiit RJ, Naser HA, Sedighi O, Shokri MR, Shuhaibar B, Wabnitz CCC, Le Quesne WJF (2021) A regional review of marine and coastal impacts of climate change on the ROPME sea area. Sustainability 13(24):13810. https://doi.org/10.3390/su132413810

Lovelock CE, Barbier E, Duarte CM (2022) Tackling the mangrove restoration challenge. PLoS Biol 20(10):e3001836. https://doi.org/10.1371/journal.pbio.3001836

Luo YY, Not C, Cannicci S (2021) Mangroves as unique but understudied traps for anthropogenic marine debris: a review of present information and the way forward. Environ Pollut 271:116291. https://doi.org/10.1016/j.envpol.2020.116291

Marshall CD, Al Ansi M, Dupont J, Warren C, Al Shaikh I, Cullen J (2018) Large dugong (Dugong dugon) aggregations persist in coastal Qatar. Mar Mammal Sci 34(4):1154–1163. https://doi.org/10.1111/mms.12497

Massetti E, Guerriero L, Baldantoni D, Gratani L (2013) Mangroves as thermo-regulators of the coastal environment: a case study in Southern Italy. iForest-Biogeosciences For 6:241. https://doi.org/10.1016/j.wse.2022.10.004

Mateos-Molina D, Lamine EB, Antonopoulou M, Burt JA, Das HS, Javed S, Giakoumi S (2021) Synthesis and evaluation of coastal and marine biodiversity spatial information in the United

Arab Emirates for ecosystem-based management. Mar Pollut Bull 167:112319. https://doi.org/10.1016/j.marpolbul.2021.112319

McCook L, Jompa J, Diaz-Pulido G (2001) Competition between corals and algae on coral reefs: a review of evidence and mechanisms. Coral Reefs 19:400–417. https://doi.org/10.1007/s00338 0000129

McLeod IM, Boström-Einarsson L, Creighton C, D'Anastasi B, Diggles B, Dwyer PG (2019) Habitat value of Sydney rock oyster (Saccostrea glomerata) reefs on soft sediments. Mar Freshw Res. https://doi.org/10.1071/MF18197

Ministry of Development Planning and Statistics (2017) Ministry of development planning and statistics. Environment Statistics Report 2017 Doha–Qatar

Mitra A, Mitra A (2020) Ecosystem services of mangroves: an overview. Mangrove forests in India: exploring ecosystem services, pp 1–32. https://doi.org/10.1007/978-3-030-20595-9_1

Moberg F, Folke C (1999) Ecological goods and services of coral reef ecosystems. Ecol Econ 29:215–233. https://doi.org/10.1016/S0921-8009(99)00009-9

Naser HA (2013) Assessment and management of heavy metal pollution in the marine environment of the Arabian Gulf: a review. Mar Pollut Bull 72(1):6–13. https://doi.org/10.1016/j.marpolbul.2013.04.030

Naser HA (2014) Marine ecosystem diversity in the Arabian Gulf: threats and conservation. In: Biodiversity–the dynamic balance of the planet, pp 297–328. https://doi.org/10.5772/57425

Nordlund LM, Jackson EL, Nakaoka M, Samper-Villarreal J, Beca-Carretero P, Creed JC (2018) Seagrass ecosystem services–what's next? Mar Pollut Bull 134:145–151. https://doi.org/10.1016/j.marpolbul.2017.09.014

Norton JD, Allen S, Abdelmajid M, Alsafran BB, Richer R (2009) An illustrated plant species checklist for Qatar. Browndown Publications, Gosport, p 96

Otero V, Lucas R, Van De Kerchove R, Satyanarayana B, Mohd-Lokman H, Dahdouh-Guebas F (2020) Spatial analysis of early mangrove regeneration in the Matang Mangrove Forest Reserve, Peninsular Malaysia, using geomatics. For Ecol Manage 472:118213. https://doi.org/10.1016/j.foreco.2020.118213

Paparella F, Xu C, Vaughan GO, Burt JA (2019) Coral bleaching in the Persian/Arabian Gulf Is modulated by summer winds. Front Mar Sci 6:205. https://doi.org/10.3389/fmars.2019.00205

Pascal N, Allenbach M, Brathwaite A, Burke L, Le Port G, Clua E (2016) Economic valuation of coral reef ecosystem service of coastal protection: a pragmatic approach. Ecosyst Serv 21(Part A):72–80. https://doi.org/10.1016/j.ecoser.2016.07.005.

Pendleton LH, Ahmadia GN, Browman HI, Thurstan RH, Kaplan DM, Bartolino V (2018) Debating the effectiveness of marine protected areas. ICES J Mar Sci 75(3):1156–1159. https://doi.org/10.1093/icesjms/fsx154

Pergent G, Romero J, Pergent-Martini C, Mateo MA, Boudouresque CF (1994) Primary production, stocks and fluxes in the Mediterranean seagrass Posidonia oceanica. Mar Ecol Prog Ser 139–146. https://www.jstor.org/stable/24844820

Pernot O, Abu-Dieyeh MH, Simon L, Al-Khyatt J, Al-Ghouti M (2017) Human and avicennia marina mangrove populations: with special reference to Qatar. Arab World Geogr 20(2–3):208–236. https://doi.org/10.5555/1480-6800.20.2.208

Perri E, Tucker ME, Słowakiewicz M, Whitaker F, Bowen L, Perrotta ID (2018) Carbonate and silicate biomineralization in a hypersaline microbial mat (Mesaieed sabkha, Qatar): roles of bacteria, extracellular polymeric substances and viruses. Sedimentology 65(4):1213–1245. https://doi.org/10.1111/sed.12419

Phillips RC (2003) The Arabian Gulf and Arabian Region. World Atlas of Seagrasses. UNEP WCMC, London, pp 74–81

Phillips A, World Conservation Union (2002) Management guidelines for IUCN category V protected areas: protected landscapes/seascapes, vol 9. IUCN--the World Conservation Union

Pilcher NJ, Al-Maslamani I, Williams J, Gasang R, Chikhi A (2015) Population structure of marine turtles in coastal waters of Qatar. Endanger Species Res 28(2):163–174. https://doi.org/10.3354/esr00688

Planning and Statistics Authority. (2021). Proportion of fish stocks within biologically sustainable levels (Indicator 14.4.1). Planning and Statistics Authority, State of Qatar. https://gis.psa.gov.qa/GISApps/SDG/Data/PDF/En/SDG14.pdf

Pittman SJ, Yates KL, Bouchet PJ, Williams S (2017) Seascape ecology: identifying research priorities for an emerging ocean sustainability science. Mar Ecol Prog Ser 574:1–19

Planning & Stastics Authority (2018) Qatar Second National Development Strategy 2018 to 2022. NDS2Final.pdf

Pountos I, Giannoudis P (2016) Is there a role of coral bone substitutes in bone repair? Injury 47(12):2606–2613. https://doi.org/10.1016/jinjury.2016.10.025

Preen A (2004) Distribution, abundance and conservation status of dugongs and dolphins in the southern and western Arabian Gulf. Biol Cons 118(2):205–218. https://doi.org/10.1016/j.biocon.2003.08.014

Price AR, Coles SL (1992) Aspects of seagrass ecology along the western Arabian Gulf coast. Hydrobiologia 234:129–141. https://doi.org/10.1007/BF00014245

Rabaoui L, Roa-Ureta RH, Yacoubi L, Lin YJ, Maneja R, Joydas TV, Qurban MA (2021) Diversity, distribution, and density of marine mammals along the Saudi waters of the Arabian gulf: update from a multi-method approach. Front Mar Sci 8:687445. https://doi.org/10.3389/fmars.2021.687445

Rice J, Moksness E, Attwood C, Brown SK, Dahle G, Gjerde KM, Westlund L (2012) The role of MPAs in reconciling fisheries management with conservation of biological diversity. Ocean Coast Manag 69:217–230. https://doi.org/10.1016/j.ocecoaman.2012.08.001

Riegl B (2002) Effects of the 1996 and 1998 positive sea-surface temperature anomalies on corals, coral diseases and fish in the Arabian Gulf (Dubai, UAE). Mar Biol 140:29–40. https://doi.org/10.1007/s002270100676

Rodrigues-Filho JL, Macêdo RL, Sarmento H, Pimenta VR, Alonso C, Teixeira CR, Pagliosa PR, Netto SA, Santos NC, Daura-Jorge FG, Rocha O (2023) From ecological functions to ecosystem services: linking coastal lagoons biodiversity with human well-being. Hydrobiologia 850:2611–2653. https://doi.org/10.1007/s10750-023-05171-0

Romañach SS, DeAngelis DL, Koh HL, Li Y, Teh SY, Barizan RSR, Zhai L (2018) Conservation and restoration of mangroves: global status, perspectives, and prognosis. Ocean Coast Manag 154:72–82. https://doi.org/10.1016/j.ocecoaman.2018.01.009

Saifullah SS (1996) Mangrove ecosystem of Saudi Arabian Red Sea coast-an overview. Mar Sci 7:263–270

Samara F, Bejarano I, Mateos-Molina D, Abouleish M, Solovieva N, Yaghmour F, Saburova M (2023) Environmental assessment of oyster beds in the northern Arabian Gulf Coast of the United Arab Emirates. Mar Pollut Bull 195:115442. https://doi.org/10.1016/j.marpolbul.2023.115442

Sánchez-Vidal A, Canals M, de Haan WP et al (2021) Seagrasses provide a novel ecosystem service by trapping marine plastics. Sci Rep 11:254. https://doi.org/10.1038/s41598-020-79370-3

Sequeira A et al (2025) Global tracking of marine megafauna space use reveals how to achieve conservation targets. Science 388(6751):1086–1097

Shark Conservation Society (2009) Qatar expedition-shark conservation society expeditions

Sheppard C, Al-Husiani M, Al-Jamali F, Al-Yamani F, Baldwin R, Bishop J, Benzoni F, Dutrieux E, Dulvy NK, Durvasula SRV, Jones DA, Loughland R, Medio D, Nithyanandan M, Pillingm GM, Polikarpov I, Price ARG, Purkis S, Riegl B, Saburova M, Samimi Namin K, Taylor O, Wilson S, Zainal Z (2010) The gulf: a young sea in decline. Mar Pollut Bull 60:13–38. https://doi.org/10.1016/j.marpolbul.2009.10.017

Sherman RE, Fahey TJ, Battles JJ (2000) Small-scale disturbance and regeneration dynamics in a neotropical mangrove forest. J Ecol 88(1):165–178. https://doi.org/10.1046/j.1365-2745.2000.00439.x

Shinn EA (1976) Coral reef recovery in Florida and the Persian Gulf. Environ Geol 1(4):241–254. https://doi.org/10.1007/BF02407510

Smyth D, Al-Maslamani I, Chatting M, Giraldes B (2016) Benthic surveys of the historic pearl oyster beds of Qatar reveal a dramatic ecological change. Mar Pollut Bull 113(1–2):147–155. https://doi.org/10.1016/j.marpolbul.2016.08.085

Stamatopoulos C, Abdallah M (2015) Standardization of fishing effort in Qatar fisheries: methodology and case studies. J Mar Sci Res Dev 5(3):1

Torquato F, Jensen HM, Range P, Bach SS, Ben-Hamadou R, Sigsgaard EE, Riera R (2017) Vertical zonation and functional diversity of fish assemblages revealed by ROV videos at oil platforms in The Gulf. J Fish Biol 91(3):947–967. https://doi.org/10.1111/jfb.13394

Ullah H, Nagelkerken I, Goldenberg SU, Fordham DA (2018) Climate change could drive marine food web collapse through altered trophic flows and cyanobacterial proliferation. PLoS Biol 16(1):e2003446. https://doi.org/10.1371/journal.pbio.2003446

UNEP (2016) GEO-6 regional assessment for West Asia. United Nations Environment Programme, Nairobi, Kenya

UNEP (2020) Out of the blue: the value of seagrasses to the environment and to people. UNEP, Nairobi. https://www.unep.org/resources/report/out-blue-value-seagrasses-environment-and-people.

UNESCO-IOC/European Commission (2021a) MSPglobal policy brief: marine spatial planning and the sustainable blue economy. Paris, UNESCO. (IOC Policy Brief n° 2). https://unesdoc.unesco.org/ark:/48223/pf0000375720

UNESCO-IOC/European Commission (2021b) MSPglobal international guide on marine/maritime spatial planning. Paris, UNESCO. (IOC Manuals and Guides n° 89)

Unsworth RK, Nordlund LM, Cullen-Unsworth LC (2019) Seagrass meadows support global fisheries production. Conserv Lett 12(1):e12566. https://doi.org/10.1111/conl.12566

Van Lavieren H, Klaus R (2013) An effective regional marine protected area network for the ROPME sea area: unrealistic vision or realistic possibility? Mar Pollut Bull 72(2):389–405. https://doi.org/10.1016/j.marpolbul.2012.09.004

Vaughan GO, Al-Mansoori N, Burt JA (2019) The Arabian Gulf. In World seas: an environmental evaluation. Academic Press, pp 1–23. https://doi.org/10.1016/B978-0-08-100853-9.00001-4

Wabnitz CC, Lam VW, Reygondeau G, Teh LC, Al-Abdulrazzak D, Khalfallah M, Cheung WW (2018) Climate change impacts on marine biodiversity, fisheries and society in the Arabian Gulf. PLoS ONE 13(5):e0194537. https://doi.org/10.1371/journal.pone.0194537

Ward-Paige CA (2017) A global overview of shark sanctuary regulations and their impact on shark fisheries. Mar Policy 82:87–97

Warren JK (2006) Evaporites: sediments. Berlin, Heidelberg Springer-Verlag, Berlin Heidelberg, Resources and Hydrocarbons

Wells AJ (1962) Recent Dolomite in the Persian Gulf: Nature 194(4825):274–275

Wells S, Addison PF, Bueno PA, Costantini M, Fontaine A, Germain L, Zorrilla MX (2016) Using the IUCN green list of protected and conserved areas to promote conservation impact through marine protected areas. Aquat Conserv Mar Freshwat Ecosyst 26:24–44. https://doi.org/10.1002/aqc.2679

Woodhead AJ, Hicks CC, Norström AV, Williams GJ, Graham NAJ (2019) Coral reef ecosystem services in the Anthropocene. Funct Ecol 33:1023–1034. https://doi.org/10.1111/1365-2435.13331

WorldBank (2023) What you need to know about blue carbon and its importance. The World Bank. https://www.worldbank.org/en/news/feature/2023/11/21/what-you-need-to-know-about-blue-carbon

Open Access This chapter is licensed under the terms of the Creative Commons Attribution 4.0 International License (http://creativecommons.org/licenses/by/4.0/) , which permits use, sharing, adaptation, distribution and reproduction in any medium or format, as long as you give appropriate credit to the original author(s) and the source, provide a link to the Creative Commons license and indicate if changes were made.

The images or other third party material in this chapter are included in the chapter's Creative Commons license, unless indicated otherwise in a credit line to the material. If material is not included in the chapter's Creative Commons license and your intended use is not permitted by statutory regulation or exceeds the permitted use, you will need to obtain permission directly from the copyright holder.

Chapter 7
Economic Valuation of Qatar's Ecology

M. Contestabile, J. Bruder, M. Ali, and N. Hamwy

Abstract Policies to protect and restore Qatar's natural environment must be informed by sufficient knowledge on how local ecosystems function. They must also be informed by an understanding of the costs of action against the potential costs of inaction. This chapter introduces the economic argument for policy intervention to protect and restore Qatar's environment and formulates recommendations for creating a sound evidence base to inform environmental policy in Qatar. The chapter begins with a brief historical account of how nature, initially considered a provider of free commodities, was gradually introduced into how we think about economic systems, and how this informs the latest perspectives on sustainable development and sustainability. The chapter then discusses the theory and practice of economic valuation of ecosystems and the services they provide, and applies these ideas to Qatar. After situating Qatar's environmental protection and restoration ambitions in the context of existing national policy, the chapter provides a review of existing knowledge on Qatar's ecosystems, the services they provide and their economic value, highlighting the substantial gaps that still exist. The chapter concludes with recommendations for how to build the evidence base for policymakers on the economic value of Qatar's environment.

Keywords Ecosystem services · Natural capital · Environmental economics · Policy valuation · Sustainable development · Economic resilience

M. Contestabile · M. Ali (✉)
Earthna, Center for a Sustainable Future, Qatar Foundation, Doha, Qatar
e-mail: muezali@qf.org.qa

M. Contestabile
e-mail: marcello.contestabile@googlemail.com

J. Bruder · N. Hamwy
Carnegie Mellon University in Qatar, Doha, Qatar
e-mail: jbruder@andrew.cmu.edu

© The Author(s) 2026

A. D. Chatziefthimiou et al. (eds.), *Pathways to Nature Conservation and Resilience in Hot and Arid Lands*, Advances in Geographical and Environmental Sciences,
https://doi.org/10.1007/978-3-032-11046-6_7

7.1 Introduction

Policies to protect and restore the natural environment of Qatar must be informed by the relevant scientific knowledge on how local ecosystems function. Decisions on policy choices must consider the costs of action against the potential economic costs of inaction. Given the trade-offs inherent to any policy making process, developing environmental policies that are both effective—can be expected to achieve the intended goal, is efficient and also achieve the said goal at the lowest possible cost—requires, among other things, combining insights from both ecology and economics.

Chapters 4, 5 and 6 in this report describe Qatar's natural environment and present the science explaining the complex functioning of its ecosystems, and the interactions between different ecosystems and between ecosystems and human activities. This chapter introduces the economic argument for policy intervention to protect and restore Qatar's environment, presents existing evidence, and sets out a policy and economic research agenda to inform future policymaking.

The chapter begins with a brief historical account of the emergence of the fields of environmental and ecological economics to illustrate how *nature*, initially considered a provider of free commodities, was gradually introduced into the field of economics, and how *nature*, as a finite natural resource, is informing the latest perspectives on sustainable development and sustainability. This is followed by a discussion of the economic value of the environment, the services it provides, and the methods used to assess the economic value of these services. Then an overview of Qatar's environment and ecosystems is presented, followed by the role of the environment in its National Vision and Development Strategy, the specificities of its environment, particularly with emphasis on its desert, coastal and marine ecosystems, and the evidence available on the value of its services. Finally, a list of recommendations for creating a sound evidence base to inform environmental policy in Qatar is provided.

7.2 The Economics of the Environment: An Overview[1]

7.2.1 *The Evolving Role of Nature in Economic Theories*

Classical economics of the eighteenth and nineteenth centuries viewed land as a factor of production, alongside capital and labour. Capital included all assets, such as machines that could be used in the production of goods and services without being consumed by it. Land, on the other hand, included all natural resources that can be found on the Earth's surface or under it. *Nature*, therefore, was characterised

---

[1] The majority of the material covered in the following section draws from the work of Dasgupta (2021) and Bukhard and Maes (2017). When other sources were used, they are cited as appropriate in the text.

in economic theory as the provider of free commodities, such as water, timber and minerals, the supply of which could not be fundamentally affected by humans.

This view of nature did not fundamentally change in the neoclassical economic theories of the late nineteenth and twentieth century, which redefined capital as any asset that can produce a stream of income over time. This reformulation introduced the idea of capital as a stock that generates a flow of goods and services which can generate income. In this framework, *nature* can also be classified as capital. More precisely, as a stock of resources producing a flow of commodities that can be used as inputs in the production of goods and services. As a free stock with uncontrollable flows of commodities, *nature* did not receive much attention by neoclassical economists either.

Neoclassical economics is centered around the idea that markets are the most efficient mechanism for resource allocation, including natural resources. Since the 1950s, neoclassical economic theory focused on the determinants of economic growth. A new paradigm developed that saw continued economic growth become a key objective of governments, with all that this entails in relation to the unrestrained exploitation of the environment.

In the 1970s, the finiteness of nature and the negative impacts of overexploitation of natural resources on the environment emerged as a counter force to the underlying assumptions in neoclassical economic theory. More specifically, the report 'Limits to Growth' (Meadows et al. 1972), explored model scenarios showing how excessive demographic and economic growth are bound to result in overshooting the Earth's carrying capacity and ultimately lead to demographic and economic collapse. In 1973, the economist E. F. Schumacher in his book 'Small is beautiful' (Schumacher 1973), introduces the term 'natural capital', acknowledging the economic value of nature. 'Ecosystem services', as both a concept and a term, was coined in the 1980s, but it did not become widely used until a decade later (Daily 1997).

Building on the growing body of evidence on the impact of environmental pollution on social and economic development, environmental economics emerged in the late 1960s as a sub-field of economics. Environmental economics, rooted in neoclassical economics, focused on studying the correct pricing of the flows of goods and services provided by *nature*—referred to today as ecosystem services—based on the observation that their market price, when they had one, did not fully reflect their value to society. In environmental economics, the difference between market price and societal value reflects the inability of markets to correctly price ecosystem services. This is referred to as an 'externality'. This called for governments to correct markets by internalizing such externalities through the introduction of suitable regulation. Ensuring the correct pricing of ecosystem services also required developing methods to estimate their real societal value—which environmental economists call 'accounting value' or 'shadow price'—thus the emergence of a new field of research called environmental valuation.

In the early 1980s, a new, related field emerged: ecological economics. Ecological economics builds on the theoretical foundations of neoclassical economics and makes use of concepts developed by environmental economists. However, while environmental economics adds a quantitative characterization of nature and its services to the neoclassical representation of the economy, ecological economics provides a fundamental change of perspective by placing the economy within the broader boundaries of *nature*. The difference is substantial. Environmental economics attaches economic value to the natural environment and studies the economic impact of environmental policies, implying that a certain extent of degradation of natural capital can be accepted, if the overall benefit to the economy exceeds the cost. This perspective is also often referred to as 'weak sustainability'. Ecological economics, on the other hand, by subordinating the economic system to nature, implies that natural capital is essential to human existence and, therefore, cannot be traded with other forms of capital. This perspective is also referred to as 'strong sustainability'.

Despite these foundational differences, both approaches are framed around natural capital and ecosystem services, concepts that have garnered much attention in economics over the last two decades. In contrast to classical economics, which only considered capital as the total stock of man-made assets, contemporary economics identifies three different types of capital: produced capital, human capital and natural capital.

The value of a nation's wealth, now measured by produced, human and natural capital, provides a measure of the economy that is complementary to Gross Domestic Product (GDP) and can be used as a measure of sustainable development. This new measure of wealth is called 'inclusive wealth' (Polasky et al. 2015). Within this framework, a necessary, but not sufficient, condition for sustainable development is that inclusive wealth increases over time. The practical use of inclusive wealth was made possible, among other things, by the development of methods for measuring the value of human and natural capital, either directly or through the value of the services they provide.

Under this framework, the three forms of capital are all essential for human prosperity. Each one contributes to economic activity through the services it provides. Natural capital, for example, provides ecosystem services. The services provided by each form of capital can also contribute to building up other forms of capital, sometimes at the expense of degrading one form of capital to produce more of another. However, no form of capital can be directly traded with another and they are not directly substitutable because they perform complementary roles. The interdependencies between the three forms of capital, the services they provide and human activities that can degrade natural capital are represented schematically in Fig. 7.1.

Natural capital is made up of both living organisms, often referred to as biodiversity, and the non-living chemical and physical parts of the environment. Together, living and non-living components of the environment form ecosystems. Biodiversity is particularly important because it is a critical indicator of the value of an ecosystem. The higher the biodiversity, the more resilient an ecosystem is to external perturbations and the higher its productivity in delivering services. A loss of biodiversity, therefore, corresponds to ecosystem degradation and loss of value. The importance of the non-living parts of ecosystems should not be understated. In the literature, they

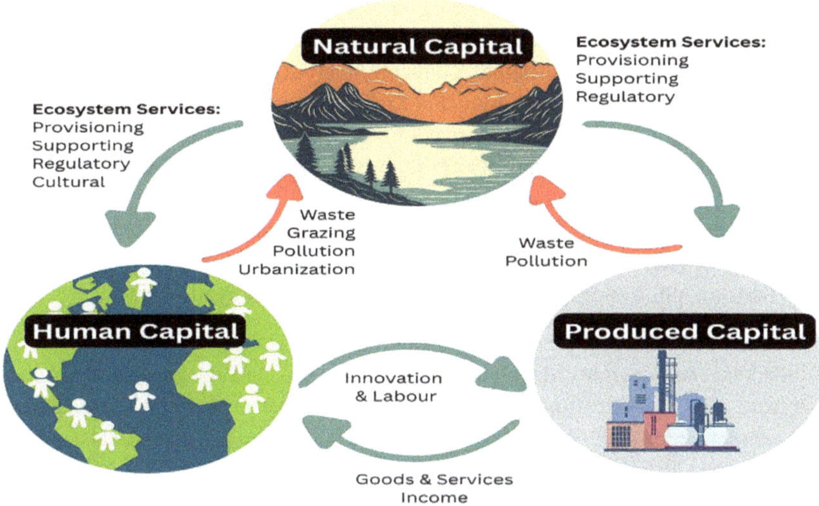

Fig. 7.1 The three forms of capital—produced, human and natural—that constitute inclusive wealth; green arrows represent the flows of goods and services between them; red arrows are strains put on natural capital (*Source* adapted from Dasgupta 2021)

are sometimes treated separately and referred to as 'geosystems', and the services they provide as 'geosystem services'.[2]

Ecosystem services are crucial for human economic activity and the development of produced and human capital. There are many definitions for ecosystem services, some emphasizing the important role they play in sustaining human life, others highlighting their value in economic activity (Costanza et al. 1997; Brauman 2015). However, all definitions identify ecosystem services as the link between natural capital and other forms of capital, and the vehicle through which the natural world contributes to human wellbeing. For example, in the report 'The economics of ecosystems and biodiversity' (TEEB 2010), ecosystem services are defined as "the direct and indirect contributions of ecosystems to human well-being".

The Common International Classification of Ecosystem Services–CICES–(Haines-Young and Potschin 2017), an adaptation of the pioneering Millennium Ecosystem Assessment (2005), is currently the most widely used classification system adopted by both practitioners and academics. CICES identifies the following four broad categories of ecosystem services:

– *Provisioning services*: the goods that humans harvest or extract from ecosystems, such as food, water, raw materials for construction and for a multitude of manufacturing processes, fuels (both fossil, such as oil, and non-fossil, such as wood) and medicines.

---

[2] For the purposes of this chapter, the term *ecosystem services* will be used to describe the services provided by both the living and non-living parts of the ecosystems. The merging of these concepts is increasingly viewed as a global best practice (Gray 2011).

- *Cultural services*: the non-material benefits that humans obtain from ecosystems through experience, interaction and spiritual enrichment, cognitive development, improved mental-health, connection with one's heritage, ethical values, eco-tourism, education, recreation and aesthetic experiences.
- *Regulating and maintenance services*: the benefits that humans obtain from nature's ability to regenerate itself, absorb waste and pollutants, and regulate the quality of air, water and soil.
- *Supporting services*: the benefits derived from the functions performed by nature that are critical to habitats for all species, including humans. For example, pollination, photosynthesis, soil formation and nutrient cycling.

Within each category there are multiple types of services. The type of service included in each category depends on the ecosystem and its geospatial location. Therefore, any classification system must be applied critically and adapted to the environment under study (for more details, see Hassan et al. 2005). A classification adapted from CICES is used in this chapter.[3]

In any given ecosystem, ecosystem services are intrinsically linked to one another and, therefore, it is not possible to increase one indefinitely without interfering with the others. Ecosystems are self-regulating but only within certain bounds. For example, over-use of provisioning services will result in physical degradation of the ecosystem—which, from an economic perspective, is a loss of natural capital—and will reduce the ecosystem's ability to provide cultural, regulating and supporting services.

7.2.2 Environmental Valuation: Theory and Practice

Understanding the role of nature in the economy—or of the economy within nature—is vital for charting a sustainable path for human and economic development. Central to this understanding is both the finiteness and value of natural capital, as opposed to *nature* being a free provider of services not affected by human activity. To gauge whether a national development strategy or an individual project is sustainable, a monetary value must be assigned to the potential impacts on natural capital that the proposed plan or project entails.

Environmental valuation is a collection of methods to assign economic value to natural capital and ecosystem services in the absence, or existence, of market prices. These methods, however, do not necessarily reflect the accounting value of natural capital. Even with the broad set of methods available to practitioners and researchers, environmental valuation is a rapidly evolving field in both theory and practice.

Despite noticeable growth in adoption and popularity of certain environmental valuation methods, pricing nature and its services remains controversial among

---

[3] Despite the popularity of CICES as a classification system, due to the circumstances under which it was developed, it under-represents the services of marine and coastal environments. This is particularly problematic for Qatar's ecosystem.

economists, especially ecological economists. One argument against pricing *nature* is that doing so implies that natural capital can be traded for other forms of capital, which overlooks certain facts. First, a large proportion of natural capital is not substitutable, and degradation of nature, and the services it provides, threatens human existence. Second, due to our limited understanding of the functioning of complex ecosystems, there is a real risk of underestimating the value of the services ecosystems provide. Ecological economists have instead been exploring non-monetary and non-market approaches to the economics of ecosystems and the goods and services they provide. On the other hand, one argument in favour of pricing the environment is that doing so, at the very least, includes nature in economic decision-making.

The need for environmental valuation is centered around the premise that markets are not effective at correctly pricing nature. First, most natural capital and ecosystem services can be accessed at no cost because they have no property rights or, when they do, these rights are difficult to enforce. This leads to a 'tragedy of the commons', whereby individual actions are inconsistent with the limits of ecosystems. Furthermore, not all services provided by the ecosystems are known to us; some services are not visible and we do not fully comprehend the complex functioning of ecosystems. Finally, it is important to keep in mind that human decision making is influenced by numerous cognitive biases, which challenge sound decision making processes, especially for complex, future scenarios. Quantitative valuation of ecosystems and their services, therefore, eases the burden of decision makers by reducing the cognitive complexity of decision making.

In principle, limited knowledge and cognitive biases can be addressed by introducing suitable policies. For example, market-based instruments, such as taxation or cap-and-trade mechanisms, or command-and-control measures, such as regulating access to resources, can be used to control consumption. However, for a policy to be both effective and efficient, the accounting value of *all* relevant ecosystems and the services they provide must be estimated.

To ensure all relevant ecosystem services are captured, it is necessary to first conduct a mapping exercise. Rigorously built and well-documented maps are an essential tool for quantifying natural capital in both physical and economic terms. The field of mapping ecosystem services has developed rapidly since the first peer-reviewed maps were published in the mid-1990s. Mapping ecosystem services is a complex exercise with no standard method. However, different approaches are possible, and the choice of methods must take into consideration the characteristics of the ecosystem and the policy questions to be addressed.[4]

The first step of mapping ecosystem services is developing a map of the ecosystem under study. Most mapping exercises are conducted using Geographic Information Systems (GIS), a powerful tool that presents multiple layers of data in a geospatially resolved fashion. To identify all major interactions between nature and humans, an

---

[4] For an in-depth discussion of mapping ecosystem services see (Burkhard and Maes 2017; Tallis et al. 2011).

ecosystem map must include the geographic features of the region, spatial information on all different components of the ecosystem—both biotic and abiotic—and a superimposed map of human presence and activities. In addition, the map should qualify and quantify each component of the ecosystem and the functions they perform. This allows us to assess how human activity could impact the ecosystem by degrading some of its functions in both quantitative and qualitative terms. Sophisticated maps of ecosystems capture the complex dynamics that link the different components and their interactions with human activity. Ecosystem maps can include time at different scales to capture daily, seasonal, and long-term impacts of human activity. Ecosystem maps vary in complexity depending on the characteristics of the ecosystem and the purpose for which they are built and are useful tools with application in environmental management and impact assessment exercises.

Mapping ecosystem services requires multidisciplinary competencies, and, despite the substantial development of the field in recent years, many challenges remain. Due to uncertainties and the resource-intensive nature of mapping ecosystem services, a tiered approach provides users with a variety of methods with incremental complexity and allows them to choose the simplest one that meets their needs (Burkhard and Maes 2017; Tallis et al. 2011; TEEB 2010). A peer-reviewed map of Qatar's ecosystem services is not yet available. Mapping Qatar's ecosystem services poses particular challenges because desert, coastal and marine ecosystem services have received comparatively little attention in the literature. Table 7.1 illustrates the uncertainties for coastal and marine ecosystems in the relationship linking ecosystems and the services they provide. While posing a challenge, this also creates an opportunity for the research community in Qatar to make a substantial contribution to this field.

Once all ecosystem services are mapped, the next step is to conduct a rigorous economic valuation of each of them in turn.[5]

The value of ecosystems is typically estimated through the value of the services they provide. In the literature, a major distinction is made between use and non-use values of the environment. Use values pertain to all those ecosystem services that are used, directly and indirectly, by the present generation. These include the values of provision services, such as agricultural products, and of cultural services, such as the enjoyment of natural landscapes. Non-use values pertain to all those services that we do not use, such as landscapes we know exist even if we never visit them, and to the preservation of ecosystems and all their services for future generations. The methods used to assess use and non-use values differ: use value can be estimated based on market prices of related goods or on costs incurred to enjoy them, while non-use value is typically assessed by asking people about their attitudes towards specific ecosystems and their services, and how much they would be willing to pay for them. The main valuation methods are presented in Table 7.2.

When estimating the total value of an ecosystem or its services using multiple methods, particular attention must be paid to avoid double-counting values that may

---

[5] Environmental valuation is a complex discipline that has been studied by environmental economists over the last few decades and, just like the case of ecosystem service mapping, in

Table 7.1 Illustration of the uncertainties in mapping of coastal and marine ecosystem services (Key: '✓' indicate areas that have been documented in the literature; '?' indicate areas where knowledge is still insufficient (adapted from Drakou et al.(2017)))

	Provisioning			Regulating and maintenance							Cultural		
	Food provision	Water storage / provision	Biotic materials / Biofuels	Water purification	Air quality regulation*	Coastal protection	Climate regulation	Ocean nourishment	Life cycle maintenance	Biological regulation*	Recreation tourism	Symbolic / Aesthetic values	Cognitive effects
Beach and dunes	✓	✓	✓	✓	✓	✓	✓	✓	✓	?	✓	✓	✓
Coastal wetland	✓	✓	✓	✓	?	✓	✓	✓	✓	?	✓	✓	✓
Estuary		✓	✓	✓	?	✓	✓	?	✓	?	✓		✓
Mangrove	✓	?	✓	✓	?	✓	✓	✓	✓	?	✓	✓	✓
Coral reef	✓	?	✓	✓	?	✓	?	✓	✓	?	✓	✓	✓
Maerl bed*	?	?	?	?	?	?	?	?	✓	?	?	?	?
Oyster reef	✓	?	?	✓	?	✓	?	✓	✓	?	?	?	?
Macroalgal bed	✓	?	?	?	?	✓	✓	?	✓	?	?	✓	?
Seagrass meadow	✓	?	?	✓	?	✓	✓	✓	✓	?	✓	?	?
Unconsolidated sediments	✓	?	?	✓	?	?	✓	✓	✓	?	?	?	?
Open ocean / pelagic	✓	✓	✓	✓	✓	?	✓	✓	✓	?	✓	✓	✓

* These habitats and Ecosystem Services are still very poorly analysed.

be captured by more than one method. Moreover, certain methods suffer structural limitations. For example, stated preference methods are prone to many biases, and the extent of the knowledge and direct experience of an ecosystem and its services can influence the results of the valuation. Finally, caution is necessary when transferring the results of valuations across ecosystems in different locations as this overlooks local specificities.

---

this chapter we will only provide a brief overview of the main methods as well as the practical difficulties that one faces when estimating the economic value of ecosystems and ecosystem services. For a more detailed introduction to valuation methods, see (Defra 2007).

Table 7.2 Overview of main valuation methods used for use and non-use values of ecosystems and ecosystem services

	Ecosystem value	
	Use value (direct and indirect)	Non-use value
Valuation method	**Revealed preference methods**: The value of an ecosystem service is based on the costs users incur **Example**: Cost of travel to visit a natural amenity **Hedonic methods**: The value of an ecosystem service is based on the price of a related market good **Example**: The price of a house overlooking a natural amenity compared to a similar house that does not	**Stated preference methods**: include *contingent valuation methods* (CVM), based on questionnaires and surveys on willingness to pay (WTP) and willingness to accept, and *choice experiments*, where respondents are required to perform repeated choices between bundles of attributes **Example**: WTP to protect a natural amenity; willingness to accept compensation for losing an amenity due to human intervention

7.3 Qatar's Environment: Policy and Ecosystem Overview

7.3.1 The Role of the Environment in Qatar's National Vision and Development Strategy

In its National Vision 2030—QNV2030—(https://www.zotero.org/google-docs/? NRR7Yt General Secretariat for Development Planning 2008) and 3rd National Development Strategy (Planning and Statistics Authority 2024), the State of Qatar articulates the need to protect its biodiversity from adverse anthropogenic impacts (;). Qatar is a signatory to the Convention on Biological Diversity (CBD). To support its terrestrial ecosystems, Qatar increased its protected area from 11 to 29%, the highest percentage relative to surface area in the world. Building on Qatar's National Biodiversity Strategy 2015–2025 (Ministry of Environment (now Ministry of Environment and Climate Change) 2014), in 2021, Qatar's Ministry of Municipality and Environment (now Ministry of Environment and Climate Change), in collaboration with UNEP, proposed plans to establish a biodiversity database to ensure progress toward meeting the objectives of the national strategy through cumulative knowledge (*Ministry of Municipality*, n.d.). This aligns with the first goal in Qatar's National Biodiversity Strategy (2015–2025).

QNV2030 emphasized that achieving a balance between environmental management and social and economic development is one of the five[6] major challenges for the future.

There are many cross-cutting themes across the future challenges highlighted in QNV2030, and the importance of environment-conscious growth is implicit in most. For example, the vision underscores the importance of creating renewable wealth from non-renewable resources to meet the needs of future generations. This is a recognition of the limits and potential negative impacts of resource-fueled economic growth. The Vision's environmental stance is embedded within Qatar's three National Development Strategies (NDS1, NDS2 and NDS3), which include general targets for reducing air pollution, enhancing waste management, natural heritage conservation and increasing environmental awareness (Table 7.3). Moreover, QNV2030 acknowledges, both directly and indirectly, that uncontrolled economic growth could lead

Table 7.3 Government targets for different environmental areas of concern in NDS1 and NDS2

Topic	Targets		
	NDS1	NDS2	NDS3
Air pollution	1. Eliminate instances of excess ozone levels through improved air quality management 2. Halve gas flaring in billion cubic meters per million tonnes of energy produced from the 2008 level	1. Reduce the levels of air pollutants in accordance with Qatar's ambient air quality standards by 2022	1. Enhance air quality (no target)
Waste	1. Establish a solid waste management plan, strongly emphasizing recycling 2. Recycle 38% of solid waste, up from the current 8% 3. Contain domestic waste generation at 1.6 kg per capita per day	1. Fix the domestic waste generation rate under 1.6 kg per capita per day during the period 2018 to 2022 2. Recycle 15% of the solid waste generated by the end of 2022	1. Programs to reduce waste through societal behavioural change (no target)

(continued)

---

[6] The other four major challenges being: balancing modernization and preservation of traditions; meeting the needs of the current generation and the needs of future generations; finding a balance between managed growth and uncontrolled expansion; and controlling the size and the quality of the expatriate labour force and the selected path of development.

Table 7.3 (continued)

Topic	Targets		
	NDS1	NDS2	NDS3
National and natural heritage conservation	1. Establish a comprehensive electronic biodiversity database 2. Expand actively managed protected areas	1. Raise awareness of the current and future status of biodiversity and create and operate a biodiversity database by the end of 2022 2. A sustainable management of nature reserves and ecosystems by 2022	1. Protect 30% of land area, 30% of sea area, and recover 30% of degraded natural habitats by 2030
Environmental awareness	1. Build an environmentally aware society 2. Appoint a well-known national champion for the environment to raise awareness and commitment through demonstration projects and conversation partnerships	1. Build an environmentally aware and supportive society for environmental sustainability	1. Programs to reduce waste through societal behavioural change (no target)

to environmental damage and the potential negative impacts of rapid urban development. One area where environment relevant policies have been overlooked is in preserving traditions in the face of globalization and interconnectedness.

QNV2030 rests on four pillars: human, social, economic and environmental development. The four pillars group the government's plans for overcoming the aforementioned challenges and achieving the country's 2030 ambitions. The environmental development pillar stresses the importance of a balance between development and environmental protection, including air, land, water and biological diversity. The government aims to achieve this through increased awareness across the population, an agile legal system that enforces environmental laws, and efficient and proactive environmental institutions. Furthermore, the government plans to play a proactive role in the region in assessing and mitigating impacts of climate change, supporting international efforts and encouraging regional cooperation in climate change mitigation, and developing an urban development plan that ensures sustainable urban expansion.

In light of Qatar's rapid urban expansion, the Ministry of Municipality and Environment created the Qatar National Development Framework (QNDF), a spatial development strategy based on QNV2030, NDS1 and NDS2. Table 7.4 shows the policies proposed by QNDF to protect and restore the natural environment in Qatar. While QNDF does not take an ecosystems services approach, the policies implicitly acknowledge the need for a systematic approach to environmental management

Table 7.4 Policies to protect the natural environment as part of Qatar National Development Framework (QNDF)

Policy	Details	Purpose	Policy actions
ENV1: Sustainable planning and development	Use the precautionary approach in the planning, assessing, construction, monitoring and on-going enforcement of development and infrastructure to ensure impacts on the natural environment are minimised	To mitigate the effects of human activity to promote and protect the natural environment and resources for the benefit of current and future generations	1. In absence of a defined sustainability assessment tool, a precautionary approach is used 2. Develop and implement a sustainability assessment tool to support EIA[a] and SEA[a] processes for projects
ENV2: Climate change management	Safeguard human life, life, public health, culture, heritage, infrastructure, development, economic activities and the natural environment of Qatar from the potential impacts of climate change in the Gulf region	To accurately assess the risk posed by climate change and to establish a strategy to minimize the risk to life, economic activities, infrastructure and the natural environment	1. Use local data to undertake a comprehensive assessment of risk posed by climate change on various outcomes 2. Develop a set of adaptation strategies to with irreversible risks 3. Develop a national GHG reporting system 4. Prepare and implement a strategy on reducing GHG emissions
ENV3: National Environmental Management Plan (NEMP)	Identify, protect and manage environmentally sensitive areas to conserve and enhance the natural environment for current and future generations	To identify and protect environmentally sensitive areas	1. Prepare a NEMP to ensure sustainable use of terrestrial, coastal and aquatic environmentally sensitive areas 2. Develop National Marine Spatial Plan

(continued)

Table 7.4 (continued)

Policy	Details	Purpose	Policy actions
ENV4: Integrated Coastal Zone Management (ICZM)	Achieve effective management and use of coastal land and marine environment to protect and sustain the nation's valuable coastal assets	To safeguard, restore and enhance the coastal environment, processes and ecosystems for existing and future generations	1. Prepare an ICZM plan to identify, monitor and manage coastal zone assets 2. Use a precautionary approach (pending ICZM plan) to identify impacts and develop mitigation measures to prevent environmental degradation in coastal areas
ENV5: Biodiversity	Safeguard biodiversity in Qatar from development pressures to preserve the natural environment for existing and future generations	To protect and enhance native flora, fauna and habitats to enhance, promote and future generations	1. Complete, implement, monitor and report on the National Biodiversity Management Plan 2. Identify areas of significant habitats of biodiversity and develop a Habitat Conservation Strategy
ENV6: Protection of groundwater resources	Safeguard groundwater resources, recharge wells and permitted groundwater abstraction facilities to ensure water security	To protect groundwater resources from excessive usage and contamination	1. Prepare and implement a National Groundwater Management Strategy 2. Conduct audit and establish inspection, enforcement and monitoring program for septic systems 3. Establish reticulated sewerage systems for all urban areas
ENV7: High impact land uses and buffer zones	Implement buffer zones between high impact land uses and adjoining land uses to safeguard the natural and build environments, whilst ensuring the continued operation of the high impact land use	To protect adjoining land uses to safeguard the natural and built environments, whilst ensuring the continued operation of the high impact activities	1. Establish buffer zones in collaboration with MME through the EIA process

(continued)

Table 7.4 (continued)

Policy	Details	Purpose	Policy actions
ENV8: National Waste Management Strategy	Establish new waste management infrastructure, facilities and systems within an integrated National Waste Management Strategy that provides sustainable waste reduction, recycling and disposal solutions for all waste	To more efficiently manage the collection, treatment and disposal of hazardous and non-hazardous waste to minismise the impact on the natural environment	1. Create and implement an integrated National Waste Management Strategy 2. Introduce waste management requirements in applications for new developments 3. Prepare and implement public awareness campaigns and education on waste management
ENV9: Sites for Waste Management Facilities	Identify and provide sites for waste management facilities in pursuit of an integrated National Waste Management Strategy, QNV2030 and international best practice		1. Identify and provide sites for waste management facilities for all urban communities and industrial areas 2. Establish standards and criteria for applications for new hazardous and non-hazardous waste projects

[a] EIA—Environmental Impact Assessment; SEA—Strategic Environmental Assessment

and preservation. Moreover, certain policies, such as ENV1 and ENV4, implicitly acknowledge the need for a framework by emphasizing that a precautionary approach to environmental management must be adopted in "the absence of a defined sustainability assessment tool" (MME 2016: 135).

Qatar's National Master Plan (QNMP), the spatial expansion plan of QNV2030, includes the Qatar National Development Framework (QNDF), Municipality Spatial Development Plan (MSDP), urban centre plans, and zoning regulations. QNMP attempts to incorporate environmental management and protection measures through zoning regulations, which are designed to help municipalities implement MSDP. The zoning regulations are divided into three levels of zoning requirements. The first level consists of 21 planning zones categorized into residential zones, industrial, logistics and large format retail zones, environmental zones, sport and recreation zones, and other zones. The environmental zones comprise two zones: environment and conservation zones, to maintain and protect areas that have ecological, natural and scenic value, and green belt zones, which provide a buffer area to control urban sprawl. The second level includes seven special planning zones that deal with developments located in areas of sensitivity and cultural and environmental significance, such as heritage sites and coastal areas. There are also special planning overlays with specific protection mandates. For example, some of the objectives of the Coastal Protection Overlay (CPO), a special planning overlay, is to identify coastal areas suited for urban expansion needs, preserve culturally and environmentally sensitive areas along the coastline, and ensure future development reflects the character of protected areas.

7.3.2 Overview of Qatar's Ecosystems

Qatar's unique and fragile ecosystems are embedded within desert, coastal and marine environments. Qatar's marine environment holds some of the most biodiverse areas in the Arabian Gulf (Fanning et al. 2021) and is home to some iconic and threatened species, such as the Dugong, Hawksbill Sea Turtle and Whale Shark. This unique environment provides a number of ecosystem services for Qatar, including food, recreation and tourism, climate regulation, transportation, shipping and offshore industry, and represents aspects related to a rich and ancient cultural heritage. Today, these ecosystems face several threats, the majority of which are caused by human activity. Serious threats to the marine environment are due to destruction of the seabed, pollution (see, for example, Burt et al. 2016; Vaughan et al. 2019) and rapid coastline development (see, for example, Burt and Bartholemew 2019). Moreover, unregulated fishing is depleting Qatar's fish stock at unsustainable rates (Al-Abdulrazzak 2013). Other studies elucidate how poaching and disused fishing nets affect threatened species, such as dugongs (Chilvers et al. 2004). Chapter 6 on marine ecology describes Qatar's marine ecosystems in detail and highlights the important role they play for Qatar and their interconnectedness with ecosystems beyond Qatar's borders.

Deserts harbor surprisingly high biodiversity (Safriel et al. 2005), including some of the most endangered species in the world (IUCN 2022). The diversity of mammalian species and plants in desert ecosystems have not been fully captured. In Qatar, systematic research is far behind and data are lacking (Kamel & Madkour 1984; Richer 2009). Desert species confer a wide range of adaptations that are suitable for the desert climate, many of them may play a role in supporting sustainability efforts. Furthermore, deserts and other dryland ecosystems are also important terrestrial global carbon stocks (Trumper et al. 2008). https://www.zotero.org/google-docs/?DwMDwp. Carbon sequestration can further increase through improved land management. Importantly, desert genetic biodiversity is key to improving dryland agricultural productivity (Darkoh 2003). Qatar's desert environment provides a number of services, including space for cities and industry, agriculture, construction materials, and several important connections to cultural heritage and recreation. However, lack of awareness and knowledge, rapid urbanization, construction, pollution and other destructive human activities, have severely diminished Qatar's terrestrial environment, its biodiversity and its beauty. Chapter 4 on terrestrial ecology describes Qatar's terrestrial ecosystems in detail and addresses threats to biodiversity in Qatar's desert, current conservation efforts and positive actions being taken, gaps, best practices and corresponding recommendations. The reader is directed to this chapter for detailed insights into Qatar's terrestrial environment.

7.4 Qatar's Ecosystem Services and Their Value

7.4.1 Ecosystem Services Mapping and Valuation in Qatar

Having introduced all the relevant concepts and background, this section critically reviews the available evidence on Qatar's ecosystem services. The methods used for ecosystem service mapping and valuation are well established and the evidence they generate is used to inform environmental policies in many countries (IPBES 2019). However, countries in the Middle East and North African region lag far behind, with only a handful of researchers contributing to this effort (see, for example, Teff-Seker and Orenstein 2019; Zurqani et al. 2019).[7]

7.4.2 Qatar's Marine Environment and Ecosystem Services

Few studies have been conducted on the ecosystem services of Qatar's marine environment. To date, only Qatar's coral reefs and seagrass beds have been mapped in some capacity (Fanning et al. 2021). Though studies on Qatar's ecosystem services are limited, the preliminary research highlights the benefits existing studies have for understanding Qatar's environment. One benefit is documenting the links between how people use and value ecosystems in an accessible way, along with details about how the ecosystem functions. Together, this information becomes more accessible and digestible for ecosystem management decisions.

In their study, Fanning et al. (2021) utilize the Ecosystems-Based Management (EBM) framework proposed by Granek et al. (2010). One strength of the EBM approach is that it provides best practices on how to engage in the participatory process, which makes it especially suitable for some of the challenges Qatar faces in assessing and understanding the services its ecosystem provides for stakeholders. Typically, in non-westernized settings, including the input of stakeholders in EBM

---

[7] In this section, we provide a review on the work conducted to assess Qatar's marine and terrestrial ecosystem services. To this end, Google scholar, Qatar National Library's Online Resources, PsychInfo, Scopus and PubMed databases were searched using the terms *Qatar, ecosystem services, ecosystem services framework, desert, marine, environment.* As no studies on ecosystem services in Qatar's desert were found, we broadened our search to neighboring geographic regions in the MENA region to gain insight about the value of applying the ecosystem services framework in (similar) desert environments. References cited in the papers retrieved from the databases were also checked for their relevance for the following review.

implementation can be a barrier as their views are often not encouraged or sought after.[8] Another strength of the EBM framework is its focus on exposing potential trade-offs in the decision-making processes about ecosystem development, while ensuring the continued functioning of the ecosystem.

Following the five-step EBM framework guidelines, stakeholder data were collected by conducting interviews and focus groups with natural scientists, local knowledge holders, government representatives, academics, fisheries, NGOs, recreational users and private citizens. The interviews focused on the functions provided by Qatar's coral reef and seagrass ecosystems for ecosystem 'consumers'. The findings shed light on a shared vision for Qatar's marine environment and highlighted an overwhelming desire from all stakeholders to protect and improve existing regulations and management, whilst elucidating opportunities for achieving the vision and principles for guiding the decision-making process. The researchers summarize the stakeholder's input, as the desire for Qatar to have "*a healthy marine environment, identified as a place showing a balance in the abundance and diversity of species and habitat, a place where the environment is incredible for recreation, having more access to beaches, and having healthy water quality for both the ecosystem and people. Corals will have the chance to grow and fish will have the chance to reproduce (pg. 8).*"

Moreover, the stakeholder engagement process elicited ideas for solutions for reaching conservation goals, which included no-take marine protected areas, increased awareness of Qatar's marine environment, enhanced transparency, accountability and inclusivity for the marine environment, and increasing conservation best-practices as tools for implementing these goals. Stakeholders stressed the importance of political will and the need for political leaders to lead the implementation of EBM in Qatar.

The stakeholder interview data are combined with natural science data on coral reefs and seagrass beds in order to determine the primary ecosystem functions. Though Qatar's coral reefs remain understudied, existing research offers crucial insights on the status and functions of coral reef ecosystems, including characterizations of the health, number and location of existing and locally extinct coral species. The findings show that shallow coral reefs are now largely dead and lack sufficient biodiversity to support their recovery; whilst deeper offshore coral reefs are stressed, they are healthier than shallow coral reefs, harboring more biodiversity. The research suggests that maintaining the coral reefs in Qatar is crucial for several reasons, including the role they may play in seeding coral reefs throughout the Arabian Gulf. Furthermore, GPS data used to track the activity of fishing boats in Qatar's waters revealed the extent of dependence of the fishing industry on the coral reefs, where up to 9 frequently fished areas were also coral reef sights, including the

---

[8] The EMB approach has been trialed in the UAE with some success (Lamine et al. 2020; Mateos-Molina et al. 2021) and has also led to positive human-nature outcomes in other countries (for example, see Arkema et al. 2015). Chilvers B, Delean S, Gales N, Holley D, Lawler I, Marsh H, Preen A (2004) Diving behaviour of dugongs, Dugong dugon. Journal of Experimental Marine Biology and Ecology 304(2): 203-224. Ministry of Municipality and Environment (MME) (2016) Qatar National Development Framework 2032. Ministry of Municipality and Environment, Doha.

most heavily fished, largest and diverse reef in Qatar, Fasht East Halul. Collectively, these results shed light on the value of coral reefs for stakeholders, the vulnerability of coral reefs and the urgent need for improved management. Furthermore, the research highlights the need to include fishermen in the ecosystem management exercises to improve their awareness of best fishing practices.

To date, only a small portion of Qatar's seagrass beds have been mapped and are likely to represent a larger coastal ecosystem than coral reefs. The existing data on seagrass beds point to several important regulating roles, such as nutrient recycling and light attenuation. Seagrass beds are strongly associated with increased biodiversity and include habitat and nurseries for many of the juvenile fish species that are crucial to the fishing industry. Some of the fish species living in the seagrass beds are unique to that ecosystem and are therefore reliant on the seagrass for their survival. Overall, conservative estimates indicate that Qatar's seagrass beds are home to more than 10 million individual fish.

The outcome of the combined stakeholder and ecosystem analysis (Step 1 in the 5-step framework) is expected to identify priority ecosystem services. In the meantime, the study highlights the most likely key ecosystem benefits. The marine environment provides provisioning services through supporting the fisheries and potential aquaculture industries. Therefore, the management of these ecosystems is essential to maintain a steady supply of fish for Qatar's growing population. Cultural ecosystem services include clean and unpolluted marine areas for recreational fishing, diving and other water sports, which together form an important component of national pride. Cultural services also include protecting species of national significance, namely the dugong and the whale shark. The regulating services provided by Qatar's marine ecosystem are numerous and include buffering the effects of pollution and climate change.

While further research is required to implement the next steps of the EBM framework, valuable knowledge has been accumulated on the benefits that can be derived from appropriate management of coral reef and seagrass ecosystems. This knowledge provides the opportunity to establish priority actions that specifically respond to the benefits valued by the various stakeholders: maintaining fish production and ensuring the recreational, cultural and aesthetic benefits provided by the ecosystems. Qatar faces many conservation challenges, including a lack of scientific understanding of the ecosystems, a need for improved conservation planning, regulation, monitoring and enforcement of conservation laws, as well as the need to increase public engagement and awareness. Adapting an EBM approach provides a common language for scientists, policy makers and stakeholders to engage in transparent communication to reach ecosystem management consensus. Therefore, it is essential to provide a scaffolded approach to ease the complexity of the conservation questions and foster dialogue among groups with different interests and beliefs to design and implement management plans that are mutually acceptable. These actions in turn will help decision makers make the most impactful decisions going forward.

Another important understudied and vulnerable marine ecosystem in Qatar are mangrove forests. The scientific data shows that Qatar's mangroves are adapted to its extreme climate and are the only mangrove species able to survive the high salinity and harsh climatic conditions (Chatting et al. 2020). Mangroves are effective carbon sinks and provide several ecosystem services, including stabilizing coastlines and act as buffers against threats of extreme weather events (Pernot et al. 2015). These are only some of the ecosystem services mangroves provide, many of which are not well understood and generally understudied (Sheppard 2000). Decisions to develop Qatar's coastline have led to deforestation and loss of mangrove ecosystems. These natural systems were replaced by hotels, residences, ports, roads and other developments. Moreover, off-road driving in Qatar, water sports, sewage run-off, pipe leaks, and pollution contribute to the destruction of mangroves and their seedlings, disrupt the soil and nutrient status, and disturb the organisms living within the mangroves (Chatting et al. 2020; Shehadi 2015). Conducting an EBM valuation of Qatar's mangroves would potentially serve to help policy makers protect and manage these fragile ecosystems in the future.

In Qatar, some preliminary research using the EBM framework has aided in understanding marine ecosystem services. This research provides insights into the ways in which stakeholders value Qatar's marine ecosystems and scientific data on ecosystem functioning. Together, this information provides important insights which can be used to guide decision making and management strategies. However, much more research and cooperation across stakeholders is needed for these efforts to have a lasting legacy (Burt et al. 2017; Fanning et al. 2021).

7.4.3 Qatar's Terrestrial Environment and Ecosystem Services

In Qatar, the terrestrial environment is already in a much altered state from its original pristine habitat. Qatar, counterintuitively, is undergoing rapid desertification, primarily due to construction, overgrazing, agriculture, recreation activities and pollution (see Richer 2009 for an overview). There is extensive biodiversity richness and abundance loss in the Qatari desert, primarily due to habitat loss and overgrazing, driving many species towards an IUCN classification as locally threatened or extinct. There is an urgent need for restoration of the terrestrial habitats, and for future decisions to protect existing resources. To date, no efforts have been made to evaluate the terrestrial ecosystem services in Qatar. In line with this, deserts tend to be underestimated in terms of their biodiversity, and desert climates tend to be both understudied and underfunded compared to other types of habitats (Durant et al. 2012). Despite these shortcomings, lessons from neighboring regions, such as Libya, Egypt and Jordan, can provide examples of the benefits of using ecosystem services as a method for desert preservation. These studies can serve as an inspiration for valuing Qatar's terrestrial environment. The following sub-sections provide insight into the benefits of establishing an ecosystem service-based approach for Qatar's terrestrial environments.

7.4.3.1 Soil

Biological soil crusts (biocrusts) cover the surface layer of soils in many desert habitats, including those in Qatar, and provide essential terrestrial ecosystem services. Biocrusts enable water retention, soil stability, development of deeper soil horizons and other supporting and regulating services crucial for the healthy functioning of desert ecosystems (Chatziefthimiou et al. 2020). The microbial community that makes up biocrusts and desert soils more broadly, is diverse and resilient, yet its bioprospecting capacity for drought resistance or nitrogen fixing, for example, genetic resources is unexplored (Powell et al. 2015; Conkey et al. 2023). A thorough assessment and accounting of Qatar's biocrusts and soils could reveal a plethora of health and climate benefits amongst other ecosystem services.

In neighboring regions, such as Libya, ecologists have used an ecosystems services approach to understand how to improve the management of soil resources (Zurqani et al. 2019). The project assessed the market and non-market value of soil uses, services and disservices. Soil was found to provide a number of vital services such as food and fuel and habitat for biodiversity. Soil also regulated water purification and retention, carbon sequestration, nutrient cycling and served in waste regulation. Culturally, soils provided recreation services as well as links to cultural heritage and opportunities for education. Disservices from anthropogenic activity counteract the benefits provided by the ecosystem services. Unchecked human urbanization, recreational and agricultural activity led to soil erosion and pollution, increasing desertification, salinization and accelerated climate change. Researchers, stakeholders and policy makers in Libya benefited from insights into the services provided by soil resources by adopting an ecosystems services approach. In turn, the gaps in knowledge about soil could be addressed and the information aided diverse stakeholders in making evidence-based decisions about the best methods for sustainable development.

7.4.3.2 Flora

Native plant species and the ecosystems they live within are also important terrestrial resources in desert environments. To date, Qatar's desert plant species have not been studied using an ecosystem services approach. There are regional experiences from which Qatar could benefit. Scientists in Egypt have conducted field surveys and collected data samples of the plant species in Egypt's desert (Bidak et al. 2015). Desert plants were found to provide crucial ecosystem services. Some of the provision services that plants provide include food, medicine and genetic material. Many of the medicinal benefits of understudied desert plants have not yet been discovered. Plants play several regulatory and maintenance roles, such as climate regulation, prevention of soil erosion, ensuring continued soil fertility, and water cycling. Egypt's cultural heritage is influenced by its local plant species, which historically played a crucial role in education. Plants were also key to promoting desert biodiversity. Due to the negative effects of anthropogenic activity, all of the ecosystem services provided by desert plans in Egypt were found to be under threat. Additionally, due to declining plant ecosystem services, biodiversity and habitat availability were also found to be impacted. Continued mismanagement of the native desert plant species in Egypt could potentially lead to a collapse in the ecosystem services. Though similar studies have not been conducted in Qatar, it is likely that local plants provide similar ecosystem services locally and suffer similar anthropogenic stressors. Like in Egypt, studying Qatar's desert plants and mapping their services would lead to novel findings, provide data for sustainable development policies, and elucidate to decision makers the importance of protecting desert plants.

7.4.3.3 Human Well-Being

Finally, the significance of the natural environment for human well-being has been demonstrated in many studies. The studies point to the important role of connections to natural environments as a means to improve mental health, enhance cognitive functioning and reduce stress (Bratman et al. 2019). Desert environments, despite providing valuable cultural services, are often excluded from these studies, which tend to focus on 'green and blue' spaces. For example, researchers in the Wadi Araba desert, which borders Jordan and Israel, examined how the desert environment impacted human well-being and assessed the desert for cultural services (Orenstein and Groner 2015). People reported finding solace and connected spiritually and physically to what they perceived as energizing and relaxing environments. The desert seemed to trigger the human imagination and offered mental and physical escapes from urban settings (Teff-Seker and Orenstein 2019). Further, the desert was found to provide space for knowledge acquisition and sharing, through the educational opportunities it afforded, especially through its rich and readily available geodiversity. Despite these positive ecosystem benefits for human well-being, lack of awareness of the fragility of ecosystems and the impact that human activity has on biodiversity has led to poor decision making on developing desert land. In Jordan's Wadi Araba desert, the positive values people hold for the desert are in conflict with poor levels of knowledge and awareness about how human activity impacts the desert environment. This reality has led to irresponsible and costly desert management decisions (Orenstein and Groner 2015). Similarly, in Qatar, locals reported high levels of connection to nature but demonstrated low levels of endorsement for conservation (Bruder et al. 2022). Taken together, despite high levels of connection to nature, low levels of awareness and understanding of the importance of the natural environment, its fragility and how to protect it might be driving poor conservation decisions. Therefore, a multidisciplinary approach is necessary to fully capture and explain the anthropogenic component in decision making. For example, social science methods can be used to increase awareness about the impacts of anthropogenic activity and influence behavior. Finally, there is a cultural need and responsibility toward future generations to engage in impactful, intentional conservation.

7.5 Conclusions and Recommendations

Policies to protect and conserve the environment must be informed by evidence obtained from multiple disciplines, including ecology, sociology, anthropology and economics. The contribution of different scientific fields to the arduous task of environmental protection cannot be overstated. This chapter highlights how nature gradually became an essential part of economic theories and how this led to the development of methods to assign economic value to the natural environment. Despite the foundational differences between competing economic approaches to the environmental question, *weak* and *strong* sustainability, both approaches acknowledge the importance of nature for human prosperity.

The inclusion of nature in economic theories and the adoption of the concepts of 'natural capital' in the 1970s and 'ecosystem services' in the 1980s set the foundations for environmental valuation methods used to design environmental policy today. Natural capital, made up of living and non-living parts of the environment, human capital and produced capital are the three forms of capital that constitute inclusive wealth. The ecosystem services provided by natural capital can be grouped into four inter-linked, broad categories: provisioning services, cultural services, regulating and maintenance services, and supporting services.

Recognizing the important role nature plays in the economy is essential for charting a sustainable path for human and economic development. Attaching an economic value to nature and the services it provides is important for measuring the impacts of large projects and national strategies. This is achieved through environmental valuation, a collection of methods to assign economic value to natural capital and ecosystem services. The valuation exercise consists of two major steps. First, a rigorous mapping exercise is carried out to ensure all ecosystems and related services are accounted for. Second, an economic valuation of the ecosystems is conducted by estimating the value of the services they provide.

In response to the anthropogenic impacts on biodiversity, Qatar introduced several measures to protect its environment, embedded in Qatar National Vision 2030, Qatar's 1st, 2nd and 3rd National Development Strategies and the National Biodiversity Strategy 2015–2025. Moreover, Qatar is a signatory to the Convention on Biological Diversity. Considering rapid urban expansion, the government created the Qatar National Development Framework, a spatial development strategy part of Qatar's National Master Plan, and implicitly acknowledges the need for a systematic approach to environmental management and preservation.

Qatar's unique and fragile ecosystems are embedded within desert, coastal and marine environments, which provide many ecosystem services. Human activity has taken a toll on most of Qatar's natural capital. Urban expansion and coastline development are threatening marine and coastal ecosystems. In the desert, terrestrial ecosystems act as carbon sinks and desert biodiversity is essential for dryland agriculture. However, human activity, including industrial activity, construction and pollution, has significantly diminished terrestrial ecosystems.

Few studies have been conducted on ecosystem services in Qatar. One attempt to map Qatar's marine ecosystem uses the Ecosystems Based Management. The findings show that most stakeholders are keen on protecting and improving existing regulations and management practices. However, the lack of political will and evidence for informed decision making have hampered efforts to conduct nation-wide, comprehensive environmental valuations. Moreover, there are technical limitations that must be overcome. For example, mapping Qatar's ecosystem is challenging because desert, coastal and marine ecosystem services have received relatively little attention globally and the tools that exist are tailored for temperate and tropical environments.

Finally, efforts to protect Qatar's marine and terrestrial ecosystems are hampered by many challenges, including the cumulative effects by intra- and inter-state developments and a general lack of awareness among citizens of the importance and diversity of Qatar's ecosystems. Overall, this review suggests that various socio-ecological and institutional barriers exist and that these barriers need to be addressed at an interdisciplinary and multi-stakeholder level.

To that end, the following steps can be taken to accelerate efforts towards environmental valuation of Qatar's ecosystems:

1. The development of a comprehensive map of Qatar's ecosystems, the functions they perform and points of interaction with human activity.

   - This process should start with a plan of how the mapping exercise will be implemented.
   - Given the size of the task, pragmatic approaches can be adopted that make use of existing methods in the field. For example, a tiered approach that identifies high-priority ecosystems in the initial planning phase.
   - Adequate resources should be made available to the mapping exercise. This presents an opportunity for the research community in Qatar to set a blueprint for environmental valuation of hot and arid climates.

- Based on the results of the mapping exercise, a rigorous economic valuation should be performed of the ecosystem services provided.

   - This exercise can make use of existing valuation methods–use value and non-use value–but adapt them to the local context.

- Acknowledging the time required to conduct ecosystems mapping, and that the evolving nature of ecosystems requires baseline assessments to be revisited, it is crucial that new information is regularly analysed to be included in decision making and environmental management discussions.
- Lessons can be learned from regions with similar climates and methodological challenges. So, a thorough review of existing studies on regions with similar climates should be conducted before the planning phase.

In addition to providing data and a framework to help decision makers understand the value of ecosystem services in the present and future, a sound ecosystem services approach can be used to raise awareness about the importance of ecosystems in a particular environment. These understandings can be used to engage the public and other stakeholders, such as fishermen, farmers and recreational users, to elicit their views on environmental management. By increasing awareness of the roles of the various stakeholders, sustainable development practices can be advanced more effectively in Qatar. This would potentially also require a coordinated effort with other countries in the Gulf region to collectively institute formal rules to sustainably manage the exploitation of resources and avert a tragedy of the commons. Only then can an ecosystem valuation be translated into effective action.

References

Al-Abdulrazzak D (2013) Total fishery extractions for Qatar: 1950–2010. Fish Cent Res Rep 21:31–37. Scopus

Arkema KK, Verutes GM, Wood SA, Clarke-Samuels C, Rosado S, Canto M, Rosenthal A, Ruckelshaus M, Guannel G, Toft J, Faries J, Silver JM, Griffin R, Guerry AD (2015) Embedding ecosystem services in coastal planning leads to better outcomes for people and nature. Proc Natl Acad Sci 112(24):7390–7395. https://doi.org/10.1073/pnas.1406483112

Bidak LM, Kamal SA, Halmy MWA, Heneidy SZ (2015) Goods and services provided by native plants in desert ecosystems: examples from the northwestern coastal desert of Egypt. Glob Ecol Conserv 3:433–447. https://doi.org/10.1016/j.gecco.2015.02.001

Bratman GN, Anderson CB, Berman MG, Cochran B, de Vries S, Flanders J, Folke C, Frumkin H, Gross JJ, Hartig T, Kahn PH, Kuo M, Lawler JJ, Levin PS, Lindahl T, Meyer-Lindenberg A, Mitchell R, Ouyang Z, Roe J, Daily GC (2019) Nature and mental health: an ecosystem service perspective. Sci Adv 5(7):eaax0903. https://doi.org/10.1126/sciadv.aax0903

Brauman KA (2015) Get on the ecosystem services bandwagon. Integr Environ Assess Manag 11(3):343–344. https://doi.org/10.1002/ieam.1654

Bruder J, Burakowski LM, Park T, Al-Haddad R, Al-Hemaidi S, Al-Korbi A, Al-Naimi A (2022) Cross-cultural awareness and attitudes toward threatened animal species. Front Psychol 31(13):898503. https://doi.org/10.3389/fpsyg.2022.898503.PMID:35712146;PMCID: PMC9194822

Burkhard B, Maes J (eds) (2017) Mapping ecosystem services. Pensoft Publishers, Sofia, Bulgaria

Burt JA, Bartholomew A (2019) Towards more sustainable coastal development in the Arabian Gulf: opportunities for ecological engineering in an urbanized seascape. Mar Pollut Bull 142:93–102. https://doi.org/10.1016/j.marpolbul.2019.03.024

Burt JA, Smith EG, Warren C, Dupont J (2016) An assessment of Qatar's coral communities in a regional context. Mar Pollut Bull 105(2):473–479. https://doi.org/10.1016/j.marpolbul.2015.09.025

Burt JA, Ben-Hamadou R, Abdel-Moati MAR, Fanning L, Kaitibie S, Al-Jamali F, Range P, Saeed S, Warren CS (2017) Improving management of future coastal development in Qatar through ecosystem-based management approaches. Ocean Coast Manag 148:171–181. https://doi.org/10.1016/j.ocecoaman.2017.08.006

Chilvers B, Delean S, Gales N, Holley D, Lawler I, Marsh H, Preen A (2004) Diving behaviour of dugongs, Dugong dugon. J Exp Mar Biol 304(2):203-224

Chatting M, LeVay L, Walton M, Skov MW, Kennedy H, Wilson S, Al-Maslamani I (2020) Mangrove carbon stocks and biomass partitioning in an extreme environment. Estuar Coast Shelf Sci 244:106940. https://doi.org/10.1016/j.ecss.2020.106940

Chatziefthimiou AD, Banack SA, Cox PA (2020) Biocrust produced cyanotoxins are found vertically in the desert soil profile. Neurotox Res 39:81–106. https://doi.org/10.1007/s12640-020-00224-x

Conkey AT, Purchase C, Richer R, Yamaguchi N (2023) Terrestrial biodiversity in arid environments: one global component of climate crisis resilience. In: Cochrane L, Al-Hababi R (eds) Sustainable Qatar: social, political and environmental perspectives. Springer Nature, pp 229–256. https://doi.org/10.1007/978-981-19-7398-7_13

Costanza R, d'Arge R, De Groot R, Farber S, Grasso M, Hannon B, Limburg K, Naeem S, O'Neill RV, Paruelo J, Raskin RG, Sutton P, Van Den Belt M (1997) The value of the world's ecosystem services and natural capital. Nature 387(6630):253–260

Dasgupta P (2021) The economics of biodiversity: the Dasgupta review. abridged version. HM Treasury, London

Daily G (1997) Nature's services: societal dependence on natural ecosystems. Island Press, Washington, DC

Darkoh MBK (2003) Regional perspectives on agriculture and biodiversity in the drylands of Africa. J Arid Environ 54:261–279. https://doi.org/10.1006/jare.2002.1089

Defra (2007) An introductory guide to valuing ecosystem services. Department for Environment, Food and Rural Affairs, London

Drakou EG, Liquete C, Beaumont N, Boon A, Viitasalo M, Agostini V (2017) Mapping ecosystem services. In: Burkhard B, Maes J (eds) Pensoft Publishers, Sofia, Bulgaria

Durant SM, Pettorelli N, Bashir S, Woodroffe R, Wacher T, De Ornellas P, Ransom C, Abáigar T, Abdelgadir M, El Alqamy H, Beddiaf M, Belbachir F, Belbachir-Bazi A, Berbash AA, Beudels-Jamar R, Boitani L, Breitenmoser C, Cano M, Chardonnet P, Baillie JEM (2012) Forgotten biodiversity in desert ecosystems. Science (New York, N.Y.) 336(6087):1379–1380. https://doi.org/10.1126/science.336.6087.1379

Fanning LM, Al-Naimi MN, Range P, Ali A-SM, Bouwmeester J, Al-Jamali F, Burt JA, Ben-Hamadou R (2021) Applying the ecosystem services—EBM framework to sustainably manage Qatar's coral reefs and seagrass beds. Ocean Coast Manag 205:105566. https://doi.org/10.1016/j.ocecoaman.2021.105566

General Secretariat for Development Planning, State of Qatar (2008) Qatar national vision 2030. Doha, July 2008. https://www.psa.gov.qa/en/qnv1/Documents/QNV2030_English_v2.pdf

Granek EF, Polasky S, Kappel CV, Reed DJ, Stoms DM, Koch EW, Kennedy CJ, Cramer LA, Hacker SD, Barbier EB, Aswani S, Ruckelshaus M, Perillo GME, Silliman BR, Muthiga N, Bael D, Wolanski E (2010) Ecosystem services as a common language for coastal ecosystem-based management. Conserv Biol 24(1):207–216. https://doi.org/10.1111/j.1523-1739.2009.01355.x

Gray M (2011) Other nature: geodiversity and geosystem services. Environ Conserv 38(3):271–274. https://doi.org/10.1017/S0376892911000117

Haines-Young R, Potschin MB (2017) Common international classification of ecosystem services (CICES) V5.1 and guidance on the application of the revised structure. www.cices.eu

Hassan R, Scholes R, Ash N, Condition M, Group T (2005) Ecosystems and human well-being: current state and trends: findings of the condition and trends working group (Millennium Ecosystem Assessment Series)

https://www.zotero.org/google-docs/?a6MXcH

IPBES (2019) Global assessment report on biodiversity and ecosystem services of the intergovernmental science-policy platform on biodiversity and ecosystem services. Zenodo. https://doi.org/10.5281/zenodo.6417333

IUCN (2022) The IUCN red list of threatened species. Version 2022–1. https://www.iucnredlist.org. Accessed 1 Dec 2022

Kamel A . [عبد. ا ك], Madkour G (1984) New records of some mammals from Qatar: insectivora, lagomorpha and rodentia. http://qspace.qu.edu.qa/handle/10576/10010

Lamine EB, Mateos-Molina D, Antonopoulou M, Burt JA, Das HS, Javed S, Muzaffar S, Giakoumi S (2020) Identifying coastal and marine priority areas for conservation in the United Arab Emirates. Biodivers Conserv 29(9):2967–2983. https://doi.org/10.1007/s10531-020-02007-4

Mateos-Molina D, Pittman SJ, Antonopoulou M, Baldwin R, Chakraborty A, García-Charton JA, Taylor OJS (2021) An integrative and participatory coastal habitat mapping framework for sustainable development actions in the United Arab Emirates. Appl Geogr 136:102568. https://doi.org/10.1016/j.apgeog.2021.102568

Meadows DH, Meadows DL, Randers J, Behrens WWI (1972) The limits to growth; a report for the Club of Rome's project on the predicament of mankind. Universe Books, New York

Millennium Ecosystem Assessment (2005) Ecosystems and human well-being synthesis. Island Press, Washington, DC. ISBN 1-59726-040-1

Ministry of Municipality and Environment (MME) (2016) Qatar National Development Framework 2032. Ministry of Municipality and Environment, Doha

Orenstein DE, Groner E (2015) Using the ecosystem services framework in a long-term socio-ecological research (LTSER) platform: lessons from the Wadi Araba Desert, Israel and Jordan. In: Rozzi R et al (eds) Earth stewardship. Ecology and ethics, vol 2. Springer, Cham. https://doi.org/10.1007/978-3-319-12133-8_18

Pernot O, Simon L, Al-Khyatt J, Al-Ghouti M, Abu-Dieyeh MH (2015) Ecosystem services and mangroves in Qatar: preservation issues 2015(5):37. https://doi.org/10.5339/qproc.2015.qulss2015.37

Planning and Statistics Authority (PSA), State of Qatar (2024) Third Qatar national development strategy 2024–2030. Doha, January 10, 2024. https://www.psa.gov.qa/en/nds1/nds3/Documents/QNDS3_EN.pdf

Polasky S, Bryant B, Hawthorne P, Johnson J, Keeler B, Pennington D (2015) Inclusive wealth as a metric of sustainable development. Annu Rev Environ Resour 40(1):445–466

Powell JT, Chatziefthimiou AD, Banack SA, Co PA, Metcalf JS (2015) Desert crust microorganisms, their environment, and human health. J Arid Environ Spec Issue Toxins Desert Environ 12(Part B):127–133. https://doi.org/10.1016/j.jaridenv.2013.11.004

Richer R (2009) Conservation in Qatar: impacts of increasing industrialization. SSRN Electron J. https://www.academia.edu/10369438/Conservation_in_Qatar_Impacts_of_Increasing_Industrialization

Safriel U et al (2005) In: Hassan RM, Scholes R, Ash N (eds) Millennium ecosystem assessment. Island Press, Washington, DC, ch 22

Schumacher EF (1973) Small is beautiful; economics as if people mattered. Harper & Row, New York

Shehadi MA (2015) Vulnerability of mangroves to sea level rise in Qatar: assessment and identification of vulnerable mangroves areas. Master Thesis. http://qspace.qu.edu.qa/handle/10576/3902

Sheppard C (2000) Seas at the millennium: an environmental evaluation. Mar Pollut Bull 41:1–4. https://doi.org/10.1016/S0025-326X(00)00098-9

Tallis H, Ricketts TH, Daily GC, Polasky S (2011) Natural capital: theory and practice of mapping ecosystem services. Oxford University Press

TEEB (2010) The economics of ecosystems and biodiversity: mainstreaming the economics of nature: a synthesis of the approach, conclusions and recommendations of TEEB

Teff-Seker Y, Orenstein DE (2019) The 'desert experience': evaluating the cultural ecosystem services of drylands through walking and focusing. People Nat 1(2):234–248. https://doi.org/10.1002/pan3.28

Trumper K, Ravilious C, Dickson B (2008) Carbon in drylands: desertification, climate change, and carbon finance. UNEP-UNDP-UNCCD Technical Note for Discussions, Istanbul, Turkey

Vaughan GO, Al-Mansoori N, Burt JA (2019) Chapter 1—the Arabian Gulf. In Sheppard C (ed) World seas: an environmental evaluation, 2nd edn, pp 1–23. Academic Press. https://doi.org/10.1016/B978-0-08-100853-9.00001-4

Zurqani HA, Mikhailova EA, Post CJ, Schlautman MA, Elhawej AR (2019) A review of libyan soil databases for use within an ecosystem services framework. Land 8(5), 5. https://doi.org/10.3390/land8050082

Open Access This chapter is licensed under the terms of the Creative Commons Attribution 4.0 International License (http://creativecommons.org/licenses/by/4.0/), which permits use, sharing, adaptation, distribution and reproduction in any medium or format, as long as you give appropriate credit to the original author(s) and the source, provide a link to the Creative Commons license and indicate if changes were made.

The images or other third party material in this chapter are included in the chapter's Creative Commons license, unless indicated otherwise in a credit line to the material. If material is not included in the chapter's Creative Commons license and your intended use is not permitted by statutory regulation or exceeds the permitted use, you will need to obtain permission directly from the copyright holder.

Chapter 8
Tourism in Qatar: Pathway Toward a Place-Based Ecotourism Policy and Sustainable Practices

A. D. Chatziefthimiou, A. S. Weber, R. Hinnawi, I. Petrovska, and M. Baeuerle

Abstract Ecotourism and Sustainable Tourism have become standard practices in international tourism development strategies, officially endorsed by such organizations as the United Nations World Tourism Organization (UNWTO) and The International Union for Conservation of Nature (IUCN). However, very few countries have succeeded in re-organizing their tourism industries for long-term sustainability. Qatar is fortunate to possess a relatively new tourism industry such that sustainable practices can be built from the ground up. However, Qatar's natural environment experiences both high natural stressors (water scarcity, high temperatures, low productivity and fragile ecosystems, etc.) as well as unique anthropogenic stressors (rapid urbanization, high CO_2 emissions, terrestrial and marine habitat loss, coastal dredging, etc.). Therefore, incorporating sustainable tourist principles into all dimensions of tourism development is imperative to prevent serious long-term

Alan S. Weber—Equally contributing first author.

A. D. Chatziefthimiou (✉) · R. Hinnawi
Earthna, Center for a Sustainable Future, Qatar Foundation, Doha, Qatar
e-mail: a.d.chatziefthimiou@gmail.com

R. Hinnawi
e-mail: rhinnawi@qf.org.qa

A. S. Weber
Weill Cornell Medicine Qatar, Doha, Qatar
e-mail: alw2010@qatar-med.cornell.edu

I. Petrovska
University American College Skopje, Skopje, Republic of Macedonia

Formerly, ARIU—University of Derby, Qatar, Doha, Qatar

I. Petrovska
e-mail: ilijanata@gmail.com

M. Baeuerle
Sustainability Hive, Doha, Qatar
e-mail: Martin@sustainabilityhive.org

© The Author(s) 2026

A. D. Chatziefthimiou et al. (eds.), *Pathways to Nature Conservation and Resilience in Hot and Arid Lands*, Advances in Geographical and Environmental Sciences,
https://doi.org/10.1007/978-3-032-11046-6_8

negative environmental, economic, social, and cultural impacts for the State of Qatar. This chapter outlines the challenges facing tourism development in Qatar, showcases the discreet steps to transition tourism operations in nature toward ecotourism, and offers suggested best practices as well as a regulatory framework for sustainability in this sector.

Keywords Sustainable tourism · Ecotourism · Environmental management · Cultural heritage · Place-based policy · Nature tourism · Qatar tourism strategy

8.1   Definitions and Background

This chapter describes the overall practices, policy, and future opportunities to develop and manage tourism in Qatar as a sustainable activity, environmentally-friendly practice, and a contributor to GDP growth as part of Qatar's economic diversification strategies. Although Qatar is far from reaching "peak oil"—the period of maximum oil and gas output—its hydrocarbon resources (which generate the majority of Qatar's GDP) are finite. Thus all of the oil-rich Gulf nations have developed economic diversification plans which often include tourism, education, research and other knowledge economy activities, to mitigate against the inevitable reduction of government revenues as natural gas and petroleum reserves decline.

A distinction should be made between the closely related terms "sustainable tourism" and "ecotourism." According to the United Nations World Tourism Organization (UNWTO), sustainable tourist development should "make optimal use of environmental resources that constitute a key element in tourism development, maintaining essential ecological processes and helping to conserve natural heritage and biodiversity... respect the socio-cultural authenticity of host communities... and ensure viable, long-term economic operations, providing socio-economic benefits to all stakeholders that are fairly distributed...." (UNEP and UNWTO 2005). Fennell defined the closely related term ecotourism as "a sustainable form of natural resource-based tourism that focuses primarily on experiencing and learning about nature, and which is ethically managed to be low-impact, non-consumptive and locally oriented. It typically occurs in natural areas and should contribute to the conservation or preservation of such areas" (Fennell 2003, p. 24). The International Union for Conservation of Nature (IUCN) simply states: "Ecotourism respects the resources of Protected Areas and the prosperity of locals" (IUCN 2023). In conclusion, ecotourism can be defined as a type of sustainable tourism practice in which the primary attraction for the tourist is the environment itself.

One of the main issues with ecotourism is the inevitable human impacts resulting from all forms of tourism, which are amplified in fragile ecosystems. The Nabataean ruins at Petra Jordan, one of the most popular tourist sites in the Middle East, and situated in a dry, rocky desert environment with many similarities to Qatar's natural environment and culture, has been the subject of several studies on over-tourism and ecotourism. In a qualitative study of experienced guides at Petra, Aloudat identified the ecotourism challenges that could be replicated in Qatar if ecotourism is not carefully managed and regulated: "crowding, stepping on the fragile rocks and plants, raising the voices…. the pollution including water bottles, the pressure on the sewage and drainage services …." (Aloudat 2022, p. 76).

Only recently have arid and hyper-arid desert regions in North Africa and the Arabian Peninsula and surrounding seas (aside from well-known attractions such at the pyramids) become attractive to tourists. Until the mid-twentieth century deserts were perceived as uninviting and dangerous areas to visit due to remoteness, "lack of infrastructure, poverty of local inhabitants, harsh and sometimes life-threatening environmental conditions and the negative misconceptions of deserts as wasteland, lifeless, and inhospitable" (Weber 2018, p. 189). However, now the exoticness of these destinations is actively promoted in the tourism industry as unique and unspoiled destinations. Innovations such as GPS, cell phone reception and service, and accurate maps have opened these areas for safe exploration and enjoyment.

8.2 Tourism in Qatar

Although Qatar currently has abundant hydrocarbon resources, tourism is growing as one of its economic pillars for economic diversification for its post-oil future. In addition to economic benefits, tourism should also provide social and environmental benefits for sustainable development. The travel and tourism sector contributed to 10.3% of Qatar's GDP and 12% to its employment in 2021 (World Travel and Tourism Council 2022).

During the last decade, Qatar's tourism sector has shown rapid growth, and it ranked 51st among 140 countries for 2019 (World Economic Forum). Despite the economic and social blockade from Gulf countries (2017–2021) and the COVID-19 pandemic in 2020, tourism in Qatar grew rapidly with a 32% increase in international arrivals in the first half of the year 2021 compared to 2020. Approximately 38% of total visitors to Qatar are from GCC (Gulf Cooperation Council) markets, following the lifting of the blockade in January 2021 (Qatar Tourism 2022). These numbers further increased with more than 1.4 million visitors during the FIFA World Cup Qatar 2022™ (Qatar 2022).

New legislation has supported the tourism sector toward opening and building Qatar as a tourist destination. The introduction of visa-free travel to 88 countries made Qatar one of the most open countries in the Middle East (UNWTO 2018). The Qatar National Tourism Sector Strategy 2030 (QNTSS 2030; QTA 2014) was launched in 2014, with the purpose of developing Qatar's tourism sector and to ensure sustainable development, in line with the national development plan entitled Qatar National Vision 2030 (QNV 2030; Qatar General Secretariat for Development Planning (2008). The QNTSS 2030 is positioning Qatar as "a world class hub with deep cultural roots", focused on its unique culture and heritage (QTA Vision Statement 2014). The strategy is initiating development of diverse tourism products, covering cultural, urban, MICE (Meetings, Incentives, Conference and Exhibitions—business events), sports, sun and beach, health and wellness, nature, and education areas. The environmental sustainability dimension of the plan is focused on pollution management, waste generation, minimizing the burden on natural resources, and preserving the country's ecosystems (QTA 2014).

There have been many successful government initiatives for tourism development since the QNTSS 2030 was launched. Overall infrastructure development is evident with new highways, metros, electrical buses, taxi networks, new ports, stadiums, museums, park areas, shopping malls, and entertainment centers. Many initiatives were taken to further develop tourism, to ease visa procedures, to increase investments in tourism, to develop HR capabilities, and to increase professional training (Qatar Chamber 2022). There have been a number of new openings in the hospitality industry, including new hotels, restaurants, cruise lines, and a few new public beach areas (Qatar Development Bank 2021). In recent years, festivals and cultural events have increased, local artists have been encouraged, and art and cultural sites have been developed, along with exhibitions and conferences (enhanced MICE sector), positioning Qatar as a destination for corporate meetings, which is challenging due to the regional competition in this area (Qatar Development Bank 2021).

Furthermore, the FIFA World Cup Qatar 2022™ placed Qatar on the world sports tourist map and provided unique experiences for visitors. Sports tourism can be further developed using the current infrastructure and experience for future Asian Games, Arab Games, Olympic Games, etc. Several marketing campaigns were run to enhance the brand identity "Destination Qatar", attracting tourists from around the world and providing mobile app and online educational platforms for tourism providers (Qatar Chamber 2022).

Qatar Tourism, the regulating body for tourism in the State of Qatar, should focus on further enhancing the underdeveloped health tourism and ecotourism sectors, following future worldwide trends (Qatar Development Bank 2021). Sustainable practices in hotel industry and ecotourism are becoming important preconditions for many worldwide passengers (Qatar Chamber 2022), and should be further implemented in every aspect of tourism in Qatar. Qatar tourism should additionally build financial services, and diversify ecotourism markets as these are two of the weaknesses in Qatar tourism initiatives in the last few years (Yap et al. 2022). As technology is entering every area of life, Qatar will by necessity need to follow and implement innovations to compete with the leading economies to enhance the tourist experience in Qatar and ensure sustainability.

8.3 Sustainable Tourism: Reducing National Environmental Impacts

Tourism, done right, can be a source for good to the State of Qatar, their people, and the environment. Yet often, it has negative impacts especially on the environment but also on people, heritage, and cultural places.

A trend, accelerated by the Covid-19 pandemic, is Slow Tourism, which typically takes sustainability practices into consideration associated with the environmental, social, and economic impacts travel has. Slow Tourism is based on traveling for a longer period of time at a reduced speed, allowing the traveler an authentic, deep, and culturally-relevant experience (Tourism teacher 2023).

8.3.1 Sustainability Management System

"A Sustainability Management System is a management system (set of interrelated elements) to establish a sustainability policy and objectives and processes to achieve those objectives. It addresses environmental, social, cultural, economic, quality, human rights, health, safety, risk and crisis management issues" (gstcouncil.org).

To achieve a sustainable tourism regime and ecotourism in the country, businesses need to be built on strong sustainable management systems, which incorporate operational policies, procedures, strategies, and action plans, including long and short-term goals, which should be reviewed periodically and readjusted to meet new developments and drive continuous improvement. These policies would minimize the negative impacts of business on the environment, especially the conservation of energy and water, waste reduction and proper recycling. They would also minimize harm to local heritage and cultural sites, inhabitants and indigenous people. Furthermore, they should be focusing on maximizing the positive impact that businesses can have on the environment, by enhancing natural environments and biodiversity, and socio-economic aspects of their surroundings. A key to enhancing socio-economic aspects of tourism is a holistic sustainable procurement practice, sourcing local and fair-trade products.

8.3.2 Water and Energy Efficiency

The implementation of water and energy efficient practices in the built environment has increased in Qatar in the last decade. This is mainly driven by government regulations, and adherence to building standards like the Global Sustainability Assessment System (GSAS) and the Leadership in Energy and Environmental Design (LEED), and secondarily as a means to reduce operational costs, by saving on energy and water bills. Another driver, particularly in the hotel sector, is that all big international hotel chains are aiming to reduce their overall carbon emissions, which is the result of international pressure and client requests (Booking.com 2023). This was reflected in the surveys the Qatar Green Building Council conducted in 2021 (Qatar Green Building Council 2021). Though most tourism establishments in the urban areas are connected to the general energy, domestic water, and Treated Sewage Effluent (TSE) supply, many remote tourism establishments are heavily relying on on-site generators to meet their operational energy supply and delivery of potable water.

8.3.3 *Waste Disposal, Recycling*

Though the FIFA World Cup Qatar 2022™ has enhanced the management of waste and recycling in Qatar, the current infrastructure of waste recycling is still underdeveloped, and businesses find it challenging to implement a full-scale waste recycling and management system. The recycling of paper and cardboard is widespread and implemented throughout the country, yet other waste streams are often a novelty. Waste materials may be easily collected, but there are very few industries who reuse these recycled materials, especially plastic and glass (Qatar Green Building council 2021).

8.3.4 *Certification, Regulation, Enforcement for Tourist Infrastructure*

Qatar Tourism has introduced environmental guidelines for the hotel sector as part of their hotel classification system in 2016 (Qatar Tourism 2017). These guidelines include mandatory criteria that establishments must meet and optional criteria to enhance the star rating of the property. Yet there are no local guidelines in place focusing on tour operators, attractions, and camp sites thus tourism businesses have turned to voluntary 3rd party certifications, like Green Key, to demonstrate their sustainability efforts and achievements and to meet requirements of international clients.

8.4 Tourism in Nature in Qatar, and the Path to Ecotourism

The term "ecotourism" was first coined by Héctor Ceballos-Lascuràin in 1987 to describe tourism in nature, with the add-on benefit of education, i.e., learning about the nature in and around one's destination (Ceballos-Lascurain 1987). "Eco" refers to ecology or ecological, which besides the scientific field of study, describes any process that is circular, self-sustained, with built-in controls to avoid waste accumulation, to prevent exhausting natural resources, and to maintain or enrich biodiversity (Velenturf and Purnell 2021).

Although ecotourism may automatically be conceptualized as an outdoor sports activity or an exploration of local nature, certain types of cultural or outdoor/public art exhibitions may qualify as ecotourism as well: if for example the exhibition is an art-meets-science collaboration, if the concept of the exhibit raises environmental awareness and promotes discourse, or if it highlights and interconnects indigenous ecological heritage to modern sustainability practices. Additionally these activities must meet the ecotourism criteria described in the paragraphs below.

As ecotourism has been refined and re-defined, "Best Practice" guidelines have been developed that, on paper, bring ecotourism closer to what it should have been in the first place: a vehicle for the conservation of nature, biodiversity, natural resources, and the heritage of indigenous communities (Drumm and Moore 2005). Conservation *is* the cornerstone of ecotourism, and for it to be safeguarded these distinct principles need to be embedded and realized in any ecotourism venture: low environmental impacts; benefit to nature and community-economic or otherwise; community involvement; educational and research programs; and place-based policy with law enforcement and dynamic regulations (Table 8.1; Samal and Dash 2023).

Despite the development of guidelines, as of today, the majority of projects and destinations offering "ecotourism" experiences, locally and globally, are still just tourism carried out in nature as they lack one or more elements of ecotourism. In

Table 8.1 Ecotourism suggested principles and descriptions

No.	Principle	Requirement(s) description
1	Low environmental impacts	EIA, as well as ecological, environmental and waste management plans carried out
		Flexibility on site selection
		Adherence to environmental sustainability practices and standards
		Responsive and dynamic management with regular monitoring to inform and enforce regulations
2	Benefit to nature and community/Community involvement	Socio-economic study as part of the EIA conducted
		Community is one of the Site's stakeholders involved in the decision-making
		Community is involved at all levels of ecotourism operations
		Economic benefits from operations trickle to the community and to restoration and nature conservation program(s)
		Community is involved in execution of restoration and nature conservation program(s)
3	Educational and research programs	Tour guide training program development
		Educational program development which incorporates traditional knowledge
		Research developed around restoration and nature conservation program(s)
		Research findings feed back into educational programs
4	Place-based policy with law enforcement and dynamic regulations	Development of policy based on the particular conditions of any given location
		Continual oversight to ensure the policy is adhered to/enforce law
		Flexibility in policy and regulations to be responsive and continually adjusted to lower environmental impacts

the section below we describe and expand on the distinct elements of ecotourism, and showcase discreet steps on how tourism in nature can be modified and turned into true ecotourism in Qatar and elsewhere. These observations then can be used as the building blocks for the synthesis of a comprehensive, site-specific, place-based ecotourism policy that is currently lacking in Qatar.

8.4.1 Low Environmental Impacts

Clearly, to ensure low environmental impacts in ecotourism, an Environmental Impact Assessment (EIA) and a monitoring program need to be carried out for each step during all stages of the development of an ecotourism initiative: namely, site selection, design, construction, operation, and end-of-life/decommission. In Qatar, EIAs are required by Environmental Protection Law No. 30 of 2002, and its subsequent amendments, for any new private or public development (State of Qatar 2002; See Chapter on "Environmental Law and Ethics"). EIAs and monitoring programs have also been identified as integral components for the success of environmental and biodiversity protection and conservation both in the Kunning-Montreal Global Biodiversity Framework 2022 (CBD 2022), and in the legally binding international High Seas Treaty (UNGA 2023).

During the EIA process, environmental impacts are identified and classified based on our capability to prevent them, and if that is not possible, then mitigation and restoration options are presented. If impacts are altogether unmitigable, then off-setting pathways and compensation schemes are drawn up to finance conservation efforts elsewhere. EIAs also consider the socio-economic aspects of the development, and this is the entry point for community involvement that is carried throughout the ecotourism project's life cycle.

Flexibility in the site selection and design stages is essential in guaranteeing that the majority of negative environmental impacts are prevented. A high-level preliminary survey helps determine which sites contain the most sensitive ecological receptors, and thus should be avoided. The listed sites can be further thinned out if they are sitting on bird migration flyways or bird and turtle nesting sites, etc.

The more extensive ecological survey of the selected site(s) records in detail the local climatic conditions, and habitats; takes inventory of the biodiversity; maps how wildlife transverses the area, and where they feed and make their burrows (home) at and around the site. These findings inform the rest of the stages of the development plan, so that the fitness and survival of any form of biological life is not harmed or lost, that hotspots of biodiversity and keystone species are protected, that habitats and resources used by wildlife are conserved, and that pollution is minimal, especially light and noise pollution.

The ecological survey's findings can also inform the design of structures to be built, so that they are bio-textured in a way to create additional habitat heterogeneity and thus introduce a net gain of biodiversity (See section below on "benefits"), on top of the desired net zero impacts (Chithra and Amritha 2015). Other sustainable development standards that must be met, and are subsequently described in the EIA report include but are not limited to: local sourcing of any materials for construction; use of energy from renewable sources; net zero carbon emissions from operations and structures, etc.

The productivity and carrying capacity of the environment on site needs also to be determined (how much stress can it withstand without losing its functionality), and as described in the EIA report, to help regulate the visitor numbers, something that is successfully done in Arizona's Grand Canyon National Park (IUCN 2018; NPS 2023a). A sensitivity matrix of the area, combining the findings of the ecological survey and the productivity and carrying capacity of the site, along with regular monitoring, can propel more efficient management decisions. In Yellowstone National Park, for example, managers practice intermittent or circular closure of trails and areas close to the tipping point, to give them a chance to (actively or passively) self-restore, while opening other pristine or already restored adjacent trails through this managerial practice (NPS 2023b).

Regulating visitor numbers and development activities is acutely important in drylands such as those of Arizona and Qatar, since their biodiversity levels are low to begin with, and recuperation after disturbance will likely take longer than in more productive temperate or tropical climate settings (Qifeng, 2004; WTO 2007). This fact raises the need for place-based policy and/or ecosystem-based policy, as opposed to importing and trying to fit or apply policy created for tropical or temperate climates to the desert climate.

8.4.2 Benefit to Nature and Community Involvement

As mentioned above, community involvement commences at the EIA, as part of the socio-economic study. Community engagement and inclusion is essential and as equally important for the success of ecotourism as it is for nature conservation more broadly (Yao et al. 2020; Quintana et al. 2021). Usually, the community is one of the stakeholders engaged in workshops to determine the fate of a development, the environmental values to preserve, and to make decisions on whether to maintain the current levels of biodiversity or to introduce a net gain, etc. It is also during that time that their level of involvement, and at which stages, is clarified. For a number of reasons, the ideal scenario is for the ecotourism operations to be run by the community. The local community has a sense of place, and thus more incentive to protect and conserve it, and the educational and training aspects of ecotourism grant them additional agency to do so.

Naturally, community-based ecotourism creates job opportunities and uplifts the economic status of the local community. In addition to the community, the economic benefits must trickle down to nature and wild species as well. Setting aside a percent of the profits to revert back to conservation and restoration efforts, is a practice that the best managed ecotourism destinations exercise (Kim et al. 2019). In Yellowstone National Park for instance, whole areas are closed off during the winter months when visitor numbers dwindle down, and these "set-aside-funds" are used to manage and catalyze recuperation of nature, or enact conservation programs such as re-introduction of at-risk species (Gray Wolf), etc. (NPS, 2023b).

8.4.3 Educational and Research Programs

Education sits at the heart of ecotourism, and it does not only target visitors, but all and any fraction of the community that is involved in any aspect of the operation (Wood 2002). Best practices necessitate that tour guides in ecotourism be required by law to receive training and become certified nature guides, as is the case for cultural tour guides (Qatar Tourism 2023a, b; State of Qatar 2023). At the time of this writing, neither laws nor training programs exist for nature guides in Qatar.

Of course, research is the basis of the educational programs and material being developed for ecotourism activities, as well as for the tour guide training programs. At its best, this material is continually fed and updated with real-time research findings as our understanding of phenomena evolves. The other element that makes a research program an essential component of ecotourism is that it serves as the knowledge base and "executor" for conservation efforts in the local context. Educational program development that incorporates indigenous practices and knowledge has a bi-directional benefit, in that the local intangible heritage is conserved at the service of natural heritage conservation and vice-versa.

Through education the Red Flag issues of tourism in nature are raised and dissipated: namely, exploitation of indigenous people, "trophy tourism" for wildlife and plants, visits to natural habitats of endangered species, or to zoos where animals are kept in captivity in tiny cages (Ananthaswamy 2004; Geffroy et al. 2015; Hariohay et al. 2018). Any practice that doesn't recognize equal rights and freedoms among all sentient species and peoples should be re-examined. It should be noted here that, unfortunately, trophy tourism is advertized as a nature loving experience, even if animals are hunted down and killed for thrill. Also, direct animal contact, through endangered-animal watching and zoo visits, is marketed as *the* means to raise environmental awareness among the visitors, and *the* way to propel them to conservation action. Clearly these are inaccurate statements, not supported by ethical codes of conduct or conservation principles (Biasetti and de Mori 2021).

Furthermore, research shows that awareness raised through knowledge acquisition is the key to inspire conservation action, not direct animal contact (Clifford-Clarke et al. 2022). Case in point, one does not need to travel to the North Pole to directly interact with polar bears to comprehend the plight they are faced with: habitat and food-source loss due to the climate crisis. Documentary footage and illustrated written reports create sensitization and incite action to equal degrees.

Educational programs and activities in ecotourism, instead, are built from the tenets that one never interferes with the life cycle of wild animals, and that awareness is raised through respect for and exposure to nature, with no need for direct exposure to wild animals (Drumm and Moore 2005; IUCN 2018). A brilliant teaching tool here is incorporation of tracking and trailing practices employed by indigenous populations. This way one discovers the wildlife of a given place through tracks and signs of activity, not direct contact, while at the same time traditional knowledge is conserved (Tom Brown Jr. Tracker School 2023). Contact with big animals can be afforded in a principled manner, still from a safe distance, if through a conservation program animals are kept in wide open spaces either for rehabilitation before they are re-introduced in the wild, or when they are permanently hosted there because for any number of reasons they can never be released back to the wild (WERC 2023).

The bright example of such conservation efforts that all the countries of the GCC region can showcase with pride is the breeding program for the survival of the Arabian Oryx (*Oryx leucoryx*) that went extinct in the wild on the Arabian peninsula in the 1960s (IUCN 2011; IALA 2023). Creating an ecotourism program around this species that most notably resides in the open areas of Ras Brooq and by Abu Samra in Qatar, will tick many boxes: education, community involvement, and conservation.

This kind of program was enacted for another charismatic, critically endangered, and protected species: the Hawksbill turtle (*Eretmochelys imbricata*). In this case, in 2022, the Ministry of Environment and Climate Change (MoECC), worked hand-in-hand with the Children's Museum Dadu, and the NGO Qatar Natural History Group, to create educational material given out during the tours in the turtle protected area, to raise awareness and to create the impetus for life-long community engagement in conservation efforts like this one (Qatar Moments 2022). It follows that the educational material was based on research that the ministry conducts on the Hawksbill turtles in collaboration with scientists from Qatar University and other research institutes. Establishing this initiative as an annual program, and further involving the community in research aspects of conservation, would make it an exemplar in ecotourism.

To come full circle, a research station or facilities on site of the ecotourism operation, would provide additional benefits to the ones mentioned above: provision of a knowledge base for education, and conservation and monitoring execution (Sarkar et al. 2019). A station provides opportunities for researchers to carry out fieldwork and experiments on site; additionally, it creates opportunities through training for operators to become stewards of the surrounding environment, and it creates opportunities to involve the school community in "service learning" projects. Involvement of school and the community at large can be harnessed to obtain additional data through citizen science initiatives as well (Hein et al. 2018).

One isolated initiative to showcase these fundamental ecotourism principles was the collaboration among MoECC, the Aquatic Fisheries and Research Center, and the kayak operator Aquasports in Al Khor (Aquasports Qatar 2022). They all brought together their resources, scientific expertise and operational skills to release into the open waters hamour (*Epinephelus tauvina*) grown in the research facility, and in this way restock the fisheries of this species. If this initiative were to be an established program, it too would become an ecotourism exemplar, similarly to the Hawksbill turtle conservation program, and could involve local community members or students.

8.4.4 Place-Based Policy with Law Enforcement and Dynamic Regulations

The cruciality of place-based policy for ecotourism has been described and substantiated above. In lieu of this analysis, since it is currently lacking, enforcement of the Environmental Protection Law No. 30 of 2002 and other pertinent ones by the relevant authorities becomes critically important as do further dynamic regulations. For this to be possible, continued oversight is required, as does sustained mutual trust between the authorities and the tourism operator. A hopeful example that showcases local cooperation is the collaboration of the authorities with kayak operators like Aquasports in Al Khor, whereby their operations are constantly monitored by the MoECC and their managerial practices are continually adjusted to lower the impacts of their operations. Most recently, one operator was required by authorities to cease using generators and bright lights during the night, to safeguard the natural rhythms of wildlife from noise and light pollution (Personal Communication, Chatziefthimiou AD, January 2023). And, as an adaptive regulation example, the authorities required Aquasports to move their camp away from the shoreline (the original location that was assigned to them) to a less ecologically sensitive area, to alleviate any disturbance to shorebirds and other wildlife (Aquasports, Personal Communication Chatziefthimiou AD, March 2023).

Another hopeful message that has become apparent from our review of local tourism operations in nature is that even in the absence of place-based policy, adherence to sustainable development laws and best practices serve and satisfy some ecotourism principles. An example is the Studio Olafur Eliasson outdoor/public art exhibit "The Curious Desert", where the National Museum of Qatar (NMoQ) and the artist's studio (SOE) worked with Applus+ ecologist and environmental engineers to keep their impacts to a minimum, in collaboration with MoECC (QM 2023; NMoQ, Personal Communication, March 2023). The exhibit consists of a series of large iron pavilions installed in a natural area of Qatar. The theme of this exhibit in itself raises awareness on sustainability and the climate crisis through a dialogue between art and natural phenomena. Another positive development springing from this exhibit is that a restoration plan during and post decommission is currently being drafted, and, along with all other environmental protocols followed during this exhibit, this plan will serve as the blueprint for all Qatar Museums (QM) outdoor/public art exhibits in the future.

And to conclude, Aquasports, which has been discussed above, is, based on our assessment, only one step away from being a full-blown ecotourism venture—the final step being the development of an educational/training program, something that the company has identified as a need as well (Aquasports, Personal Communication, March 2023). Besides this, they satisfy all other requirements of best practices ecotourism: collaboration with the authorities, stewardship of the surrounding nature, camp hardscape built from refurbished material which also reflects the local culture (Bedouin tents), zero pollutant emissions, energy obtained from renewable resources, and taking part in conservation programs.

8.5 Regulation of the Tourism Sector in Qatar

Qatar has set ambitious goals to diversify its economy and transition it from an oil and gas-based economy to a knowledge-based economy as per Qatar's strategic development plan, Qatar National Vision 2030. Tourism is a sector that has been identified to play a major role in this transition. The county has put forth rigorous efforts to attract tourists from all over the world. In 2021 and in preparation for the FIFA World Cup Qatar 2022™, Qatar promised to offer visitors the ultimate experience therefore, Qatar Tourism launched the country's largest tourism campaign, "Experience a World Beyond." The campaign aims to attract 10 million visitors by 2030. The global campaign is catered to international travelers who are looking for unique, authentic experiences blending cosmopolitan modernity with Arabic tradition.

The unique natural environment of Qatar and its biodiversity were strongly present in the campaign in a direct effort to encourage ecotourism. Discover Qatar announced natural and cultural destinations, strategies, and regulations to meet the goal. Although work is in progress, more government attention and collaboration among various ministries needs to culminate in a holistic national strategy to protect Qatar's unique ecology and nature and support regeneration developments.

According to The Tourism Sustainable Development Index (TSDI), data collection is vital to drive strategy formulation for resilient and green destinations (Murmuration 2020). Data can be collected to generate a green index and a human index. The green index is related to CO_2 emissions per capita per year, water and air quality, urbanization and forest cover, and urban development proximity to protected areas. The human index is related to education, international tourist numbers, and life expectancy. In order to advocate for Qatar as a sustainable tourism destination offering eco-destinations, collaborations among the Ministry of Municipalities, Ministry of Environment and Climate Change, Qatar Tourism, Discover Qatar, and Qatar Museums along with ecologists and the private sector are very important to coordinate efforts for a comprehensive approach toward a national strategy for responsible ecotourism and greening tourism destinations based on data and best practices.

The authors recommend that an Ecotourism Committee be formed, with members representing the stakeholders listed above for the creation of a holistic national "ecotourism strategy." The role of the committee would be to develop strategies, guidelines, and regulations for implementing sustainable ecotourism practices—in other words, to act more as a legislative body. Other stakeholders that should be included and/or serve as advisors to the Committee, include universities, tourism organizations, hospitality establishments, funding institutions, local community groups, NGOs, and media during the development phases and at the time of dissemination.

8.6 Recommendations and Moving Forward

For any developed strategy and campaign for ecotourism in Qatar to be successful, it should include the following elements:

1. **Promote cultural heritage and cultural sites preservation**. Create community initiatives that empower locals to lead the design of unique tourism opportunities that reflect the Qatari traditions and heritage.
2. **Create awareness** among the public and private sectors and consumers on ecotourism's important role for sustainable communities and in consideration of the ecosystems' carrying capacity. Plans and campaigns would create interest in nature among the public and change attitudes toward protecting natural areas.
3. **Identify urban and rural destinations**, map routes and trails based on the seasons, and the biodiversity status of habitats and species, and the activities allowed.
4. **Regulate infrastructure and developments on sites** with fragile habitats and species-at-risk.
5. **Create guidelines with best practices** and tools for developing, regulating, managing, and monitoring of the ecotourism destinations and activities to promote low impacts on the environment, and build long-term ecosystem sustainability.
6. **Create partnerships** to build capacity and training for both private and public sectors to manage the operation of offered services and grant the proper implementation of developed guidelines.

Promoting Cultural Heritage and Cultural Preservation Sites The State of Qatar has been proactive in identifying, developing and preserving cultural and heritage sites (natural and built heritage). For example, Msheireb Museums have refurbished as museum spaces in downtown Doha the Bin Jelmood House, the Company House, Radwani House, and Mohammed Bin Jassim House, which chronicle Qatar's history related to pearl diving, general trade, the oil industry, and the slave trade. However, some cultural sites are unprotected: the Kassite-period Murex dye factory in Al Khor excavated in 1976 by Jacques Tixier on Purple Island is not protected by fencing or signage.

Creating Awareness for Sustainable Communities As the Qatari government considers ecotourism as a tool to diversify the national economy, it is important to present ecotourism to the community as a means of fostering cultural identity and to showcase the desert as a living environment. Awareness campaigns need to firstly, represent and revive the important role of the desert, and marine and coastal ecology as a source of life, inspired by the knowledge of the original inhabitants of Qatar. Secondly, campaigns should advocate for the potential of regenerating these ecosystems to support the sustainability of modern communities today, especially with the rising environmental challenges Qatar is facing such as extreme heat events, water scarcity, and urbanization.

In Qatar the local populations maintain a strong connection to the desert and sea. Therefore, it is crucial to involve the community and engage them in any national conservation plan and campaign. The MoECC has made efforts to raise awareness around Qatar's unique ecosystems on their social media platforms, events, and publications. The following are additional general recommendations to take into consideration when designing future awareness campaigns to engage public and private sectors and the community for a greater impact:

- Bridge the gap between environmental sciences, marine and terrestrial ecology and public outdoor activities through the development of well-designed educational excursions to Qatar's unique ecosystems, in order to offer visitors learning opportunities while enjoying leisure and entertaining activities in nature.
- Design initiatives on a yearly basis and at a nation-wide level. The campaign would be designed in coordination with various public and private sectors. To highlight the importance of each ecosystem, and to offer educational resources, maps with the identification of controlled routes and best practices guidelines for site visitors would be prepared.
- Identify nation-wide ecosystem restoration projects that allow visitors the opportunity to contribute positively to environmental protection and allow for carbon offset opportunities for the private sector.

Identifying Urban and Rural Destinations

Key urban and rural marine and terrestrial ecosystems should be identified to act as "Learning Ecosystems" (Wise 2019). Sites are to be provided with educational infrastructure and centers that offer eco-activities and learning opportunities on a per season basis. Nature maps under the supervision of the proposed Ecotourism Committee should be developed in partnership with government ministries, Qatar Tourism, Qatar Museums, ecologists, tourist agencies, educational institutes, and developers. Maps would clearly identify coordinates, routes and trails, authorized means of transportation, and landmarks. Furthermore, educational materials along with the maps would create public awareness and community interest in exploring and protecting nature. It is important to identify here the role of Qatar Museums and other national attractions such as exhibitions and parks to act as urban ecological destinations where they provide the community with opportunities to learn and to increase biodiversity within urban areas.

Regulating Infrastructure and Developments on Sites with Fragile Habitats and At-Risk-Species

There are many ecolabels that focus on greening the hospitality sector from an operational and building perspective such as Green Key, LEED, and GSAS. These ecolabels drive sustainable changes and grant continuous progress at the strategy and policy management levels, in addition to incorporating global standards among various existing hospitality establishments. In addition to these, for any new hospitality development to be deemed sustainable, it must be designed and constructed in a manner to reduce the negative impact on the environment, and take into consideration the site's history, cultural heritage and ecological importance to revive its role within the context based on low impact development principles (Regenesis Group). Low impact developments address many strategies and principles such as passive design principles, the use of local materials, the preservation and restoration of natural landscapes with native and locally-adapted plants, water sensitive site design, etc. The Environmental Impact Assessment is a widely used tool to implement these goals (See Section 8.4.1). Additionally, the sustainable development code should address and regulate development and the types of activities that may be allowed in proximity to inland and coastal marine Protected Areas, and to areas with sensitive receptors with endemic and endangered animal and plant species (species-at-risk).

The above suggested course of actions can be turned into legislative tools, and measures by the Green Development and Environmental Sustainability Department at MoECC, which is tasked to promote sustainability in these ways.

Creating Guidelines with Best Practices and Tools for Ecotourism Destinations

When identifying ecotourism destinations, it is crucial to develop best practice guidelines, codes of conduct and educational material for those sites. This information can be developed on a per age group manner, and made available via digital platforms such as apps, QR codes, and through on-site signage. In recent years, Qatar Museums has made great strides in such educational on-site signage across restored heritage sites in nature. Equally important is to monitor activities and their impact on the environment. A regular monitoring program will support policy development, allow for dynamic land management and regulating the number of visitors. The following are some ideas that may be considered:

For urban areas

- Maps of activity routes and spots in green areas for biking, bird watching, hiking, etc.
- Best practices to protect the environment and opportunity offerings to the community to contribute to ministirial restoration efforts.
- Information about the ecosystem ecology and its importance to the health of Qatar's environment.

For rural areas

- Maps of trails for each site to limit site disturbance and negative environmental impacts,
- Clear identification and demarcation of Protected Areas,
- Site-specific guidelines for activities allowed,
- Increased numbers of trained environmental rangers to control and oversee activities,
- Infrastructure and services to control pollution within sites such as trash bins, campfire allowed zones with pits, etc.,
- Identification of allowed camping sites and provision of green criteria for public camping sites and hotel operated camps. Criteria need to tackle onsite water and

energy generation, use and re-use, activities allowed, material used and waste handling.

Creating Partnerships to Build Capacity and Training

Building capacity means building high quality guiding services by trained and educated people on environmental sustainability, cultural and ecological issues for any sustainable destination. The following are some ideas to build capacity among researchers, tour operators, the public, stakeholders, small and medium sized businesses, and service providers:

- Create an educational center to train and certify tour operators, rangers, and guides, which can also serve as a visitor's welcoming center and/or research field station facility,
- Create digital platforms such as websites and apps for tour operators, rangers, guides and the public to be able to access information related to the Protected Areas and other nature destinations. The website is to be updated regularly and provide laws, guidelines, best practices, maps, and show the services provided and activities allowed as per seasons,
- Develop a clear annual agenda for environmental awareness events, such as conferences to create networking opportunities among academics, operators, various stakeholders and service providers. Also to create a platform to share information, data, updates on national plans and to share international best practices,
- Create management committees for Protected Areas, collaborations between ministry rangers and local communities,
- Train businesses and service providers on principles of green and sustainable operations and how to provide benefits for the local community and contribute positively to the biodiversity and habitats within their sites.

References

Aloudat AS (2022) Overtourism in petra protected area: tour guides' perspectives. In: Fennell DA (ed) Routledge handbook of ecotourism, pp 70–82

Anathaswany A (2004) Massive growth of ecotourism worries biologists. New Scientist. https://www.newscientist.com/article/dn4733-massive-growth-of-ecotourism-worries-biologists/

Aquasports Qatar (2022) Qatar Environment Day 2022. https://www.instagram.com/p/CaoSgZVs7Dq/?utm_medium=copy_link

Biasetti P, de Mori B (2021) The ethical matrix as a tool for decision-making process in conservation. Front Environ Sci 9. https://www.frontiersin.org/articles/; https://doi.org/10.3389/fenvs.2021.584636

booking.com-sustainable-travel-report 2023 (2023) https://globalnews.booking.com/download/31767dc7-3d6a-4108-9900-ab5d11e0a808/booking.com-sustainable-travel-report2023.pdf

CBD (Convention of Biological Diversity) (2022) Convention of the parties to the convention of biological diversity. Decision Adopted: 15/4. Kunming-Montreal Global Biodiversity Framework. https://www.cbd.int/doc/decisions/cop-15/cop-15-dec-04-en.pdf

Ceballos-Lascurain H (1987) Estudio de Perfectibilidad Socioeconómica del Turismo Ecológico y Anteproyecto arquitectónico y urbanístico del Centro de Turismo Ecológico de Slan Kalan, Quintana Roo, México. México: Sedue

Chithra K, Amritha Krishnan K (2015) BIOTECTURE—a new framework to approach buildings and structures for green campus design. In: Leal Filho W, Muthu N, Edwin G, Sima M (eds) Implementing campus greening initiatives. World sustainability series. Springer, Cham. https://doi.org/10.1007/978-3-319-11961-8_10

Clifford-Clarke MM, Whitehouse-Tedd K, Ellis CF (2022) Conservation education impacts of animal ambassadors in zoos. J Zool Bot Gard 3(1):1–18. https://doi.org/10.3390/jzbg3010001

Drumm A, Moore A (2005) Ecotourism development—a manual for conservation planners and managers. Volume I—an introduction to ecotourism planning, 2nd ed. The Nature Conservancy, Arlington, VA, USA.

Fennell DA (2003) Ecotourism.(2nd edition). London, UK: Routledge

Geffroy B, Samia DSM, Bessa E, Blumstein DT (2015) How nature-based tourism might increase prey vulnerability to predators. Trends Ecol Evol 30:755–765. https://doi.org/10.1016/j.tree.2015.09.010

Global Sustainable Tourism Council (2022) GSTC sustainable tourism glossary. Available at gstcouncil.org

Hariohay K, Jackson GR, Fyumagwa RD, Roskaft E (2018) Trophy hunting versus ecotourism as a conservation model? Assessing the impacts on ungulate behaviour and demographics in the Ruaha-Rungwa ecosystem, Central Tanzania. Environ Nat Resour Res 8(2). https://doi.org/10.5539/enrr.v8n2p33

Hein MY, Couture F, Scott CM (2018) Ecotourism and coral reef restoration. Coral reefs: tourism, conservation and management. Earthscan Oceans, Oxford

IALA (Institute of Animal Law of Asia) (2023) Arabia Oryx: threats and conservation efforts. https://www.ialasia.org/projects/arabian-oryx-threats-and-conservation-efforts

IUCN (2023) Sustainable tourism. https://www.iucn.org/our-work/region/mediterranean/our-work/ecosystem-resilience-and-spatial-planning/sustainable-tourism. Accessed July 19 2023

IUCN (2011) A grain of hope in the desert. https://www.iucn.org/content/a-grain-hope-desert

IUCN (2018) Tourism and visitor management in protected areas: guidelines for sustainability. Best Practice Protected Area Guidelines Series. Monographic Series no. 27. https://doi.org/10.2305/IUCN.CH.2018.PAG.27.en

Kim M, Xie Y, Cirella GT (2019) Sustainable transformative economy: community-based ecotourism. Sustainability 11(18):4977. https://doi.org/10.3390/su11184977

Murmuration (2020) TSDI index. https://www.tourism-sdi.org/wp-content/uploads/2021/02/Methodology-TSDI.pdf

NPS (National Park Service) (2023a) Grand Canyon, National Park Arizona: backcountry rules and regulations. https://www.nps.gov/grca/planyourvisit/backcountry-regs.htm

NPS (National Park Service) (2023b) Yellowstone National Park: state of the park 2023. https://www.nps.gov/yell/learn/management/upload/State-of-the-Park-2023-Web.pdf

Qatar Chamber of Commerce and Industry (2022) Study on the reality of Qatar tourism sector in the time of Covid-19 (Challenges and Solutions). Research and Studies Department, Doha

Qatar Development Bank (2021) Tourism sector in Qatar: current state assessment series. Doha, Qatar. https://www.qdb.qa/en/Documents/Tourism-Sector-in-Qatar-Current-State-Assessment-Series.pdf

Qatar General Secretariat for Development Planning (2008) Qatar national vision 2030. https://www.npc.qa/en/QNV/Documents/QNV2030_English_v2.pdf

Qatar Green Building Council (2021) Qatar green building council: green hotels challenges and pathways

Qatar Moments (2022) Dadu invites culture pass family members to watch turtles hatching. https://www.qatarmoments.com/dadu-invites-culture-pass-family-members-to-watch-turtles-hatching-481131.html

Qatar Tourism (2017) Hotel_Classification_Manual_for_2017. https://www.qatartourism.com/content/dam/qatar-tourism/industry-resources/classification/Hotel_Classification_Manual_for_2017.pdf

Qatar Tourism (2022) Qatar now. Issue 1, Lusail, Qatar

Qatar Tourism (2023a) https://www.qatartourism.com/

Qatar Tourism (2023b) Tour guide training. https://www.qatartourism.com/en/licensing-e-services/e-services/tour-guides

Qatar2022 (2022) FIFA World Cup Qatar 2022. Qatar hosts more than 1.4 million visitors during FIFA World Cup™. https://www.qatar2022.qa/en/news/qatar-hosts-more-than-one-million-visitors-during-fifa-world-cup#:~:text=More%20than%201.4%20million%20fans,FIFA%20World%20Cup%202022%E2%84%A2.

Qifeng G (2004) Slow recovery in desert perennial vegetation following prolonged human disturbance. J Veg Sci 15:757–762

QM (Qatar Museums) (2023) Olafur Eliasson's artistic practice in Qatar. https://qm.org.qa/en/stories/all-stories/interview-with-olafur-eliasson/

QTA (2014) Qatar national tourism sector strategy 2030. Qatar Tourism Authority, Doha, Qatar. www.qatartourism.gov.qa

Quintana ACE, Giron-Nava Alfredo, Urmy S, Cramer AN, Domínguez-Sánchez S, Rodríguez-Van Dyck S, Aburto-Oropeza O, Basurto X, Weaver AH (2021) Positive social-ecological feedbacks in community-based conservation. Front Mar Sci 8. https://www.frontiersin.org/articles/; https://doi.org/10.3389/fmars.2021.652318

Regenesis Products (2023) https://regenesisgroup.com/projects/

Samal R, Madhusmita DM (2023) Ecotourism, biodiversity conservation and livelihoods: understanding the convergence and divergence. Int J Geoheritage Park 11(1):1–20. https://doi.org/10.1016/j.ijgeop.2022.11.001

Sarkar D, Chapman CA, Valenta K, Angom SA, Kagoro W, Sengupta R (2019) A tiered analysis of community benefits and conservation engagement from the makerere university biological field station, Uganda. Prof Geogr 71(3):422–436. https://doi.org/10.1080/00330124.2018.1547976

State of Qatar (2002) Law No. 30 of 2002 promulgating the law of the environment protection. https://almeezan.qa/LawPage.aspx?id=4114&language=en

State of Qatar (2023) Al Meezan Qatar legal portal. Law No (6) of 2012. Tourism Law. https://www.almeezan.qa/LawPage.aspx?id=4785&language=en

Tom Brown Jr. Tracker School (2023) Courses. https://www.trackerschool.com/

Tourism Teacher (2023) What is slow tourism and why is it so good? https://tourismteacher.com/slow-tourism/

UNEP and UNWTO (2005) Making tourism more sustainable-a guide for policy makers, pp 11–12. https://www.unwto.org/sustainable-development

UNGA (United Nations General Assembly) (2023) Agreement under the United Nations Convention on the Law of the Sea on the conservation and sustainable use of marine biological diversity of areas beyond national jurisdiction. https://documents-dds-ny.un.org/doc/UNDOC/GEN/N23/177/28/PDF/N2317728.pdf?OpenElement

UNWTO (2018) UNWTO: Qatar among 10 most open visa countries in the world. https://www.unwto.org/middle-east/press-release/2018-09-03/unwto-qatar-among-10-most-open-visa-countries-world. Accessed 14 Jan 2022

Velenturf APM, Purnell P (2021) Principles for a sustainable circular economy. Sustain Prod Consum 27:1437–1457. https://doi.org/10.1016/j.spc.2021.02.018

Weber AS (2018) Desert landscapes and tourism in the Middle East and North Africa. In: Timothy D (ed) Routledge handbook on tourism in the Middle East and North Africa, 1st edn. Routledge, pp 189–198

WERC (Wildlife Education and Rehabilitation Center) (2023) Wildlife education. https://www.werc-ca.org/wildlifeeducation

Wise (2019) Learning ecosystems. https://www.wise-qatar.org/2019-wise-research-learning-ecosystems-innovation-unit/

Wood M (2002) Ecotourism: principles, practices and policies for sustainability. UNEP Publications Collections. https://wedocs.unep.org/20.500.11822/9045

World Economic Forum (2019) The travel and tourism competitiveness report 2019. Travel and Tourism at a Tipping Point. ISBN-13: 978-2-940631-01-8. https://www3.weforum.org/docs/WEF_TTCR_2019.pdf

World Travel and Tourism Council (2022) Qatar 2022 annual research: key highlights. https://wttc.org/Research/Economic-Impact

WTO (World Tourism Organization) (2007) Sustainable development of tourism in deserts–guide for decision makers. https://www.e-unwto.org/doi/book/; https://doi.org/10.18111/9789284411931

Yao X, He J, Bao C (2020) Public participation modes in China's environmental impact assessment process: an analytical framework based on participation extent and conflict level. Environ Impact Assess Rev 84:106400. https://doi.org/10.1016/j.eiar.2020.106400

Yap G, Saha S, Alsowaidi SS (2022) The competitiveness of Qatari tourism: a comparative and SWOT analysis

Open Access This chapter is licensed under the terms of the Creative Commons Attribution 4.0 International License (http://creativecommons.org/licenses/by/4.0/), which permits use, sharing, adaptation, distribution and reproduction in any medium or format, as long as you give appropriate credit to the original author(s) and the source, provide a link to the Creative Commons license and indicate if changes were made.

The images or other third party material in this chapter are included in the chapter's Creative Commons license, unless indicated otherwise in a credit line to the material. If material is not included in the chapter's Creative Commons license and your intended use is not permitted by statutory regulation or exceeds the permitted use, you will need to obtain permission directly from the copyright holder.

Chapter 9
Education for Sustainability

V. L. Miller, G. Haddad, R. Hinnawi, S. D. Sever, C. C. H. Lim, N. Al Salem, F. Hassan, and A. D. Chatziefthimiou

Abstract The educational environment in Qatar is characterized by its richness, diversity, and complexity, involving various formal and non-formal educational entities. This chapter presents an introductory overview of the existing educational landscape and highlights the key policies that drive the advancement of Education for Sustainability in Qatar. The current state of Education for Sustainability practice is highlighted through a series of vignettes from both formal and non-formal contexts. Each vignette offers valuable insights and recommendations for propelling Education for Sustainability forward. Currently, Qatar has a variety of policies in place, both at the national and organizational levels, which establish sustainability as a

V. L. Miller (✉)
Qatar Academy Msheireb, Qatar Foundation, Doha, Qatar
e-mail: vmiller@qf.org.qa

G. Haddad
Arab International Academy—Lusail, Doha, Qatar
e-mail: g.maalouf@aialusail.qa

R. Hinnawi · A. D. Chatziefthimiou
Earthna, Center for a Sustainable Future, Qatar Foundation, Doha, Qatar
e-mail: rhinnawi@qf.org.qa

A. D. Chatziefthimiou
e-mail: a.d.chatziefthimiou@gmail.com

S. D. Sever
Carnegie Mellon University in Qatar, Doha, Qatar
e-mail: duygusever@cmu.edu

C. C. H. Lim
Dadu Children's Museum of Qatar, Doha, Qatar
e-mail: clim@qm.org.qa

N. Al Salem
National Museum of Qatar, Doha, Qatar
e-mail: nalsalem@qm.org.qa

F. Hassan
Friends of the Environment, Doha, Qatar
e-mail: flowereachspring@gmail.com

© The Author(s) 2026

A. D. Chatziefthimiou et al. (eds.), *Pathways to Nature Conservation and Resilience in Hot and Arid Lands*, Advances in Geographical and Environmental Sciences, https://doi.org/10.1007/978-3-032-11046-6_9

priority. In many of the examples provided in this chapter, a theme emerges of sustainability as a disposition that aligns to local values, culture, and heritage. Qatar Foundation is in the beginning stages of prioritizing education for sustainability at the strategic level and is working toward coherent alignment of policy, curriculum, and teaching and learning practices. Non-formal education entities are increasing in number and placing sustainability at the forefront of their educational enrichment programs. The chapter concludes with a list of recommendations derived from the vignettes. In the formal education sector, these recommendations include fostering learning experiences that drive relevant and authentic action and adopting progressive, and inquiry-driven pedagogies that prepare students with the skills they need to tackle complex sustainability issues. Systems for forging meaningful partnerships between formal and non-formal stakeholders and increasing students' access to outdoor learning experiences are lacking. What emerges from this intricate network of stakeholders and organizations is the crucial need for a coherent and synergistic educational ecosystem—one that transcends isolated pockets of effective practices and fosters a community actively engaged in driving change.

Keywords Sustainability education · Qatar foundation · Environmental awareness · Curriculum development · Youth engagement · Eco-schools · Informal learning

9.1   Introduction

Throughout the chapter, a variety of education contexts are introduced including university, pre-university (pre-school-grade 12), museums, and outdoor learning settings. In addition to varied contexts, the chapter also explores educational programs. This discussion requires an understanding of where each of these contexts and programs fall in terms of formal, non-formal, and informal learning. Table 9.1 provides an overview of the key characteristics of each. To further clarify, the perspectives presented in the vignettes fall into the following categories: 1. Qatar Foundation Schools, Multiversity/University and the Ministry of Education and Higher Education are formal learning contexts; 2. Eco-Schools is a non-formal learning program that may be connected and aligned with the formal school curriculum. Eco-Schools also develops educational resources and promotes learning beyond the formal school context and may include aspects of informal learning. 3. Dadu Children's Museum of Qatar and A Flower Each Spring are non-formal learning contexts that develop educational programs and resources. Because these organizations often work with families outside of the formal learning contexts, they also encourage informal learning. It is important to note that the delineations between the categories are not hard fixed. Formal learning contexts, especially those that adopt inquiry-based and progressive pedagogies, may encourage non-formal and informal learning as part of the curriculum.

Table 9.1 Key characteristics of learning contexts (adapted from Johnson and Majewska 2022)

Formal learning	Non-formal learning	Informal learning
Learning is structured (e.g., linear objectives)	Learning **may be** structured	Learning is not structured
Learning is promoted through direct teaching behaviors	Learning is promoted through indirect teaching behaviors	
Learning is intended (by educator and learner)	Learning is intended by the **learner**	Learning may not be intended by the learner
Learning is recognized by the learner and educator	Learning is recognized by the **Learner**	Learning may not be recognized by the learner
Motivation for learning may be extrinsic to the learner		Motivation for learning is intrinsic to the learner
Learning takes place in educational institutions	Learning **can** take place in educational institutions	Learning can take place anywhere
Learning has a mandated dimension	Learning has a voluntary dimension	
Learning may be recognized or measured through qualifications		Learning is not recognized or measured through Qualifications
Learning may primarily focus on propositional knowledge	Learning may focus on both propositional and procedural knowledge	
Learning tends to have a cognitive emphasis	Learning involves cognitive, emotional, social and behavioral elements	

The term "curriculum" also presents ambiguities. In formal learning, the written, taught, and assessed curriculum is intentional, coherent, and horizontally and vertically aligned. In the case of Qatar Foundation Schools, the written curriculum is also aligned to the Cloud Institute's Education for Sustainability (EfS) Standards and select Ministry of Education Standards. Non-formal educational programs often develop supplemental curriculum resources, which they sometimes refer to as curriculum. However, these resources are extracurricular and sit outside of the formal learning context and written, taught, and assessed curriculum (Fig. 9.1).

This brief introduction provides a framework to connect the various entities and stakeholders represented in the chapter and their respective roles in the complex educational landscape of formal, non-formal, and informal education and in curriculum and learning.

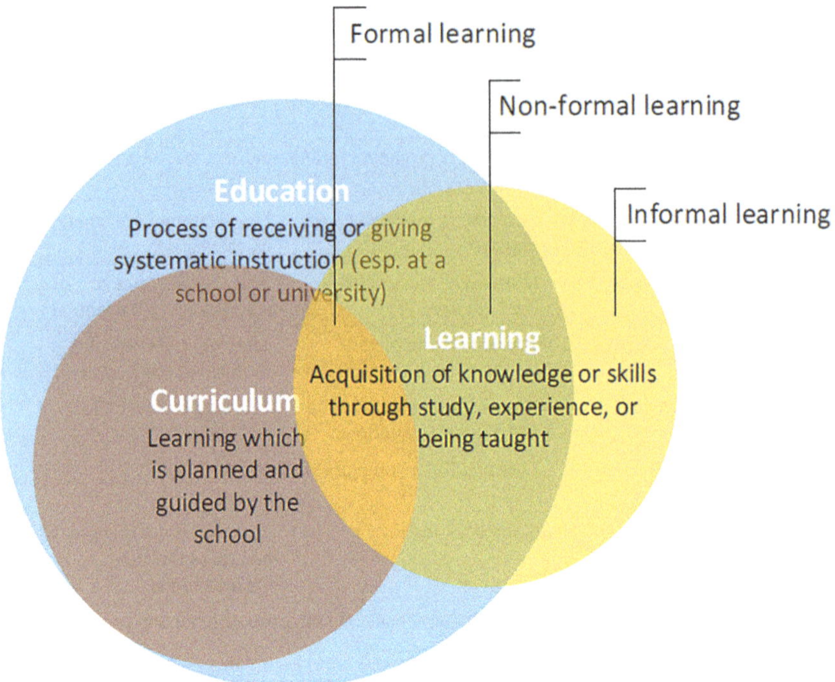

Fig. 9.1 Illustration of the overlap and interplay between formal, non-formal, and informal learning as well as education, curriculum, and learning (adapted from Johnson and Majewska, p. 6, 2022)

9.2 Local Context and Educational Landscape for Sustainability

Nestled on the Arabian Peninsula, Qatar is a compact nation that grapples with the heat and aridity of its landscape. As the effects of climate change emerge, the country finds itself facing challenges that necessitate embedding sustainability at the heart of its national vision (Table 9.2).

Qatar National Vision 2030 assigns significant emphasis on the role of education to foster sustainability and capacity building (Qatar General Secretariat for Development Planning 2008, 2). Quality education emerges as the key system for developing youth with the skills and dispositions needed to tackle the current and future needs of Qatar (Qatar General Secretariat for Development Planning 2011, 139). By promoting innovative teaching and learning practices, developing quality educational resources, and providing students with authentic and relevant experiences connected to local sustainability issues, Qatar shows an increasing commitment to incorporating sustainability into the education system.

Table 9.2 Qatar National Vision overview of major changes and pillars and their connection to education (Qatar National Vision 2030, 2008)

Qatar National Vision 2023
"Qatar National Vision 2030 builds a bridge between the present and the future. It envisages a vibrant and prosperous country in which there is economic and social justice for all, and in which nature and man are in harmony…Strong Islamic and family values will provide our moral and ethical compass… The welfare of our children, and of our children yet to be born, demands that we use our resource-wealth wisely. Qatar must continue to invest in its people so that all can participate fully in economic, social and political life"

Five major challenges

Modernization and preservation of traditions	The needs of this generation and the needs of future generations	Managed growth and uncontrolled expansion	The size and the quality of the expatriate labor force and the selected path of development	Economic growth, social development, and environmental management

Four pillars

	Human development	Social development	Economic development	Environmental development

A national network of formal and non-formal educational programs that equip Qatari children and youth with the skills and motivation to contribute to society, fostering
– A solid grounding in Qatari moral and ethical values, traditions, and cultural heritage
– A strong sense of belonging and citizenship
– Innovation and creativity
– Participation in a wide variety of cultural and sports activities

Subsequent National Development Policies underline the need to promote environmental awareness and encourage more sustainable behaviors and mindsets (Planning and Statistics Authority 2018, 318). The Qatar National Vision 2030 highlights the connection between sustainability and national values, traditions, and cultural heritage, and emphasizes the importance of creating environmentally conscious citizens (Qatar General Secretariat for Development Planning 2008, 2). Other policy commitments include support for an increase in environmental studies programs in schools, embedding principles of sustainability into the curriculum, and linking local and international education initiatives (Qatar General Secretariat for Development Planning 2011, 245–252).

The implementation of education for sustainability in the formal education system in Qatar does not follow a standardized approach. Rather, it takes place in a complex context that involves a combination of top-down initiatives, global frameworks, programs, and extracurricular activities in schools. This is partly due to the complex and diverse school system, which consists of public (Ministry of Education) schools and private schools. Qatar has a total of 1,004 schools, with 318 public schools that cater 126,256 students from pre-primary to secondary levels, and 686 private schools that serve approximately 200,241 students from different nationalities (Ministry of Education and Higher Education 2022). Schools can be further divided into community schools, international schools, and Arabic schools and vary in terms of their nationalities, curricula, and practices related to education for sustainability.

The Qatari Government recognizes the importance of developing educational curricula and training programs that meet the current and future needs of the country. The Ministry of Education and Higher Education (MOEHE) sets the standards to support high-quality learning opportunities across all educational levels. The emphasis on twenty-first-century skills, values, and attitudes in the national curriculum framework aligns with the global perspective on the significance of education for sustainability. The official national curriculum framework refers to sustainability as a core value of the Qatari education system, along with peace, tolerance, mutual respect, and cultural understanding, and encourages significant engagement in education for sustainability (MOEHE 2022b).

It is crucial to underline that the formal education system is not the only space for education for sustainability. Teaching and learning about sustainability goes beyond the classroom to translate curriculum-based knowledge into behavioral change through connecting the learners with nature, and creating first-hand experiences. To this end, programs, initiatives, and tools related to education for sustainable development emerge from a variety of channels, formal, non-formal, and informal.

Qatar's educational ecosystem is characterized by a diverse array of such initiatives, projects, and non-institutional actors, which are involved in a range of activities and partnerships that support education for sustainability. A recent study mapping these stakeholders identified that Qatar is home to over 400 actors, campaigns, and initiatives, both online and on-site, that serve teaching and learning on sustainability (Sever and Tok 2023). These include diverse sources such as schools, universities, non-profit organizations, government entities, community-based environmental conservation initiatives and awareness campaigns, as well as structured extracurricular efforts. Some of these efforts take place within the formal school system, such as programs and clubs initiated by school leadership, staff and students, and extracurricular activities led by teachers. Formal education stakeholders are also often connected with international initiatives which translate global agenda on sustainability into school agenda, such as UNESCO Associated Schools Network, or international programs which lead the schools to connect with the global efforts by also localizing their projects, such as the Eco-Schools Programs. As another example, Qatar Foundation stands out as an entity at the intersection of formal and informal efforts for education for sustainable development by integrating the concept of sustainability in the education programs, as well as leading community activities through awareness programs and its research centers.

The same mapping research reveals that initiatives taking place through K-12 schools, higher education institutions, or student-led efforts make only around 44% of the entire education for sustainability ecosystem in Qatar (Sever and Tok 2023), showcasing a rich existence of non-formal stakeholders. Government entities, private sector, local awareness campaigns, and key social figures, who showcase leadership, expertise, and efforts for sustainable development, play a critical role in offering opportunities to increase awareness on sustainability, to ensure the protection of environment and to promote a culture of responsible consumption, along with the formal education system.

9.3 Ministry of Education and Higher Education

9.3.1 Sustainable Development in the National Education Curriculum in the State of Qatar

The General Framework is the educational constitution for the Qatar national curriculum and is comprised of four fundamental components: values, goals, principles, competencies, and common issues. The environment and sustainability are emphasized in the common issues curriculum component. Instilling values that drive positive behavior and developing student competencies necessary for dealing with sustainability issues are primary goals of the curriculum.

During the process of developing subject standards for curricula, special consideration was taken to include the issue of sustainability in theory, learning outcomes, and practical application. Various topics, concepts, and activities related to sustainability were introduced gradually and cumulatively, in line with the age groups of the students, the nature of the subject, and its educational content. Care was also taken to design lessons in a manner consistent with the specifics of the study subjects and the type of activity, to enhance the educational competencies of the students and provide them with the required knowledge and skills, and guide them behaviorally to adopt a positive and active attitude toward the environment, climate, and social values.

During the 2022–2023 academic year, the Ministry of Education and Higher Education introduced the subject "Earth and Environment Sciences" as an optional subject for the eleventh and twelfth grades in the scientific and technological tracks. This subject was introduced to raise learners' awareness of the importance of environmental issues, current local and global challenges, and to raise awareness of environmental hazards and the importance of sustainability. The aim of all of these efforts is to develop a generation of students in Qatar who have a deep understanding of sustainability issues and who are able to think critically and make informed decisions to contribute to building a sustainable future for their country and the world at large.

9.3.2   Qatar Education Forum 2021: Education and Sustainable Development for a Prosperous Future

The Ministry of Education and Higher Education in the State of Qatar organized the Education Forum in 2021 under the title "Education and Sustainable Development." The objective of this important virtual forum was to bring together experts and specialists in the field of education to discuss and exchange ideas about the role of education in achieving sustainable development. The forum focused on exchanging experiences and sharing local and international efforts to develop positive education practices that contribute to sustainable development. The forum included the following topics:

- Enhancing the role of school curricula in achieving the goals of sustainable development.
- Examining ways of harmonizing earning outcomes with the requirements of the labor market and sustainable development.
- Finding new insights into early childhood education.
- Highlighting the role of educational supervision and leadership in promoting sustainable professional development for teachers.
- Fostering community partnerships to achieve education for sustainable development.

This forum was a vital occasion to enhance dialogue and cooperation between experts and practitioners in the field of education, with the aim of improving the quality of education and enhancing its role in achieving sustainable development and creating a prosperous future for society.

9.3.3   Community Partnerships and Activities Supporting Sustainability: Building a Sustainable Future

The Ministry of Education and Higher Education is a pioneer in adopting the principle of partnership with ministries, governmental, and non-governmental agencies to promote sustainable development projects. The Ministry seeks to enhance awareness and develop skills related to sustainability through projects such as the Green Computer Project, robot activities and competitions, artificial intelligence, digital manufacturing, electronic circuits, and the Hackathon project. The Ministry also invests in national and international events to raise awareness of environmental and sustainability issues such as Qatar Environment Day, World Environment Day, Earth Day, and Tree Week.

Within the framework of community partnership, many classroom and extracurricular activities are organized that promote concepts of sustainability at all levels of education. These activities include recycling, methods of reducing e-waste, using renewable and clean energy, digital transformation, programming robots to detect stains and pollution in the environment, designing posters, and collecting and analyzing information through electronic forms. Through these community partnerships and supportive activities, the Ministry of Education and Higher Education seeks to build an informed and responsible generation that contributes to a sustainable future and works to preserve our environment and our cultural heritage for future generations.

Examples of community partnerships and collaborations in the field of sustainability:

- "My School is My Farm" project: The Environmental Awareness and Education Department of the Ministry of Environment and Climate Change cooperated with a number of public and private schools to launch this project, which aims to educate students about the importance of agriculture and encourage them to grow crops in the vicinity of their schools.
- "A Flower Every Spring" program: The Qur'anic Botanic Garden program promotes environmental awareness among school students by introducing them to local plants and encouraging them to plant trees in various regions.
- "Goal 22" program: This school program, sponsored by UNESCO, aims to build a sustainable legacy for the Qatar World Cup 2022, by promoting social responsibility and raising a creative generation that contributes to change through teaching life skills and sustainable development.
- "Morouj Qatar" initiative: This initiative, launched by Qatar University in cooperation with Qatari farm owners, aims to promote practices on agriculture and reduce consumption in schools through events and workshops.
- Eco-Schools Program: The Qatar Green Building Council formerly and now Earthna Center for a Sustainable Future, in cooperation with the Ministry of Education and Higher Education, is implementing this program to promote environmental culture in schools and encourage environmentally sustainable practices.
- The "Tarsheed 22" project is a distinguished example of partnerships that aim to raise awareness in schools about the importance of energy conservation by using football as an effective way to reach to young people. The National Program for Conservation and Energy Efficiency "Tarsheed" was launched in cooperation with the Qatar General Electricity and Water Corporation. Kahramaa and the Supreme Committee for Delivery and Legacy, the project was implemented in multiple stages. The first phase began with the participation of 1,500 students from 22 schools in the local communities surrounding the stadiums that hosted the FIFA World Cup matches. The project achieved three GWH reduction in electricity consumption annually, 750k QR annual savings and reduction of 1,600 tons of CO_2e annually. The schools replaced the inefficient lighting and plumbing devices with energy-saving ones, which contributes to achieving these positive results. The project aims to expand its scope to include all schools in Qatar by 2022 with the aim of building a conscious generation that understands the importance of preserving the country's wealth in the future.

The Ministry is keen on exchanging knowledge and experiences between students from different schools and educational levels, by organizing research competitions annually. The participation of schools contributes to enhancing the creative and innovative spirit of students and strengthening their role in achieving sustainable development in society. These activities and competitions aim to inculcate the values of caring for the environment and sustainability in students' hearts and develop their environmental awareness.

9.3.4 The Role of Scientific Research in Achieving Sustainable Development and Promoting Environmental and Social Responsibility

Scientific research is one of the main requirements for achieving sustainability in education, as it contributes to the development of knowledge and technology and encourages the spirit of innovation in order to find innovative solutions to environmental, economic, and social challenges. The Ministry of Education and Higher Education pays great attention to encouraging students to conduct scientific research that focuses on sustainable development issues such as using renewable energy, managing natural resources effectively, and preserving biodiversity.

The Ministry also seeks to cooperate and establish partnerships with the authorities concerned with scientific and environmental research, which contributes to enabling students to link theoretical and applied scientific aspects. There are many examples of topics that have been addressed through students' research projects in the field development and sustainability such as.

- Designing a PMD device to convert carbon monoxide gas into a gas used in extinguishing fires.
- Using cooking oils as an environmentally friendly fuel.
- The effect of *Prosopis juliflora* trees on the diversity of vegetation in the Qatari environment.
- The effect of using limestone and sand in the Qatari environment for the purification of consumed water.
- Using solar energy to produce green hydrogen to achieve sustainable development in the State of Qatar.
- Using solar energy to design a smart self-irrigation system that reduces the amount of water consumed in irrigation and lighting public parks in the State of Qatar.

These topics reflect the Ministry's commitment to promoting scientific research and stimulating innovation among students with the aim of promoting sustainable development and environmental awareness.

9.3.5 The Role of the Qatar National Commission for Education, Culture, and Science in Promoting Sustainable Development and Global Citizenship

Science cooperates closely with the Ministry of Education and Higher Education to achieve the goals of international organizations that are compatible with the goals of education in the State of Qatar. These efforts are supported by the joining of Qatari schools to the global network of Associated Schools of UNESCO.

Since the joining of the first school in 1983 (Doha Secondary School for Boys) and until 2020, about 80 schools for boys and girls of all educational levels have joined the network. Schools joining the global network focus on implementing activities and projects that focus on protecting and preserving the environment, promoting mutual learning between cultures, promoting principles of respect for human rights, democracy and just peace, openness to and acceptance of new ideas, and implementing pioneering projects in these fields.

Among the programs and activities implemented in UNESCO affiliated schools under the supervision of the Qatar National Commission for Education, Culture, and Science include the following:

- UNESCO Ambassadors Program: Students are appointed to represent Qatar at local and international student events.
- Green Schools Program: It aims to transform schools into green environments and promote environmental awareness and sustainable development.
- Intercultural Dialogue Program: Works to enhance communication and cultural interaction between students and the local community.
- School Twinning Project: Cooperative programs are being implemented between Qatar schools and schools in other countries such as Kuwait, Oman, and Jordan.
- Media Awareness and National Affiliation Program: It aims to enable students to prepare media programs that promote issues of the homeland and local culture.

These schools also celebrate global events such as the International Day for Cultural Diversity, the World Cultural Heritage Day, the International Family Day, and the International Earth Day, with the aim of spreading awareness of the culture of sustainable development and promoting national identity and local cultural values.

9.3.6 Conclusion

The Ministry of Education and Higher Education prioritizes building an educational system that contributes to achieving the goals of education for sustainable development. These efforts are achieved through reformulation and development of school curricula to ensure coverage of various sustainability concepts and to improve the quality of the curriculum. The capabilities of the teaching staff, considered the main pillar in achieving success, are also strengthened to ensure their ability to implementing sustainable education processes.

The Ministry also focuses on linking curricula and education programs to the development of the local community, enhancing national belonging among young people and enhancing their awareness of their role as global citizens. As a result of these efforts, the principles of sustainable education have been integrated into the national education system, in both formal and non-formal settings, with the aim of promoting sustainable thinking, promoting community participation, and preserving natural resources and the environment.

Through these measures, the Ministry seeks to build an educated generation that possesses the knowledge and skills necessary to interact with the challenges of the changing world and to contribute to building a sustainable and prosperous future for the country.

9.4 Qatar Foundation Schools

Qatar Foundation (QF) is a not-for-profit organization comprising over 50 education, research, and community development entities. Created in 1995 by the sitting Emir, Sheikh Hamad Bin Khalifa Al Thani, and his wife, Sheikha Moza bint Nasser, QF is the legacy of the royal family's vision of prioritizing education as a national commodity. Education City deemed the Foundation's flagship initiative encompasses at least eight university campuses, six pre-university campuses, a national library, and various other research and community centers (Building a World-Class Education Ecosystem in Qatar n.d).

In 2022, Qatar Foundation released a comprehensive strategy entitled "Cultivating Multiversity: Qatar Foundation Strategy Refresh. 2022–2032." This 81-page document articulates a clear vision, mission, and strategic direction to steer the organization for the next 10 years. The Refresh further aligns Qatar Foundation with Qatar's National Vision 2030 (Table 9.3).

The theme of sustainability is defined as "protection and restoration of the environment while balancing economic development and well-being" and is divided into five subthemes: sustainable energy, resource security and management, environment protection and restoration, sustainable/circular economy, and well-being. Social responsibility, under the theme of social progress, gives priority to "providing opportunities for youth and the community to be empowered, connected, and engaged to contribute toward shaping society in meaningful ways." These themes and subthemes within the strategy refresh place sustainability at the forefront and as a priority for all organizational initiatives.

The Pre-University Education (PUE) division of Qatar Foundation encompasses 13 schools serving students from early childhood to year 12. The Strategy Refresh offered PUE leaders an opportunity for reflection and re-envisioning. PUE has undertaken several steps to revise education policies and curriculum to align with the new vision and created a plan of action toward education for sustainability. The QF

Table 9.3 Summary of Qatar Foundation's strategic areas of focus (Qatar Foundation 2022)

Cultivating Multiversity: Qatar Foundation Strategy Refresh 2022–2032

Themes: strategic areas of focus via which QF can jointly achieve its strategic objectives and intended impact

Progressive education	**Sustainability**	**Artificial intelligence**	**Precision health**	**Social progress**
Innovatively re-imagining and transforming education to unlock the potential of learners and prepare them for the world of tomorrow by encouraging creativity and student agency	Protection and restoration of the environment while balancing economic development and well-being	The advancement of society and improvement of the quality of human life by focusing on big data, algorithms, and cognitive computing	Individualized healthcare and disease prevention driven by emerging research, clinical approach, environment, and lifestyle	A thriving and value-based society rooted in Qatar's cultural heritage and identity

(Qatar Foundation 2022)

Curriculum Framework and Review Cycle establishes a common vision of progressive pedagogies and cultivating learners with a set of defined traits: 1. Multilingual and biliterate; 2. Socially engaged; 3. Academically competent; 4. Locally rooted, globally minded; and 5. Critical thinkers. Through a 5-year review cycle, leaders and educators will work to align the written, taught, assessed, and hidden curriculum with the QF Graduate Profile.

Each of the 11 schools within PUE is unique, and the variety of existing curricula, ranging from six International Baccalaureate schools to other highly specialized schools, reflect that individuality. Key to developing the profile are common standards that can work within the individual ethos of those varied schools. During the Curriculum Review cycle, QF Schools will work toward aligning their existing curriculum with The Cloud Institute's "Education for Sustainability Standards (EfS)." The EfS standards are organized under the areas of Cultural Preservation and Transformation, Responsible Local and Global Citizenship, The Dynamics of Systems and Change, Sustainable Economics, Healthy Commons, Natural Laws and Ecological Principles, Inventing and Affecting the Future, Multiple Perspectives, and Strong Sense of Place. The EfS standards align to the QF Strategy Refresh and provide a roadmap for schools to reimagine their curriculum and actualize the vision through action.

Together, the QF Graduate Profile and alignment to the EfS standards outline the "What" of the PUE vision for change. The "How?" encompasses a variety of pathways in the form of resources, tools, and programs available to the schools. Some of the pathways involve collaboration within QF PUE entities, such as the Educational Development Institute (EDI). EDI provides a variety of support services such as Curriculum Development and Professional Learning. Over the next 5 years, the EDI Curriculum Development Department will provide extensive and ongoing support to schools to align their written, taught, assessed, and hidden curriculum with the Graduate Profiles and the EfS. EDI will offer a variety of professional learning opportunities in the form of conferences, workshops, and in-school inquiries aimed at building educator capacity and lift teaching and learning practices. Additionally, EDI facilitates a professional learning community (PLC) for school leaders and will lead and support the curriculum review process through bi-monthly meetings. The theme chosen by school directors for EDI's annual Teaching and Learning Forum was "Learning in Service and Action: Creating meaningful connections, developing local and global learning communities." This event brought together a variety of community organizations with local educators for the purpose of forming meaningful partnerships and inspiring a culture of service learning and sustainability action initiatives.

Jamie Cloud, working in collaboration with EDI, traveled to Qatar to facilitate several workshops around the EfS Standards, offered curriculum coaching to individual schools, and conducted a train the trainer workshop. Train the trainer programs equip participating educators with the theoretical and pedagogical expertise to train other educators, therefore creating a multiplier effect and spreading the desired practice. Every QF School now has at least two representatives who can work as "sustainability champions," building capacity and deepening collective understanding across the family of schools.

Qatar Foundation's Strategy refresh further outlines a vision for "Cultivating Multiversity" among all QF Entities and encourages a dynamic ecosystem seeking to create synergy among its stakeholders. One such collaboration is the partnership with PUE Schools and Eco-Schools. Eco-Schools, under the umbrella of Qatar Foundation's Earthna (discussed in detail later in the chapter), provides another pathway for schools to create sustainable communities. All PUE schools will be required to become green flag certified within 5 years. The green flag grants schools' access to Eco-Schools curriculum resources. Pockets of strong practice are already emerging among the schools, with two already awarded the green flag.

Challenges

Change is hard. While the QF Strategy Refresh, The QF Curriculum Framework and Review Cycle, and adoption of the Education for Sustainability Standards provide a clear destination for meaningful change toward education for sustainability, the implementation of change requires the buy-in of leaders and educators within the PUE family of schools. While evidence of early adopters is already apparent, true reformation of the written, taught, and assessed curriculum will require dedication and diligence as well as collective understanding of intended outcomes. EDI established a variety of support mechanisms such as curriculum development and professional learning. However, educators often find it difficult to manage competing demands and engaging with curricular change can feel daunting.

Furthermore, each individual school is required to adhere to a variety of accreditation organizations and evaluation measures including, but not limited to, the Ministry of Education, International Baccalaureate, CIS (Council of International Schools), and MARQ (Motivation Accreditation Review Qatar). If schools are to be successful in adopting a new vision for change, then they will need support aligning their multiple program requirements and recommendations.

To develop the competencies encompassed in the QF Graduate Profile, PUE learners will need to engage in meaningful and authentic sustainable action projects. This will require tapping not only into the QF Multiversity, but also into other organizations within Qatar. Forming such partnerships requires a great deal of communication and dedication from both educators and members of the greater community. A persistent barrier to this work has been a resource directory for connecting organizations to schools. Efforts are currently underway to create a resource guide for collaborators. However, the ever-changing landscape and transient nature of Qatar presents challenges keeping the resource current.

One pathway for schools to reach the destination of Education for Sustainability is through robust outdoor education and co-curricular opportunities. Such opportunities have yet to be established, as stated previously, forming meaningful partnerships presents challenges. Creating a culture where the outdoor environment is viewed as an extension classroom will require a mindset shift among PUE learning communities including educators, learners, and families.

To fully grasp the urgency and need around sustainability, educators and learners need access to the Qatar scientific community and relevant research being conducted in Qatar. Learners would greatly benefit from engaging in citizen science programs involving them in meaningful and authentic research. Currently, these collaborations do not exist.

The concept of Indigenous knowledge is specifically addressed in the EfS standards as "Cultural Preservation and Transformation." This standard states.

> The preservation of cultural histories and heritages, and the transformation of cultural identities and practices contribute to sustain communities. Students will develop the ability to discern with others what to preserve and what to change in order for future generations to thrive.

There is a need for further examination of this concept and how it aligns to the local cultural context and fits into the broader theme of Qatari culture, heritage, and identity.

Currently, there is a lack of authentic literature and resources dealing with topics pertaining to sustainability in the Arabic language. The QF Graduate Profile requires learners to be multilingual and biliterate, "to responsibly use both English and Arabic to communicate their understanding of the world. They produce and make meaning of oral, visual, and written texts" (Curriculum Development Unit, p. 7). To achieve this level of competency, learners will need greater access to Arabic text.

9.5 Education for Sustainability in the International Baccalaureate (IB)

Description of the Current State of Practice

How does the International Baccalaureate (IB) endorse Education for Sustainability (EfS)? The IB plays a significant role in educating for sustainability because of its mission, standards, and practices, the structure and content of its programs and its orientation as a future looking organization. In this part of the chapter, we are going to shed some light on how the IB is aligned with the principles and practices of EfS without claiming that this light is a comprehensive and all-encompassing one.

IB Mission

"The International Baccalaureate aims to develop inquiring, knowledgeable and caring young people who help to create a better and more peaceful world through intercultural understanding and respect. Through its programs, the IB encourages students across the world to become active, compassionate and lifelong learners who understand that other people, with their differences, can also be right" (IBO 2021).

This better and more peaceful world is environmentally, socially, and economically sustainable. In addition, those active, compassionate, and lifelong learners are learners who have the will and skills to act, care about their local and global environments, and are continuously wondering about how to be more knowledgeable to solve more problems and have an impact as active citizens of the world, as guardians of the planet. With the principle of shared guardianship of the planet, where everyone is responsible for protecting the environment and preserving it for future generations. This responsibility extends to local, national, and global levels.

IB Learner Profile

The IB learner profile—the IB mission in action—describes a broad range of human capacities and responsibilities that go beyond academic success and exemplifies how IB learners are responsible members of local, national, and global communities. The ten profile attributes viewed from the education for sustainability perspective can lead to (Table 9.4).

International-Mindedness

The IB places a strong emphasis on developing international-mindedness in its students, which includes an understanding of global issues and a commitment to promoting global citizenship. By encouraging students to appreciate and value different cultures, beliefs, and perspectives and including the study of different languages, cultural awareness, and global issues, the IB encourages students to think beyond their own local contexts and consider the global impacts of environmental, cultural, and economic issues.

Table 9.4 Connections between the IB learner profile and sustainability

Inquirers	Learners initiate independent and collaborative inquiries into real-world local and global issues through questioning, problem and project-based learning and research projects
Knowledgeable	Learners who understand and utilize sustainability concepts with confidence to demonstrate an understanding of current problems and solutions to reach a sustainable decision
Thinkers	Learners who think critically and creatively to take responsible action and solve complex problems through making reasoned and ethical decisions
Communicators	Learners who are able to listen to different perspectives, express themselves in many ways to collaboratively find solutions to problems around them, including sustainability
Principled	Learners who have a strong sense of fairness, who respect the rights of others while taking responsibility for their own actions. They act with integrity and honesty to make the world a better place
Open-minded	Learners who understand that sustainability of their own culture and that of others happens when they appreciate the values and traditions of everyone around them
Caring	Learners who are committed to service and who, through empathy, compassion, and respect, have a positive impact on the lives of others and on the world
Risk-takers	Learners who are resilient, resourceful, determined, and work collaboratively with others to continuously find innovative ideas and strategies to change the world around them into a better place
Balanced	Learners who understand how to maintain balance in their life and take care of their well-being as well as that of others. They know that the world, others, and themselves are interdependent and hence have to keep that balance
Reflective	Learners who reflect on their strengths and weaknesses, their experience, and development as a tool to analyze their actions and make a more positive impact on others and the world around them

IB Standards and Practices

In "Circular economy: A call to action for IB World Schools" (IBO 2021), the IB standards and practices, as a holistic framework to a school ecosystem, are compared to the Ellen MacArthur Foundation's circular economy. A quote from Sustainable schools, 2008 summarizes how the holistic framework, ensured by the standards and practices, puts forth education for sustainability.

A sustainable school prepares young people for a lifetime of sustainable living, through its teaching, fabric and its day-to-day practices" (Sustainable schools, 2008).

These standards (Purpose, Environment, Culture, and Learning) and practices, because of their holistic nature and the connections among each other, allow IB schools to integrate circular economy in the fabric of the school.

IB Education

An IB education includes approaches to teaching and learning which facilitate, in an integrated and meaningful way, education for sustainability. Those approaches to learning and teaching align well with the key traits of education for sustainability. Below are some examples of how these approaches are aligned with education for sustainability.

- **Approaches to learning**: The IB's Approaches to Learning (ATL) skills are thinking, research, communication, self-management, and social skills. These skills are particularly relevant to sustainability, as they help students to analyze complex environmental issues and to communicate their findings effectively as well as provide students with the essential skills needed to understand, analyze, and address sustainability challenges.
- **Approaches to teaching**: The IB's approaches to teaching (ATT) are conceptual and assessment driven; emphasize inquiry; encourage collaborative learning; promote equity, diversity, and inclusion; and utilize a global contexts approach to teaching that emphasizes the importance of exploring real-world issues. The ATTs emphasize relevant and action-oriented learning, striving to prepare youth for the challenges of an ever-changing future.
- **Agency, action, and service**: The IB places a strong emphasis on agency, action, and service as part of its curriculum. This encourages students to take action on issues they care about and to make a positive difference in their communities. All IB programs include a service-learning component which encourages students to engage in community service projects. Many of these projects are focused on promoting environmental sustainability, such as organizing beach cleanups or planting community gardens. When students are involved in service learning, they develop the "will" to act; this will and its connected action contribute to a better and more sustainable society.
- **Transdisciplinarity**: The Primary Years Program (PYP) emphasizes a transdisciplinary approach to learning, which encourages students to make connections between and beyond different subjects and disciplines to explore themes that deal with human commonalities. Among these human commonalities are sustainability issues; hence, this allows students to explore the complex interrelationships between environmental, social, and economic factors and deepen their understanding of sustainability.
- **Interdisciplinarity**. The interdisciplinary approach in the Middle Years Programme (MYP) allows students to see the connections between different subjects, encourages students to explore issues and topics from multiple angles, and invites them to explore how they can work together to address sustainability issues. The interdisciplinary approach draws upon a range of disciplines, perspectives, and knowledge integrating them into an interdisciplinary unit which encourages students to explore issues from multiple perspectives. Students understand the interconnectedness of different systems and see how different factors can influence and impact sustainability.

Theory of knowledge (TOK): The TOK course in the DP promotes critical thinking, asks students to consider different perspectives, and fosters ethical reflection. These skills and attitudes can help students to become more informed and engaged local and global citizens who make positive contributions as active citizens of the world to lead to a more sustainable future. When, through TOK, students critically evaluate the knowledge claims that they encounter in different areas of knowledge and sharpen their critical lens to information, they develop a deeper understanding of the impact that different forms of knowledge can have on the environment and society. Additionally, students reflect on the ethical implications of knowledge and action helping them to develop a sense of responsibility and agency in relation to sustainability issues. They become mindful active citizens that critically read the word to lead to a better world.

9.6 Multiversity/University Education

Multiversity is identified as Qatar Foundation's "unique strength," for creating a dynamic ecosystem which brings together world-class education institutes at all stages of education, research entities, and incubators. This model provides an integrated, multi-dimensional (e.g., sectors, disciplines, players, etc.) network in a single location: Education City, wherein an expanded, innovative, and hands-on learning experience is offered to the students. The multiversity approach does not only increase the impact of education, it also renders Education City as a hub for knowledge, and a test-bed for leading progressive approaches to teaching and learning (Qatar Foundation 2019; Qatar Foundation 2021).

The multiversity initiative targets to achieve (1) academic integration through joint programs, (2) collaborations for developing interactive learning methods, (3) joint research projects, (4) a shared Education City experience and identity for the university students, and (5) interactions with the local industry to support national projects and national capacity development. To this end, Qatar Foundation's education ecosystem already includes some examples for teaching, research, engagement, and outreach which support multiversity:

Cluster Project: SDG Education and Global Citizenship in Qatar: Enhancing Qatar's Nested Power in the Global Arena (Further information on the Project n.d).

This is a project funded by a 5-year (2021–2026) Qatar National Priorities Research Program Grant from Qatar National Research Fund. The project is motivated by Her Highness Sheikha Moza bint Nasser's reappointment as an Advocate of the Sustainable Development Goals (SDGs). The cluster addresses the Qatari education system's policies, guidelines, current knowledge, and practices related to SDGs and focuses on the question of how the Global Citizenship Education (GCED) and Sustainable Development Goals Education (SDGED) integrated with ethical, moral, and localized educational dimensions can be increasingly absorbed into Qatar's education system. The cluster follows a multi-disciplinary approach and is designed in eight sub-projects that examine existing initiatives, policies, knowledge, and curricula in Qatar and review global best practices related to education and advancement of the SDGs. By collaborating with multiple and diverse stakeholders, this cluster aims to provide a holistic profile of the best educational practices that can be adapted and indigenized to appropriately meet the needs of Qatar's education system, in a way that respects Qatari tradition, culture, Islamic values, and heritage.

Hosted at College of Public Policy, Hamad Bin Khalifa University (HBKU), the project is a concrete example of multiversity as it is conducted with the partnership of Carnegie Mellon University, Qatar; Qatar University; Emirates College for Advanced Education; and Qatar Ministry of Education and Higher Education. The project's collaborations with multiple QF and non-QF partners aim to bring synergies and amplify impact in order to achieve progressive education, sustainability, and social progress.

Product X: Innovating for Future Economies Boot Camp

"Product X: Innovating for Future Economies Boot Camp" is another example of multiversity initiatives organized in 2022, by a team of faculty from across multiple partner universities in Education City. The Product X Bootcamp was designed in a way to bring together undergraduate and graduate students from 13 different fields of study, from HBKU, Carnegie Mellon University in Qatar, Georgetown University Qatar, Texas A&M University at Qatar (TAMUQ), and Virginia Commonwealth University School of the Arts (Qatar Foundation 2022). The target was to increase students' engagement in sustainable development and to encourage them to come up with innovative responses to complex sustainability issues, with practical real-life implications. To illustrate, the participants worked on developing a sustainable food container, tackling every stage of product design. From sourcing materials, to accessing relevant environmental laws and sustainability policy issues, to business aspects, waste management, and consumer behavior.

Joint Research Projects and Case Studies

Conducting joint research projects and case studies is another growing sphere of collaboration within the scope of multiversity principles. A recent example is the case study titled "Fostering a reading culture: evidence from Qatar Reads" (Cochrane et al. 2022), which investigated "the role of Qatar Reads in the national vision's aspiration to encourage lifelong learning, strongly enabled by fostering a reading culture" (HBKU CPP 2022). The study has been conducted by a multiversity initiative involving co-authors from HBKU, Qatar Foundation, and Qatar University.

Teaching and Learning Experiences

Education City also hosts other multiversity initiatives that offer multi-disciplinary teaching and learning experiences. To illustrate, TAMUQ offers diverse experiences to its students in order to cultivate holistic learning and multi-dimensional skills beyond the traditional classroom. These include opportunities such as cross registration or joint academic programs and courses (TAMUQ 2022).

9.7 Eco-Schools

Established in 1992, the Foundation for Environmental Education's (FEE) Eco-Schools program is the largest sustainable environmental program worldwide. As of today, it engages approximately 13 million students from around 54,749 establishments in 82 countries. Eco-Schools is a non-formal education program that schools choose to participate in. It is considered a tool that guides schools to integrate sustainability in the whole school system, both at the educational and operational levels. Therefore, sustainability is not an activity or an initiative, it is a systematic school approach. Its core mission is to engage youth in taking positive actions that transform their lives as they work on environmental issues, beginning in their schools, then radiating outward into their homes, local community, and beyond. Eco-Schools equips students with environmental knowledge and practical hands-on/heads-on activities, allowing them to experience the direct positive impact of their actions. As opposed to scenarios of doom and gloom that create anxiety and powerlessness, this pro-active approach empowers students as active citizens capable of taking responsible action for the planet. As an international collective, individual schools are aware that daily actions at the local level are occurring every day in numerous Eco-Schools around the world. With small, local actions, individual schools continuously participate in a large, global movement. The Eco-Schools' comprehensive approach strengthens the student–community–nature relationship, and integrates sustainability in the daily campus life and decision-making process.

Through its seven-step framework and 12 themes, the program places students at the center of environmental action, creating positive change-makers. The seven-step process for becoming an Eco-School includes the following: step 1: form an eco-committee, 2: carry out a sustainability audit, 3: action plan, 4: monitor and evaluate, 5: curriculum work, 6: inform and involve, and 7: produce an eco-code. Eco-Schools provides comprehensive learning tools and resources that align to the 12 themes. These resources can be integrated, adapted, and aligned with schools' existing curricular, providing authentic and contextual learning engagements. For successful implementation and greater impact, Eco-Schools provides comprehensive learning tools and resources that align to the 12 themes. These resources can be integrated, adapted, and aligned with schools' existing curricular, providing authentic and contextual learning engagements (Table 9.5).

In 2018, Earthna, Center for a Sustainable Future started Eco-Schools in Qatar, and since then the program has gained momentum among schools, students, and teachers. Qatar Foundation requires Pre-University schools to earn the Eco-Schools green flag. The Eco-Schools framework aligns with the Qatar National Vision and National Biodiversity Strategy and Action Plan. Additionally, schools view Eco-Schools as a tool for preparing students to embrace real-life challenges and contribute positively to the well-being and sustainability of their communities. The flexible framework integrates sustainability and environmental education into the existing curriculum rather than imposing an extra and disconnected subject. Eco-Schools also encourages students to activate and "sustainablize" their physical school environment. School

Table 9.5 Eco-Schools themes (https://www.Eco-Schools.global/themes)

Biodiversity and nature	Climate change	Energy	Food
Examines the flora and fauna present in the school environment and suggests ways to increase the levels of biodiversity around the school and raises the pupils' awareness of biodiversity and nature	Examines the impacts we have on the climate through our lifestyles and how our actions can influence the situation in a positive way	Suggests ways in which all members of the school can work together to increase awareness of energy issues and to improve energy efficiency within the school	Encourages young people, their parents, and the whole community to take responsible food-related choices and actions that protect the environment, promote human rights, and improve the well-being of society—every day
Global citizenship	**Health and well-being**	**Litter**	**Marine and coast**
Examines what our rights and responsibilities are on a National, European, and Global scale and encourages staff, students, and parents to look at the impacts our consumption habits have on other parts of the world	Encourages schools to promote the health and well-being of young people and the wider community and to make environmental connections to health and safety	Examines the impact of litter on the environment and explores practical means for reducing and minimizing the amount of litter produced by the school	Teaches children about local and/or global coastal and marine habitats, how people are affecting these habitats and what we can do to protect them
School grounds	**Transport**	**Waste**	**Water**
Encourages schools to introduce children to the natural environment and to biodiversity in a practical way by offering a safe and potentially exciting facility for outdoor education that can complement classroom-based activities	Suggests ways for pupils, staff, and local government to work together to raise awareness of transport issues and come up with practical solutions that will make a real difference to pupils' everyday lives	Examines the impact of waste on the environment and explores actions to minimize the amount of waste that we produce and dispose of on a daily basis	Provides an introduction to the importance of water both locally and globally and raises awareness of how simple actions can substantially cut down water use

forest projects, herbal gardens, artificial ponds, greenhouses with fruit and vegetable production, and food waste composting are commonly shared practices among Eco-Schools. These initiatives transform the school into a localized action space, where students can immediately see the impact of their positive behaviors and actions, when actively contributing to reducing their schools' footprint.

Glocalized Learning Resources

Eco-schools Qatar places great value in contextualizing the Eco-Schools themes and developing "glocalized" curriculum resources. "Glocalized" learning blends and connects local and global concepts and contexts. Learners benefit from glocalized resources because they are better able to relate to the concepts and content through familiar examples (Patel and Lynch 2013). Qatar Eco-Schools recognized the need to go beyond isolated experiences, such as beach clean ups, to a more comprehensive approach to tackling Eco-Schools themes such as Biodiversity and Marine and Coast. They developed six learning resources for the unique Qatar marine ecosystems: coral reef, intertidal, subtidal and beach, mangrove forest, open Gulf, sabkha, and seagrass beds (https://www.earthna.qa/node/3466). These place-based engagements foster a sense of identity and responsibility among students toward their local environment. They include many local resources that provide teachers and students with thorough knowledge of the ecosystems' biodiversity with prudently designed in-classroom as well as nature-based activities. The resources offer students the conceptual knowledge needed to understand the importance of these ecosystems and their vital role to the well-being of Qatari community and sustainable behavior of a nation. With meticulously prepared maps of the six unique marine ecosystems, the program offers leaning engagements that encourage the use of natural environment as a classroom environment. Overtime, these learning opportunities 1. develop environmental literacy, 2. inspire respect and appreciation for the local environment in Qatar, and 3. build skills that lead to protection and conservation practices.

Glocalized learning resources enrich the existing curriculum with a contextualized, action-based program and integrate sustainability in the educational ecosystem. A program that is sensitive to local dynamic systems, such as ecological, economic, political, and social, empowers young adults and communities with new ways of thinking. Students need to commit individually and collectively to actions toward innovative regenerative solutions for a societal transformation where the environment is at the center. The flexibility of the Eco-Schools framework encourages contextualizing learning, increasing outdoor learning experiences and opportunities for authentic problem-finding and problem-solving.

As co-learners, educators and students equally benefit from increased knowledge of local environments and ecosystems. Teachers need this knowledge to create glocalized curriculum and opportunities to empower students to take action in the local context and to build the needed dispositions for future change-makers to foster identity, ethics, and values such as empathy, responsibility, and respect to all creatures that we share earth with and that our existence and well-being relies heavily on.

Eco-Schools Journey Impact and Way Forward

Earthna administered a feedback survey in June 2022, using a questionnaire with Likert-scaled and open-ended questions, which was distributed to Qatar Eco-Schools teachers and school staff members. The purpose of the survey was to evaluate the program's strengths, opportunities for growth, and impact on student learning outcomes. As per the survey findings, the Eco-Schools program succeeded in achieving these learning outcomes: knowledge, competencies, and responsible behavior. Responses of one of the questions indicate that a total of 97% teachers agree that, when environmental education is embedded into the curriculum, it enhances students' appreciation and respect toward nature, and increases their motivation and intention to act. Moreover, a significant majority, 91% of the teachers in the sample, also indicate that a curriculum which incorporates environmental education increases students' competencies and skills, such as analysis, investigation, and action plan developments. Furthermore, the majority 93% view indicates the belief that embedded environmental education results in more responsible behavior toward sustainability. However, it is important to develop reliable tools or methodology to measure the effectiveness of the Eco-Schools program and other sustainability interventions that are introduced to students. Developing such measurement mechanism enables policy-makers to recommend the most effective integrative framework for education for sustainability in Qatar. It is important to develop evidence-based policy recommendations that promote to the development of students' environmental knowledge, ecological identity, and attitudes.

9.8 The Outdoor Classroom: A Flower Each Spring

Her Highness Shaikha Moza Bin Nasser launched A Flower Each Spring program in 1999. In 2016, A Flower Each Spring Program became a member of Qatar Foundation by joining Qur'anic Botanic Garden, combining their similar objectives in botany, environmental education, and cultural conservation. The program conducts many activities and projects that target the diverse population of Qatari residents as well as international audiences through virtual activities, currently reaching more than 40 countries.

The physical base of A Flower Each Spring, a campsite next to the Purple Island, Al Khor, provides dedicated space for protecting the environment and biodiversity of Qatar and for increasing community awareness of native flora, its benefits, and role in beautifying the local Qatari environment. The Flower Each Spring campsite facilitates annual fun-filled scientific field trips hosting up to 300 visitors. These trips aim to increase students' environmental awareness, interest, and involvement in scientific field trips and engage schools and other visitors with the local environment, horticulture, recycling, insects, birds, nature sports, marine life, and embracing the pearl diving heritage. The education program provides an outdoor classroom, a space offering unique experiences away from the regular school environment. The outdoor classroom creates connection with nature and fosters a sense of balance and harmony between individual/society and the natural world. This connection promotes students' well-being, provides an engaging learning environment, promotes sustainable behaviors and mindsets, and inspires action. The outdoor classroom appeals to varied learning styles, providing visual, auditory, or kinesthetic experiences. Learning in an open environment encourages exploration and inquiry, and it evokes a sense of freedom. De-briefing learning experiences through discussions promotes critical thinking, collaborative learning, and cultivates positive attitudes, personal growth, and moral development. Through the use of these practices, A Flower Each Spring aspires to shift human behavior through inquiry-based learning and reflective thinking.

A Flower Each Spring's learning experiences are designed to inspire students to make a difference and take action toward tackling environmental challenges in the world we live in. Raising awareness about personal responsibility in protecting the environment and encouraging children to adopt the principle of reducing waste, reusing, and recycling resources and products introduces the concept of circular economy and authentically connects it to daily actions, working toward a sustainable zero waste future transaction with nature. Participants of the program often express feelings of enthusiasm, inspiration, and a renewed commitment toward sustainability.

During the COVID-19 pandemic, A Flower Each Spring employed virtual classrooms as a mechanism to reconnect with nature from a remote and fresh perspective. Creating an easy-going environment and finding fun ways to engage children with sustainable actions was challenging, yet the program provided children from various countries shared experiences and a space to build relationships and learn from each other. This vital action helped ensure hope for a prosperous and greener future for this generation and generations to come.

Environmental preservation is a collective responsibility, and A Flower Each Spring is proud to share that value and encourage everyone to be part of positive environmental change. In the next few years, the program aspires in succeeding to achieve that objective and inspire others, especially the youth, to take further steps to promote environmental sustainability—from households, to communities, to schools, and beyond. The program hopes to play a key role in Qatar's continued regional leadership in environmental sustainability policies.

9.9 National Museum of Qatar Initiatives

The National Museum of Qatar (NMoQ) is dedicated to delivering visitors of all ages and backgrounds beneficial educational experiences. Environmental education is a priority among NMoQ's overall educational programming and activities. NMoQ seeks to increase awareness, advance environmental sustainability, and build a strong bond between visitors and the environment through a number of programs.

NMoQ's Learning and Outreach Department trained educators and museum guides organize engaging activities and tours that target visitors of all ages and abilities. These programs use interpretive speeches, guided tours, and practical activities to spread key environmental ideas, inspiring visitors to think about their part in protecting the environment and motivating them to act in their everyday lives. In addition, the museum's Learning Studio hosts workshops and demonstrations related to environmental issues. These hands-on sessions provide visitors, especially students and families, with an opportunity to learn and practice different techniques under the guidance of experts. Programs could include activities such as workshops on recycling and sustainability. Examples of previous and current activities offered by NMoQ include the planting the native thumam (*Panicum turgidum*) in collaboration with Friends of the Environment Center, designing bookmarks using dried local plants, and offering thematic tours about the natural environment in our heritage garden.

Learning and Outreach Department sponsors lectures and workshops delivered on environmental protection. These lectures focus on certain subjects including biodiversity preservation, sustainable living, and renewable energy. In addition, the museum's dedication to environmental education extends to its specially designed programs for student groups. These programs offer an immersive learning environment that is in line with curricular requirements. Students may look around the museum's exhibits and have conversations about environmental problems. The educational programs offered by NMoQ are essential in raising the next generation's awareness about environmental issues because they promote a feeling of environmental responsibility.

Moreover, storytelling occupies a basic role in spotting the light on plastic pollution, an environmental threat especially pronounced in the Qatari coastline. The Department of Learning and Outreach, primary mission is to raise youth's environmental awareness and knowledge. One such example was the development and publishing of an appealing story named "Sirenia; A Little Dugong's Tale." The story focuses on the importance of preserving and protecting our waters from pollution and other threats. The focus on environmental issues and threats in our programming is influenced profoundly by the content of the galleries.

NMoQ's first two galleries, The Formation of Qatar and Qatar's Natural Environment, showcase a captivating journey through the rich environmental history of the peninsula. The galleries feature films, fossils, and life size models of Qatar's wildlife. The collection on display shows animals, plants, and habitats from different geological periods, ending the exploration journey on a biodiversity wall and in-depth interactive screens. Inside the relevant theme family exhibit, visitors are encouraged to take a pledge on protecting the environment and expressing their ideas on similar issues.

NMoQ's programming aligns with these curricular requirements: Science and Biology, Geography and Earth Science, and Social Studies and Civics. Inside the galleries students may study the links between living beings and their environment, witness environmental events, and comprehend scientific ideas. In addition, the exhibits can shed light on the locations, resources, weather patterns, and effects of people on the environment. They can aid students in understanding the processes, landscapes, and effects of human activity on the world more thoroughly. Furthermore, the museum visit emphasizes concerns of environmental justice, indigenous viewpoints, international collaboration, and the contribution of governments, non-governmental organizations, and individuals to long-lasting change. Students can examine active citizenship techniques and discuss the societal ramifications.

One significant challenge faced by NMoQ's educators is encouraging schools to consider the museum as a place for learning. Overcoming challenges as this necessitates effective communication, collaboration with school administrators, and the development of engaging and curriculum-aligned programs that showcase the museum's educational value. By highlighting the unique opportunities for experiential and interdisciplinary learning within the museum's walls, we can gradually reshape the perception of museums as integral spaces for education.

Currently, the Learn Page at NMoQ' s website is being updated, and populated with online resources to broaden surfers' awareness concerning educational programs in recognition of the significance of digital platforms. NMoQ gives a larger audience accessible and interesting information through virtual exhibitions, educational movies, and interactive websites. These tools enable people to research environmental issues, pick up knowledge at their own speed, and self-develop into knowledgeable environmental champions.

Regarding NMoQ's future plans, the Learning and Outreach Department is in the process of forging a partnership with the Environmental Department and Science Faculty of Qatar University (QU). This alliance intends to use the knowledge of and resources from both institutions to increase public awareness on environmental issues, inspire sustainable practices, and foster a deeper understanding of the complex interrelationship between human civilization and the natural world. Finally, NMoQ intends to launch collaborative research projects and programs to investigate the interaction between culture, history, and the environment.

In conclusion, the Learning and Outreach Department of NMoQ has a notable dedication to environmental education. NMoQ encourages visitors to get a deeper awareness of environmental issues and act for change by including environmental themes into its exhibitions, interpretive programs, seminars, school programs, community engagement projects, and digital resources. NMoQ makes a substantial contribution to the development of a sustainable and environmentally conscious society through its educational activities.

9.10 Dadu Children's Museum of Qatar Initiatives

Dadu's, educational programming stimulates children to contribute confidently to a sustainable future by nurturing the development of their creativity, compassion, generosity, and trustworthiness. We believe that educating children on the ways they can make a difference will help to raise a generation of adults, even more aware and environmentally conscious than the generation before them.

Two of our ten strategic learning aims relate directly to nature conservation and environmental sustainability. They are derived from the social and environmental development pillars of the Qatar National Vision 2030 and are drafted in a developmentally appropriate context for our young audiences.

Value the Environment

We encourage a conservation ethos with a "head and heart and hand" approach, building a love for Qatar's wonderful natural heritage, connecting children with nature, and encouraging children to take action to protect it.

Sustainability

We develop children's awareness and knowledge of sustainable development and lifestyles, and cultivate values, behaviors, and habits that align with responsible and respectful use of resources.

Dadu recognizes studies showing that encouraging an interest in the outdoors sets the foundation for future interest in natural history and environmental issues. Spending time in nature, digging soil, observing insects, and connecting with the elements foster an appreciation for the environment and set up confidence, positive, and happy associations with the environment, building emotional connection to nature and a sense of morality toward environmental issues.

Dadu Gardens opened in November 2022 as the first phase of the museum is a space for children and families in Qatar to do all this and more. Children can discover the principles underpinning the practice of permaculture and witness the beginning of a long multi-year project to create Doha's first city-center Food Forest. At the Community Garden section of the Dadu Gardens, children work together to sow seeds and tend to the plots and the vermicompost. Produce from the Dadu Edible Gardens are turned into salads, smoothies, teas, and more at the Outdoor Kitchen. All of these activities are available to children of Qatar in school or family groups, positioning nature as a tool for personal and societal well-being.

Dadu also creates opportunities beyond the Dadu Gardens. Understanding the power of authentic localized learning experiences, Dadu works with partners across Qatar. One of the most powerful experiences is a result of partnership between the Ministry of Environment and Climate Change, supported by the Qatar Natural History Group to deliver Turtle Encounters tours to the Qatar Turtle Conservation Project exclusively for educators and children. For those unable to attend the tours, a complimentary workshop on Turtle's Hatching is available for school bookings and is seasonally open for public booking.

At Dadu, children are active partners rather than passive beneficiaries. The precious plastics initiative at the Dadu Gardens opens up every step of the plastic recycling process to children. They collect, sort, shred, and then mold recycled plastic into plant pots for the Dadu Gardens, experiencing first-hand the concept of circular economy. Dadu's commitment is to a longitudinal and dynamic plan based on a dual approach of instilling a deep-set appreciation and motivation to care for nature, which together with the knowledge, skills, and strategies draw-in and enable children to be stewards of our planet here and now.

9.11 Recommendations

Education for sustainability is the only way to move forward toward meaningful and lasting change. As outlined earlier in the discussion, Qatar Foundation and subsequently PUE has pursued a systematic approach for working toward a culture of sustainability permeated throughout all teaching and learning. While still in its early stages of implementation, this initiative has identified several opportunities for further development of policy and practice.

These recommendations include the following:

- Education organizations and ministries should place sustainability at the forefront of strategic planning and guiding statements. As exemplified by QF's Strategy Refresh, clearly embedding sustainability at the strategic level creates a mandate for all stakeholders to align initiatives toward sustainable actions and key performance indicators.
- Education organizations need to systematically incorporate sustainability at all levels of the written, taught, and assessed curriculum. Revising teaching and learning policies to reflect this shift is an essential step toward creating common expectations and priorities. Adopting The Cloud Institute's Education for Sustainability Standards provides a road map for schools to begin the work of creating a curriculum for sustainability.
- Successful implementation of education for sustainability requires consistent and ongoing professional learning for all stakeholders. Continuous professional inquiry and reflection is essential for building collective understanding, efficacy, and buy-in. Growing leadership capacity to support and drive a change initiative is also key to the implementation process.
- Education organizations should strive to build sustainability into their co- and extracurricular offerings, further expanding opportunities for enrichment and action.
- Educators and partner organizations need to work collaboratively to activate the natural environment as an extension to the classroom. The outdoor classroom provides limitless opportunities to connect young learners to their local context and the diversity of local ecosystems. Outdoor classrooms include school gardens, urban parks, organic farms, man-made wetlands, or pristine natural ecosystems. Through connection to these ecosystems, learners are compelled to act in more sustainable ways.
- Creating meaningful partnerships between the scientific community and educators allows young people essential opportunities to engage in meaningful and authentic sustainability learning engagements and action projects. This will require scientists conducting local research to identify areas for potential citizen science engagements.
- Engaging parents and families and enlisting them as sustainability partners is an area for development. Increasing cultural and social capital for sustainability requires investment beyond the school experience. Consistency of sustainability values, aspirations, and actions across school and home accelerates meaningful behavioral and cultural change.
- Creating meaningful community partnerships with organizations such as Flower Each Spring, DADU, NMoQ, and the multiple QF Multiversity entities further enriches the curriculum and extends the opportunity for authentic sustainability experiences.

- Eco-Schools provides a comprehensive framework for learning organizations to critically examine sustainability practices within the physical school and home environments. Organizations aspiring to create sustainable communities are recommended to work toward achieving the Eco-Schools green flag award.
- Identifying existing authentic Arabic literature around sustainability topics and creating reference lists would be a vital resource for enriching and elevating the curriculum. Currently, there is a need for more sustainability literature in the Arabic language.
- Further research is needed to explore and understand the concept of Indigenous knowledge and how it is understood and enacted in the local context, and more specifically how might be leveraged in current efforts to preserve, sustain, and promote Qatari culture, heritage, and identity.
- The majority of the recommendations require a great deal of synergy between and among varying educators, scientists, and organizations. One challenge consistently identified as a barrier to achieving this synergy is a resource platform for identifying key entities in the local context. Establishing a common shared platform for sharing key focal points including descriptions of their work with sustainability and points for potential collaboration is an urgent need.

MOE Recommendations

- Formation of a specialized national committee to follow up the implementation of the fourth goal in the sustainable development plan, with the aim of leading efforts, developing plans and implementing initiatives, and supporting the competent authorities in achieving sustainable development in Qatar.
- Developing a high-quality educational system in the professional and academic tracks, with the aim of achieving influential research outputs that contribute to supporting the knowledge economy and promoting economic development in the country.
- Building the capacities of teachers in the field of sustainable development, and training them to activate learning resources and implement relevant curricular and extracurricular activities, in cooperation and coordination with the Educational Training and Development Center, to enable them to achieve sustainable education for students.
- Launching several educational initiatives and competitions, such as the award for scientific research in the field of climate change and the sustainable school, in addition to the initiative to create websites that promote awareness of the Qatari environment and climate issues. Environmental clubs are also activated to promote environmental awareness and encourage participation in sustainability initiatives.

IB Recommendations for Growth and Improvement

- Create more urgency for education for sustainability by writing an IB position paper explicitly asking IB schools to educate for sustainability as being the only available option.
- Develop an IB education (continuum) workshop around education for sustainability in the IB.

- Develop program-specific workshops related to education for sustainability.
- Highlight the integration of education for sustainability in every program through a curriculum revision targeted to education for sustainability.
- Create more teacher support material around education for sustainability.

Acknowledgements Dr. Seda Duygu Sever Mehmetoglu is supported by NPRP grant #12C-0804-190009, entitled SDG Education and Global Citizenship in Qatar: Enhancing Qatar's Nested Power in the Global Arena, from the Qatar National Research Fund (a member of the Qatar Foundation).

References

Cochrane L, Ozturk O, Khat6ee H, Al-Hababi R, Al-Malki F, Nourin H (2022) Fostering a reading culture: evidence from Qatar Reads. Dev Stud Res 9(1):82–94

HBKU CPP (2022) CPP faculty co-author open access case study on Qatar reads. HBKU CPP

IBO (2021) Circular economy: a call to action for IB World Schools. International Baccalaureate Organization

Planning and Statistics Authority (2018) Qatar Statistical Yearbook. Doha: Planning and Statistics Authority, State of Qatar

Qatar Foundation (2019) Her excellency Sheikha Hind speaks at the education world forum in London

Qatar Foundation (2021) Higher education in Qatar foundation: preparing global leaders driving Qatar's success

Qatar Foundation (2022) QF universities organize bootcamp to encourage innovation. https://www.qf.org.qa/stories/qf-universities-organize-bootcamp-to-encourage-innovation

Qatar National Vision (2030) General Secretariat for Development Planning (July 2008). Doha, Qatar

Qatar General Secretariat for Development Planning (2008) Qatar national vision 2030

Qatar General Secretariat for Development Planning (2011) Qatar national development strategy 2011–2016: towards Qatar national vision 2030

Ministry of Education and Higher Education (2022b) General framework for national education of MoEHE https://www.edu.gov.qa/Documents/Publications/CurriculumStandards/GeneralFrameworkNationalEducation.pdf?csf=1&e=PfpExd

Ministry of Education and Higher Education (2022a) Annual statistics of education in the State of Qatar. https://www.edu.gov.qa/Documents/HigherEdTracks/Annual%20Statistics%202020-2021_%20معتمد[2083].pdf

Sever SD, Evren Tok M (2023) Education for sustainable development ecostsytem in Qatar. Research conducted under the cluster Project on SDG Education and Global Citizenship in Qatar

TAMUQ (2022) Multiversity

Curriculum Development Unit (2022) QF curriculum framework & review cycle. Qatar Foundation

Education for Sustainability (EfS) Standards and corresponding Performance Indicators (n.d) The cloud institute for sustainability education. https://cloudinstitute.org/cloud-efs-standards. Accessed 24 Mar 2023

https://www.Eco-Schools.global/

https://qbg.org.qa/qbg_programs/flower-each-spring-program/

Diploma Programme: From principles into practice (2015) International Baccalaureate Organization (UK) Ltd

Middle Years Programme: From principles into practice (2015) International Baccalaureate Organization (UK) Ltd

Primary Years Programme: From principles into practice (2018) International Baccalaureate Organization (UK) Ltd

What is an IB education? (2019) International Baccalaureate Organization (UK) Ltd

Johnson M, Majewska D (2022) Formal, non-formal, and informal learning: what are they, and how can we research them? Cambridge University Press & Assessment Research Report

Patel F, Lynch H Glocalization as an alternative to internationalization in higher education

Embedding Positive Glocal Learning Perspectives (2013) Int J Teach Learn Higher Educ 25(2)

Further information on the Project is available at: https://www.hbku.edu.qa/en/cis/maker-majlis/sdg-education

Cohen S, Horm-Wingerd D (1993) Children and the environment: ecological awareness among preschool children. Environ Behav 25(1):103–120. https://doi.org/10.1177/0013916593251005

Davis J (1998) Young children, environmental education, and the future. Early Childhood Educ J 26. https://doi.org/10.1023/A:1022911631454

James JJ, Bixler RD, Vadala CE (2010) From play in nature, to recreation then vocation: a developmental model for natural history-oriented environmental professionals. Child Youth Environ 20(1):231–256

Collado S, Rosa CD, Corraliza JA (2020) The effect of a nature-based environmental education program on children's environmental attitudes and behaviors: a randomized experiment with primary schools. Sustainability 12(17):6817. https://doi.org/10.3390/su12176817

MOE References

The strategic plan of the Ministry of Education and Higher Education

Education for Sustainable Development Post-2019 Framework-UNESCO Digital Library

Ministers of Energy and Education honor award-winners nominated d-22

Qatar is the first in the Arab world and the Gulf in the 2021 Premier League exams

The education system in Qatar is of international standards

Education for Sustainable Development (unesco.org)

Qatar National Vision 2030. July 2008. General Secretariat for Development Planning. Doha, Qatar. IBO, 2021. Circular economy: A call to action for IB World Schools. International Baccalaureate Organization.

Open Access This chapter is licensed under the terms of the Creative Commons Attribution 4.0 International License (http://creativecommons.org/licenses/by/4.0/), which permits use, sharing, adaptation, distribution and reproduction in any medium or format, as long as you give appropriate credit to the original author(s) and the source, provide a link to the Creative Commons license and indicate if changes were made.

The images or other third party material in this chapter are included in the chapter's Creative Commons license, unless indicated otherwise in a credit line to the material. If material is not included in the chapter's Creative Commons license and your intended use is not permitted by statutory regulation or exceeds the permitted use, you will need to obtain permission directly from the copyright holder.

GPSR Compliance
The European Union's (EU) General Product Safety Regulation (GPSR) is a set
of rules that requires consumer products to be safe and our obligations to
ensure this.

If you have any concerns about our products, you can contact us on

ProductSafety@springernature.com

In case Publisher is established outside the EU, the EU authorized
representative is:

Springer Nature Customer Service Center GmbH
Europaplatz 3
69115 Heidelberg, Germany

www.ingramcontent.com/pod-product-compliance
Ingram Content Group UK Ltd.
Pitfield, Milton Keynes, MK11 3LW, UK
UKHW021012080726
473054UK00003B/85